Rational Action

Transformations: Studies in the History of Science and Technology

Jed Z. Buchwald, general editor

Dolores L. Augustine, *Red Prometheus: Engineering and Dictatorship in East Germany, 1945–1990*

Lawrence Badash, *A Nuclear Winter's Tale: Science and Politics in the 1980s*

Lino Camprubí, *Engineers and the Making of the Francoist Regime*

Mordechai Feingold, editor, *Jesuit Science and the Republic of Letters*

Larrie D. Ferreiro, *Ships and Science: The Birth of Naval Architecture in the Scientific Revolution, 1600–1800*

Gabriel Finkelstein, *Emil du Bois-Reymond: Neuroscience, Self, and Society in Nineteenth-Century Germany*

Kostas Gavroglu and Ana Isabel da Silva Araújo Simões, *Neither Physics nor Chemistry: A History of Quantum Chemistry*

Sander Gliboff, *H. G. Bronn, Ernst Haeckel, and the Origins of German Darwinism: A Study in Translation and Transformation*

Niccolò Guicciardini, *Isaac Newton on Mathematical Certainty and Method*

Kristine Harper, *Weather by the Numbers: The Genesis of Modern Meteorology*

Sungook Hong, *Wireless: From Marconi's Black-Box to the Audion*

Jeff Horn, *The Path Not Taken: French Industrialization in the Age of Revolution, 1750–1830*

Alexandra Hui, *The Psychophysical Ear: Musical Experiments, Experimental Sounds, 1840–1910*

Myles W. Jackson, *Harmonious Triads: Physicists, Musicians, and Instrument Makers in Nineteenth-Century Germany*

Myles W. Jackson, *Spectrum of Belief: Joseph von Fraunhofer and the Craft of Precision Optics*

Paul R. Josephson, *Lenin's Laureate: Zhores Alferov's Life in Communist Science*

Mi Gyung Kim, *Affinity, That Elusive Dream: A Genealogy of the Chemical Revolution*

Ursula Klein and Wolfgang Lefèvre, *Materials in Eighteenth-Century Science: A Historical Ontology*

John Krige, *American Hegemony and the Postwar Reconstruction of Science in Europe*

Janis Langins, *Conserving the Enlightenment: French Military Engineering from Vauban to the Revolution*

Wolfgang Lefèvre, editor, *Picturing Machines 1400–1700*

Staffan Müller-Wille and Hans-Jörg Rheinberger, editors, *Heredity Produced: At the Crossroads of Biology, Politics, and Culture, 1500–1870*

William R. Newman and Anthony Grafton, editors, *Secrets of Nature: Astrology and Alchemy in Early Modern Europe*

Naomi Oreskes and John Krige, editors, *Science and Technology in the Global Cold War*

Gianna Pomata and Nancy G. Siraisi, editors, *Historia: Empiricism and Erudition in Early Modern Europe*

Alan J. Rocke, *Nationalizing Science: Adolphe Wurtz and the Battle for French Chemistry*

George Saliba, *Islamic Science and the Making of the European Renaissance*

Suman Seth, *Crafting the Quantum: Arnold Sommerfeld and the Practice of Theory, 1890–1926*

William Thomas, *Rational Action: The Sciences of Policy in Britain and America, 1940–1960*

Leslie Tomory, *Progressive Enlightenment: The Origins of the Gaslight Industry 1780–1820*

Nicolás Wey Gómez, *The Tropics of Empire: Why Columbus Sailed South to the Indies*

Rational Action

The Sciences of Policy in Britain and America, 1940–1960

William Thomas

The MIT Press
Cambridge, Massachusetts
London, England

MIT Press books may be purchased at special quantity discounts for business or sales promotional use. For information, please email special_sales@mitpress.mit.edu.

This book was set in ITC Stone Serif by Toppan Best-set Premedia Limited, Hong Kong. Printed and bound in the United States of America.

Library of Congress Cataloging-in-Publication Data

Thomas, William, 1979– author.
Rational action : the sciences of policy in Britain and America, 1940–1960 / William Thomas.
    p.   cm. — (Transformations : studies in the history of science and technology)
Includes bibliographical references and index.
ISBN 978-0-262-02850-9 (hardcover : alk. paper)
1. Science and state—Great Britain—History—20th century. 2. Science and state—United States—History—20th century. I. Title.
Q127.G4T48   2015
338.941'0609044—dc23
2014029657

10 9 8 7 6 5 4 3 2 1

for Jeanne T. Walter

Contents

Acknowledgments

A Graduate Research Fellowship from the National Science Foundation and a Junior Research Fellowship from Imperial College London supported the research and writing of this book. Further financial assistance was provided by an undergraduate research grant from Northwestern University and by the Department of the History of Science at Harvard University. I also benefited from resources made available to me while I was a visitor in the spring of 2006 at the Centre for the History of Science, Technology, and Medicine (CHoSTM) at Imperial College, and at the Department of the History and Philosophy of Science at Cambridge University. My employment from 2007 to 2010 as associate historian at the Center for History of Physics of the American Institute of Physics gave me time and freedom to work on the manuscript. I thank Spencer Weart and Greg Good for their crucial support as directors of the Center. I also thank everyone at CHoSTM for their hospitality and collegiality while I was working there from 2010 to 2013.

This book originated with a suggestion by Ken Alder for a topic for my undergraduate senior thesis. The material on MIT originated with a suggestion made by David Kaiser for a term paper topic during my graduate studies. It is a special pleasure to acknowledge the great influence and support of David Edgerton over a ten-year timespan. His broad knowledge, his keen sense of criticism, and his commitment to working closely with younger scholars greatly improved this book. I also thank all the historians who have offered comments at various points in time. I particularly want to acknowledge Paul Erickson and Judy Klein for illuminating conversations about decision theory. I have learned a great deal about general historiographical issues from my many chats with Chris Donohue, and from our work together on the *Ether Wave Propaganda* blog. Naturally, all errors of fact and interpretation in this book are my responsibility.

I acknowledge as well the interest and help of operations researchers, past and present, especially Freeman Dyson, John D. C. Little, John Magee, Mark Eisner, Jonathan Rosenhead, and the late Russell Ackoff and the late Saul Gass. I also want to thank the Institute for Operations Research and the Management Sciences (INFORMS) for making me a member of its History and Traditions Committee in the winter of 2013.

This book benefited from access to a number of archival collections, including the California Institute of Technology Archives, the Cambridge University Archives, the Churchill Archives Centre, the Imperial War Museum, the Johns Hopkins University Ferdinand Hamburger Archives, the MIT Archives and Special Collections, the National Academies archives, the National Air and Space Museum archives, the National Archives of the UK, the library and archives of the RAND Corporation, the Rockefeller Foundation archives, the Royal Society Library, and the U.S. National Archives at College Park, Maryland, and Waltham, Massachusetts. I particularly want to thank Nora Murphy of MIT and Jim Stimpert of Johns Hopkins for their responsive long-distance help, and Vivian Arterbury, Ann Horn, and other staff at RAND for accommodating visiting scholars. I also thank the CNA Corporation and its librarian Greg Kaminski for making wartime operations research reports available, and historian Erik Rau for his long negotiations with CNA on this count.

I want to recognize the National Portrait Gallery of the UK, the U.S. National Archives and Records Administration, the Ferdinand Hamburger Archives, the Econometric Society, the Institute of Mathematical Statistics, INFORMS, Susannah Burn, and Roberta Herman for their permission, whether by policy or specially granted, to reproduce visual materials free of charge. Solo Syndication generously agreed to reduce its usual fee. Much of chapter 25 appeared in the *Business History Review* in 2012, and is reprinted with permission from Cambridge University Press. Elements from chapter 24 appeared in the annual supplement of *History of Political Economy* in 2014, and are reprinted with permission from Duke University Press.

Scholarly research and writing are arduous, and I believe that a better professional and intellectual system could alleviate many of the more unnecessary difficulties, delays, and indignities. But, until that happens, all scholars know the profundity of the debts we owe to the people who sustain us in an undertaking that is often frustrating, thankless, and, indeed, for many of us, financially perilous. I have fared better than many. In my journeys, many friends have lent a hand and kept me tethered. I would like to express my great thanks to my parents, Jennifer and Scott

Bradford, and sister, Elizabeth Frederickson. My wife's parents, Bill and Betty Jordan, have offered continual assistance to us as we traveled the academic trail. My wife Caitlin has endured a lot and risked much, and has been an unparalleled source of strength and love. In the last couple of years, our daughter Claire has been a wonderful little friend, and our second daughter Caroline arrived just as this particular journey was reaching its conclusion. This book is dedicated to my grandmother, Jeanne Walter. She is responsible for whatever judiciousness, persistence, and humor I might have that aided me in the writing of this history.

Introduction

At the end of World War II, there was a widespread interest in what the future of science, technology, and governance would hold. To many commentators the war served as clear evidence that humanity's capacity for invention was paired with a deep irrationality that turned the products of genius inexorably to destruction. They urged that, if the human race were to survive, the world's nations would have to learn to conduct their affairs with a newfound wisdom. At the same time, such wariness was often tempered by a belief that if the work of scientists and engineers were harnessed by more enlightened policies, civilization could enter an era of unprecedented peace and prosperity.

In this same period, practical developments in the conduct of the war demonstrated to those who were privy to them that immediate and concrete benefits could be realized by improving the orchestration of research, engineering, management, and policymaking. Technologies and the methods of using them could be made more effective by designing them in tandem. A wider variety of experts could contribute to the formulation of new plans and policies, and different expert perspectives could be better harmonized. The rationales underlying practical decisions could be placed on firmer empirical foundations, and, in some cases, their logic could be fruitfully subjected to formal mathematical analysis.

Naturally, the exact implications of all these developments were a matter of dispute. J. D. Bernal, a British crystallographer and a well-known Marxist intellectual, thought that their consequences would be epoch-making. In a November 1945 lecture entitled "Lessons of the War for Science," he explained to a London audience that the wartime successes of scientists engaged in an activity called "operational research" showed that henceforth it would be possible to rationally coordinate scientific research with the needs of industry and society. To him, this prospect signaled nothing less than the dawn of a new phase of human history, in which "scientific, conscious social organization" would replace the "unplanned interaction of human wills" as a driver of progress. The moment was as important as when the advent of civilization had supplanted biological evolution. Bernal mocked anyone who might "shrink from this opportunity and the immense responsibility which it places on man for the conscious direction of his own future." Such people were, he reckoned, "in the position of the wild men in the woods of the previous transformation, who preferred to fend for themselves as had their animal forebears rather than mix in the dangerous and disturbing affairs of human society."[1]

The scale of Bernal's vision was almost singular, but his interest in planning scientific work to address social needs was widely shared. Others, however, regarded this sort of planning as a threat to the independence of academic inquiry. Warren Weaver, for one, was deeply opposed to the idea. In September 1945 the American mathematician and influential Rockefeller Foundation administrator wrote a long letter to the *New York Times* explaining that to suppose the war's technological successes validated the intelligent direction of science was to draw the wrong lessons. In November he republished his letter as a pamphlet for the Society for Freedom in Science, an organization founded in Britain in 1941 to oppose Bernal and likeminded others.[2] Yet, Weaver, in his own way, was also deeply impressed by wartime developments. In December he circulated to colleagues a draft of his chapter of the final report on the activities of the Applied Mathematics Panel, a wartime U.S. government body he had led. In it he described a "Tactical-Strategic Computer," into which could be input equations and variables describing every possible condition bearing upon the various interrelated choices that senior military officers might have to make. Within the limits of available information, the computer would work through all possible combinations of choices. When it was finished, it would select the most rational combination and a dial would light up displaying the quantitative value of that combination's "Military Worth." Weaver did not suppose the computer to be a realistic prospect. Rather, he viewed it as an idealized illustration of the analytical rigor that should henceforth inform military decision making.[3]

The Sciences of Policy

The extravagances of Bernal's and Weaver's very distinct visions drew heavily on the sense of momentousness that pervaded their moment in history. Yet neither vision was a sheer fantasy. During the war both men had been deeply involved in work that helped to improve the designs of weapons and equipment, and the planning of combat operations. Their postwar ideas were essentially extrapolations from what they felt were the principles responsible for wartime successes. Needless to say, their grandest aspirations did not come to pass. But, in the following decades, the ideas underlying those aspirations would in fact have far-reaching consequences. These included the proliferation of organizations for policy analysis, the foundation of major new professions dedicated to studying industrial and managerial problems, and the development of new kinds of mathematical models that would permeate work in fields ranging from engineering to academic social science. For convenience's sake, we can refer to this complex of developments as comprising a set of new postwar sciences of policy. Readers interested solely in the history of these sciences may turn past the next section of this introduction.[4] However, all readers should be aware that there is also an important larger story at work in this book.

In the postwar period, the sciences of policy served as a locus that interlinked a startlingly complex array of ideas about science, mathematics, philosophy, engineering, computation, expert advice, and executive decision making. By tracing the roots of these sciences in World War II, and the ways that they succeeded, failed, and evolved alongside each other in the postwar period, this book will establish much more clearly than ever before what the ideas driving their rise were. It will analyze why some of these ideas were ineffectual, while others proved powerful and enduring. Most importantly, this book will explain what people at that time actually meant when they asserted it would henceforth be possible to act more rationally. In doing so, it will show how many of their ideas we, in fact, share with them, however foreign their more outlandish expressions may seem.

Powerful Nations, Ideology, and the Concept of "Science"

Traditionally, our understanding of mid-twentieth-century ideas about science, technology, rationality, and governance has been restricted by influential conventions governing how we talk about these subjects. These conventions are inherited in large part from the postwar period, and, ironically, from many of the same historical actors whose more intricate ideas are most obscured by them.[5] The conventions are most clearly characterized by the distinction they draw between "scientific" and nonscientific ways of thinking, and by their development of a history of the evolution of the relations between those ways of thinking. While the specific subject and contents of this history can vary substantially from case to case, it always follows one of two basic narratives.

The first narrative is about the halting, eternally incomplete progress of "science" as a force of economic and political enrichment. In this narrative, science is associated closely with technology, but also with rationality, which, in turn, is implicitly defined as the virtue underlying any sound policy. The story is, of course, about science's ongoing effort to make itself more useful to society. But it is also about scientists' struggle to overcome others' neglect of, and resistance to, their work. Postwar proponents of the new sciences of policy unsurprisingly favored a version of this story. By characterizing their advocacy as a continuation of the general contribution of "science" to the war effort, they could portray their particular postwar ambitions—whatever they happened to be—as natural, clearly beneficial steps in the progress of science, both intellectually and as a social and political force.[6]

As the historian David Edgerton has argued, this story also had a major influence on the political and intellectual discourse of twentieth-century Britain, and on the subsequent historiography of that nation. Throughout the century, intellectuals, political leaders, and historians often argued that Britain was in a decline, which better-informed governance, combined with a proper investment in science and technology, might yet reverse.[7] The sciences of policy fit easily into this overarching political narrative. In fact, as in the case of J. D. Bernal, specific wartime successes such as operational research (OR) were initially taken as harbingers not of new sciences of policy, but of a much larger sea change in science–state relations. Later, when this anticipated change failed to materialize, these same wartime experiences were reduced to episodes in a longer, less optimistic history of British science. This history was replete with missed opportunities for a better government and society, which, according to the moral of the story, would require ongoing work to bring about.[8] For example, this narrative patterned a 1965 lecture given by Solly Zuckerman, the British government's chief scientific adviser, who had worked with Bernal during the war. In it Zuckerman was able to trace the history of an "uneasy alliance" between "science" and "the state" all the way from ancient Archimedes through wartime operational research to the creation of his top-level office the previous year.[9]

The second narrative is similar to the first in that it relates the concept of "science" both to technology and rationality. However, it differs crucially in that, first, it is about the rising dominance, rather than the struggle, of technology and rationality in social and political life; and, second, it does not suppose rationality to be an unalloyed virtue in policy. Some versions of this story associate the ascendancy of science and rationality with the coming of an age of "modernity" or the widespread influence of "modernist" ideology. This version of the story typically traces the origin point of modernity and modernism to the eighteenth and nineteenth centuries, and is crucial to the works of influential mid-twentieth-century intellectuals such as Lewis Mumford, Jacques Ellul, Max Horkheimer, Theodor Adorno, Herbert Marcuse, and Jürgen Habermas.[10] It remains influential among scholars and critics who purport to build their ideas on "postmodern" foundations.

Other versions of the story focus more specifically on the United States. These versions are popular among journalistic commentators, historians, and, in a rather different form, among conservative intellectuals. In their accounts, America is a place where science, technology, and rationality, including the sciences of policy, became unusually effective instruments of political power, as well as central elements of a peculiarly American form of "liberal" ideology.[11] This liberalism, while often traced to Progressive politics and the New Deal, is taken to culminate with the politics and policies of the Cold War period. It is often linked, especially via "defense intellectuals," to the arms race and the Vietnam War.[12] It is linked, especially via "modernization theorists," to aggressive international interventionism and economic development programs.[13] It is linked, especially via a new breed of policy intellectuals, to Lyndon Johnson's ambitious domestic agenda.[14] More ethereally, it is linked to a relentless instrumentality inhabiting America's postwar capitalist and consumerist culture.[15] Among historians of science, a great deal of attention has similarly been directed toward an inverse influence of "Cold War" institutions and ideology on American science policy, and on a variety of academic fields, including cognitive science, psychology, sociology, political science, and economics.[16]

Within both narratives, an emphasis on failure—and the tacit role of the narrator as a diagnostician of failure—plays a crucial role in the selection and description of the episodes populating the stories.[17] Precisely what failures are featured varies markedly from story to story. They might be flagging businesses, stagnating economies, military weaknesses, ineffective government programs, escalating international hostility, deteriorating value systems, morally repugnant policies, proliferation of vacuous and dangerous ideas, or outbreaks of intellectual or political dissent. These failures, in turn, are not taken to result merely from bad decisions or faulty policies; instead they are cast as logical outcomes of the ignorance or naiveté residing in powerful actors' deep-seated ideological outlook. Depending upon which basic narrative is being invoked, this outlook will involve either actors' irrational adherence to tradition in the face of new alternatives, or their unjustified enthusiasm for ideas they deem rational or progressive. In both cases, though, linking failure to ideology creates an important role for the narrator, who, as an intellectual with privileged access to the historical record, can claim the power to diagnose the presence of the ideology, its causes, and the nature of its pathological effects. Because it is generally assumed the present-day audience is still burdened by these pathologies in one way or another, such narratives function as morality tales. And, of course, it is the narrative's moral that lends the author's work its purpose and cogency.[18]

The history in this book has no particular moral, except the one just outlined, which is intended mainly for historians. Rather, I take the cogency of this history to rest in its ability to enhance our understanding of an important set of historical episodes, and the ideas at play in them. On the basis of existing works, it would be possible to synthesize a reasonably useful portrait of the histories of the various sciences of policy. However, such a portrait would depict these sciences as more or less interchangeable branches of a general movement to introduce newly "scientific" methods into various public and private institutions, and into different professions and academic fields. In doing so, it would not convey a good sense of the differences between these various sciences, nor would it give much sense of their proponents' thinking concerning how those sciences properly related not only to policy but also to each other. By bringing out just such issues, this book will tell a more complex, but also a more coherent story about the sciences of policy than we have heretofore possessed. Moreover, in doing so, it will substantially enhance our understanding of the intellectual, institutional, political, and cultural contexts that shaped those sciences' development and granted them (or withheld) the legitimacy they sought. Within this account, the overarching concept of "science" will be seen to have held little special authority or significance for historical actors.[19] Their thinking was too rich for their evaluations of others' ideas to be swayed, one way or the other, by any simple scientific gloss such ideas might possess. Here, then, the role of the concept of "science" is reduced to the impact it had on historical actors' rhetoric, and its capacity to link otherwise tenuously related historical developments closely to each other.

The Foundations and Evolution of the Sciences of Policy

While this book necessarily concentrates on people directly involved with the sciences of policy, it emphasizes that those people could have wildly different goals without necessarily being at odds with each other. For instance, some were deeply interested in improving the rationality of practical decision making, while others were more interested in developing abstract theories defining what it meant to make a rational decision. This book also emphasizes that, among those figures who did seek to influence practical decisions, such individuals generally regarded more traditional authorities as experienced intellectual partners, rather than as people whose outmoded ideas had to be replaced. Very often, once scientific outsiders proved their worth, traditional authorities regarded them in a similar light. The questions permeating their relations, therefore, did not so much concern *whether* plans and policies should be made more rational, but what particular reforms should be made, and by what means they could best be devised. Although scientific figures regarded this collaborative approach to decision making as crucial to the quality of their work, this attitude could also be obscured by elements of their rhetoric emphasizing the scientific nature of their contributions. Because recovering the deeper ideas so obscured is a complex task, this book has been organized into an unusually large number of shorter chapters to call closer attention to a large number of particular aspects of these ideas and their history. However, these shorter chapters are grouped into seven parts of generally chronological and thematic homogeneity.

Part I of this book is a prologue that focuses on the period spanning World War I and World War II. Rather than develop a particular history, it serves to introduce some of the cultural, intellectual, and institutional themes that help the subsequent history to make sense. Chapter 1 concerns British physiologist A. V. Hill's work developing anti-aircraft gunnery methods during World War I. It emphasizes that his work was essentially continuous with a long tradition of developments in ballistics theory and gunnery technique. However, it also emphasizes his belief that, as an outsider to that tradition, he was able to substantially increase its pace and quality, as well as his concern that, following the war, government institutions would be incapable of maintaining wartime standards. Chapter 2 is an examination of the interwar cartoon figure Colonel Blimp, which places Hill's concerns in the context of much broader cultural anxieties about the ability of British industry and the British government to act progressively. Chapter 3 discusses British and American military efforts to improve bombing technique in World War II, and illustrates the military's methods of improving its operations both with and without external or "scientific" assistance.

With these themes established, parts II and III discuss which particular new developments were able to find traction during World War II, including "operations" (in the United States) or "operational" (in Britain, Canada, and Australia) research. OR has often been singled out as a signature contribution of "science" to the war effort. Thus, the history of OR has generally been regarded as one of its progression from studies of the use of technologies, particularly radar, in combat operations, to studies of operational planning and military policy more generally. Here, we will deal with OR and operational planning, but we will also examine the technological studies as more than just steppingstones to policy studies. In this context, OR will be seen as one of several novel wartime developments directed at the problem of integrating technology design with its use, which also included the augmentation of channels for technical liaison, the deployment of "expert consultants" to combat theaters, the establishment of "field laboratories," and the creation of a branch of engineering theory called "warfare analysis." Part III will establish in further detail what these activities actually entailed, how they differed from each other, why they were considered valuable, and how they were understood to relate both to each other, and to the more traditional work of engineering and military planning.[20]

Part IV examines the immediate postwar years, when the broader implications of wartime activities remained unclear and were widely discussed. It will follow the perpetuation and expansion of wartime activities in the military, focusing on the issues involved in moving from the rapid analysis of combat conditions to the analysis of problems involved in maintaining preparedness during peace. Part IV will also discuss how the experience of wartime OR was sometimes singled out as signaling a moment of progress for "science" in the affairs of policy and management. While that discourse had little immediate effect, it did result in the creation of a new OR profession focused on civilian, and especially industrial, problems. Although the profession never became especially visible, it did prove highly successful. In 2012, according to the United States Bureau of Labor Statistics, there were 73,200 people working in the country as "operations research analysts," which compares very favorably with the 16,900 "economists," although it is still a much lower number than the 718,700 "management analysts."[21]

Part V examines the postwar development of new mathematical theories of rational decision and resource allocation, and their relationship to the advisory and professional activities discussed in part IV. Part V emphasizes that such theories flourished in settings situated at the confluence of practical managerial decision making and academic leisure. By and large, theoreticians did not expect their work to translate effortlessly into practical action by the simple virtue of its intellectual sophistication. Direct applications were largely restricted to well-structured technical problems. Otherwise, these theories were mainly valued for their more academic ability to clarify concepts, to probe the logical implications of suppositions, and to open up new abstract mathematical spaces for study. The theories were novel, and gave postwar proponents of the new field of OR traction in otherwise crowded professional and academic landscapes. However, the adoption of a theoretical canon disappointed more ambitious proponents who did not wish to confine themselves to a mere technical specialty.[22]

Part VI discusses how relations between theory and practice, and between the different branches of the sciences of policy were managed. Such management depended on sophisticated conceptions of intellectual legitimacy, professional obligation, and proper institutional structure. Even with this sophistication, though, the forging of the full postwar constellation of new sciences of policy was a far from frictionless affair. The content and practice of fields such as OR, "systems analysis," and "management science" shifted radically over the course of the 1950s. Individual and institutional goals had to be constantly reconsidered. Although these realignments became less radical in the 1960s, they never settled into a true point of equilibrium. A final chapter, constituting part VII, illustrates how a polemical rhetoric developed circa 1960 within the sciences of policy, which cast disagreements over very particular points as the product of opponents' fundamental misunderstanding of the nature of science and its relationship to rationality and policy. The conclusion of this book then reflects on how those polemics have shaded historians' subsequent portraits of the role that "science" has historically played in rationalizing ideas and legitimizing actions, not only in the twentieth century, but also across all modern history.

I Prelude

1 A. V. Hill in World War I, and His Complaints Afterward

The origins of the sciences of policy discussed in this book can be clearly traced to World War II. However, as we will see, those sciences were, in many ways, simply extrapolations of well-established practices in engineering and military planning. The fact that they were recognized as new and significant derived from the novel conditions brought by the war, which necessitated the augmentation of existing practices by scientists who were not ordinarily associated with the military. It also depended crucially on a sense that such augmentation was a historically noteworthy event, worthy of formalization and perpetuation. However, the sense that outside scientific participation was noteworthy did not originate in that war. In fact, it may have no clear origin point at all. But, it was certainly evident during World War I, when academic scientists found a number of opportunities to lend their skills to that war effort.

One important area where academics lent such skills was in anti-aircraft gunnery. Firing a gun at a moving target is to a certain extent a matter of training and instinct, and to another extent a matter of deliberate calculation. Firing rifles accurately mainly employs the shooter's acquired sense of where to aim and when to fire. Accurate artillery gunnery is more complex. Estimates of the location and movement of targets and of other factors such as wind form the basis of often-elaborate calculations that determine how the gun should be aimed, and when it should be fired. By World War I, field and naval gunnery were already well-developed sciences, with calculations encapsulated into firing tables, slide rules, and, in the case of large ships, calculating machines called fire-control devices. Anti-aircraft gunnery, in contrast, was a new field. Firing tables existed, permitting gun angles and fuse settings to be rapidly calculated based on measurements of range, height, course, and speed of the target. However, calculations could not be properly refined because positions of mid-air shell bursts at testing ranges could not be easily measured and compared to calculations. Thus, anti-aircraft gunnery largely remained a question of the skill of individual artillery crews, which remained inadequate even against massive, slow-moving Zeppelins.[1]

In late 1915 Horace Darwin proposed a device to the British Ministry of Munitions that would help train anti-aircraft crews' instincts. Darwin was co-proprietor of the Cambridge Scientific Instrument Company, and the youngest son of the late naturalist Charles Darwin, and had numerous connections to both British industry and academic science.[2] At the beginning of 1916, he recruited the commissioned Cambridge physiologist A. V. Hill into the Ministry's Munitions Inventions Department to help design the device. The device was supposed to work by giving anti-aircraft crews experience aiming and firing against a flying target. However, rather than firing a shell, it would feed information to a person who calculated where a shell would explode with respect to the target, and gave that information back to the trainee. Darwin's and Hill's labors ultimately resulted in the development of the "mirror position finder" (figure 1.1), which allowed targets to be tracked in three dimensions. Because no plotting method proved rapid enough, the finder was never used for its intended purpose of training gun crews. However, it did prove useful for plotting the location of bursts from test shells, allowing predicted explosion locations to be compared with experimental results.

To exploit this advance, Hill gathered a number of scientists, largely Cambridge colleagues, into a group called the Anti-Aircraft Experimental Section (AAES), or, informally, "Hill's Brigands."[3] The AAES was able to draw verifiable comparisons between the performance of different guns, shells, and fuses, and to calibrate particular guns against variations between shell lots, and variations in the performance of the gun itself. Most importantly, they used their instrument to develop newer and better ballistics tables, which took into account corrections for factors such as wind and variations in atmospheric density with altitude. They also developed slide rules, such as the "parallax correction instrument" (figure 1.2), which was used to correct for the offset of a central observing post from a gun.[4] These studies permitted theories of anti-aircraft gunnery to rival the sophistication of more established forms of artillery gunnery.[5] In the mid-1920s results from the AAES's work were collected into a *Text Book of Anti-Aircraft Gunnery*.[6]

Although anti-aircraft artillery never became an effective means of repelling enemy attack during the war, Hill's AAES was considered a success. Hill, however, felt that the quality of his wartime section's work might not survive the war's end. For this reason, at the end of the war, he drafted,

Figure 1.1
The mirror position finder. From War Office, 1924–1925, vol. 1.

but never completed, a memoir of his wartime experiences, which he intended to use to incite change in the military's research establishments. In it he wrote that wartime advances in anti-aircraft gunnery had been made on the backs of "able scientific men … gradually collected together from the Universities and let loose on a job." But, he urged, "The spirit which has led to this [work] has been bred by the war, and it cannot be expected that in days to come scientific developments will be able to go on so quickly and so easily without employing considerably larger staff and without expending considerably greater sums on salaries and maintenance." More broadly, he believed that the government might not be able

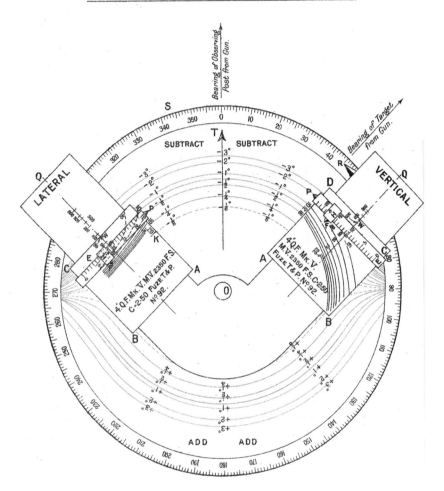

Figure 1.2
Diagram of the parallax correction instrument; the lines on the device were color coded. From War Office 1924–1925, vol. 2.

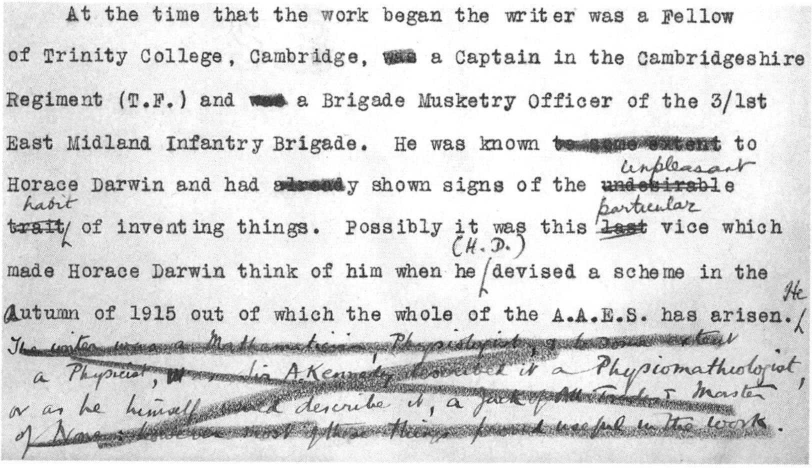

At the time that the work began the writer was a Fellow of Trinity College, Cambridge, ~~was~~ a Captain in the Cambridgeshire Regiment (T.F.) and ~~was~~ a Brigade Musketry Officer of the 3/1st East Midland Infantry Brigade. He was known ~~to some extent~~ to Horace Darwin and had ~~already~~ shown signs of the ~~undesirable~~ *unpleasant* *particular* ~~trait~~ *habit* of inventing things. Possibly it was this ~~last~~ *(H. D.)* vice which made Horace Darwin think of him when he /devised a scheme in the *He* Autumn of 1915 out of which the whole of the A.A.E.S. has arisen. /

~~The writer was a Mathematician, Physiologist ... ...~~ ~~a Physicist, ... in A Kennedy ... ... it a Physiomatheologist, or as he himself ... describe it, a Jack ... ... Master, of None ... ... useful in the work.~~

Figure 1.3

An extract from A. V. Hill's unfinished war memoir, "The Anti-Aircraft Experimental Section (1916–1918)," located at CAC AVHL I 1/37. Reproduced with the permission of Susannah Burn.

to organize technical work appropriately. His tale was mainly intended to "record … the difficulties and obstructions" that too often forced him and his colleagues to use "various improper channels" in pursuing their work.[7]

Hill also regarded his own involvement in anti-aircraft gunnery, which sparked the work of the AAES, as the product of chance and personal connections. He allowed that he had an unlikely background for a potential expert in the subject. In a passage added in and then crossed out (figure 1.3), he described his prewar work in the physiology of muscles as an eccentric combination of mathematics, physiology, and "to some extent" physics, which made him, according to one his colleagues, a "physiomatheologist." More bitingly, he described his own penchant for "inventing things" first as an "undesirable trait" and then, crossing that out, an "unpleasant habit." Fortunately, the well-placed Darwin saw fit to introduce a character exhibiting such an inventive "vice" into the Munitions Inventions Department.

As a rising academic physiologist, Hill could afford to make light of his unclassifiable talent, and of others' appreciation of it. However, his account was intended to be a serious critique of the government and military bureaucracy's ability to identify and marshal such talent to its own advantage. He did not give any indication as to what sort of system could identify useful expertise amid a national pool of possibly useful expertise. And he would certainly have been aware of the challenges involved, having served on a wartime committee tasked with reviewing—and almost always rejecting—outsider proposals for anti-aircraft-related inventions.[8] But his underlying concern was not in how to design a bureaucracy; it was in making an existing bureaucracy aware that complacency might someday lead to catastrophic failure. That theme would continue to occupy his attention in subsequent decades.

2 Colonel Blimp: Tradition, Authority, and the Difficulty of Progress

A. V. Hill's concerns about bureaucratic complacency in many ways reflected a broader anxiety that permeated intellectual life in Britain in the early twentieth century. Many commentators worried that the country was too steeped in tradition, and, for this reason, failed to adopt necessary progressive changes.[1] These worries found a rich expression in the satirical cartoons of David Low, and in particular his most famous creation, Colonel Blimp. Blimp appeared regularly as a part of the "Low's Topical Budget" feature in the British *Evening Standard* newspaper beginning in 1934. An opinionated fat man, Colonel Blimp was almost always shown enjoying the leisure of the Turkish bath. Prefacing his comments with an indignant "Gad, sir!" he held strong views on a wide range of issues, usually in accord with some authority, such as the fictional Lord Flop, or the owner of the *Evening Standard*, Lord Beaverbrook. He was usually accompanied by a cartoon version of Low who listened silently with a half-baffled, half-intimidated look on his face.

Blimp's opinions were mainly satires of conservative, ruling-class dogmas: he was militant toward foreigners (except fascists), and callous toward the less advantaged. More broadly, he could be counted on to represent whatever barrier stood in the way of a better future. In 1937 Low explained, "Without Blimps history would be a barren record of progress, with no tales of how Blimp the Great burnt the cakes, or how Sir Walter Blimp played bowls before the Spanish Armada." Blimp's ideas were rooted in the intellectually and technologically impoverished past. As Low wrote, "Blimp is Tradition. What was good enough for William the Conqueror is good enough for him. To the problems of the motor age he is apt to apply the technique of equitation; and he solves the problem of inconvenient plenty by using the economic principles of scarcity." This backwardness did not, however, make him irrelevant in times of obvious technological breakthrough. Blimps were as likely to be held responsible for ham-fisted

Figure 2.1
The Colonel Blimp panel from David Low, "Low's Topical Budget," *Evening Standard*, July 13, 1935. © Solo Syndication / Associated Newspapers Ltd. Reproduced with permission.

or inhumane uses of technology as they were for technological stagnation and ineptitude. The past, rather, was most deeply associated with a dearth of reason. Low also asserted that Blimp did not represent conservatives, specifically, but "stupidity" in general.[2]

Inasmuch as "stupidity" is a catch-all criticism, Low's satirical strength rested in his ability to adapt the Blimp character to portray positions on a whole variety of issues, which always came off as "Blimpish." In a 1935 cartoon (figure 2.1), Blimp was used to criticize opposition to the establishment of a new practice range for Royal Air Force (RAF) bomber crews. Accurate bombing, like anti-aircraft gunnery, required a substantial investment both in equipment and in the skills and techniques necessary to use that equipment. While the ostensible source of resistance to a progressive national defense in the cartoon is the Royal Society for the Prevention of Cruelty to Animals, from the mouth of Blimp it becomes a shortsighted failure to make an important investment in technical skill. Resorting to the "principles of scarcity," Blimp insists that all bombs should be reserved for their intended human targets—a position absurdly contrary to the humane concerns of the position with which he is ostensibly in agreement. Thus, the cartoon managed to serve as a polemic against antiquated military-style thinking, even though it was the RAF that was actually pressing for the range to be built. During World War II, the Fleet Lagoon near

Abbotsbury Swannery would serve as one of the ranges where prototypes of the "bouncing bomb" were tested. These tests would go on to be depicted in the highly popular 1954 film *The Dam Busters*, a celebration of not only British wartime invention and the bravery and professionalism of bomber crews, but also of precision bombing technique and the marshaling of a daring project through a skeptical bureaucracy. A bouncing bomb prototype is still incongruously on display at the Swannery today.

3   Military Heuristics and the Allied Bombing Campaigns

Following World War I, the mechanized slaughter of trench warfare became the enduring emblem of how human barbarity could turn the products of ingenuity to apocalyptic destruction. The contemporaneous advent of aerial bombardment simply promised that this mass death would soon be carried to civilian populations as well. These developments created problems for the concept of rationality insofar as it applied to warfare. On the one hand, militaries had to commit the rational faculties of their officers and engineers to the constant improvement of weapons, equipment, and tactics, lest they be vulnerable to an advanced, well-armed aggressor. On the other hand, the consequences of war were becoming so horrific that it began to seem inappropriate to regard the architects of war as being rational at all.

When mass bombardment became a reality during World War II, the rationality of bombing became a matter of intense contention. British scientists criticized the Royal Air Force's area bombing campaign against Germany on the basis that it was pursued more out of a faith in its efficacy than out of any real evidence of its value. They were convinced that more responsible scientific advice might have led to the reallocation of bombing efforts to more effective uses. In response to this line of criticism, postwar debates about the Allied bombing campaigns focused as much on their efficacy and strategic necessity as on their morality.[1] However, other postwar criticisms of strategic bombing simply diagnosed it, alongside the atomic bombing of Hiroshima and Nagasaki, as the grim culmination of the relationship between science and rationalized industry and warfare.[2]

It would be simple, then, to take the history of the origins of the sciences of policy in World War II to be essentially about either their unevenly successful attempts to bring rationality to warfare, or, conversely, their role in its ultimate stage of rationalization. Some prominent figures involved in the early history of the sciences of policy were among the key scientist critics of Britain's strategic bombing policy, both during and after the war. Their recollections of their experiences have tended to reinforce the importance of the question of whether, in some scenario, proper studies of bombing might have led to a change in the military's view of its value.[3] The question is worth asking, but it is unreflective of the history of the bulk of the work of scientific figures involved in the wartime strategic bombing campaigns, which focused instead on bombing tactics and techniques. The purpose of this chapter is to illustrate the military's own commitment, with and without its employment of scientists, to the improvement of tactics and techniques of both strategic and tactical bombing through institutional reform and intellectual debate. This heuristic culture, applied both to bombing and other kinds of combat operations, provided the basic framework on which the wartime sciences of policy were built.

Edgar Ludlow-Hewitt and the Early Phases of the War

Few individuals were more exemplary of the RAF's heuristic culture than Air Chief Marshal Edgar Ludlow-Hewitt, who was the Commander-in-Chief of the RAF's Bomber Command between 1937 and 1940. He is not well known now, but he was a well-respected officer in his time. In his memoirs, Air Chief Marshall Arthur Harris, who led Bomber Command's most devastating attacks on Germany between 1942 and 1945, lauded Ludlow-Hewitt's prewar efforts to improve the command's readiness. Harris pointed to his "immense technical ability and practical knowledge," and singled him out as "far and away the most brilliant officer" he had ever met.[4]

From the time that Ludlow-Hewitt arrived at the command, he committed himself to marshaling the knowledge and expertise that surrounded him into the development of more effective plans and tactical doctrines. Prior to the war, a major priority was accident prevention. By 1931 the predecessor organization to Bomber Command had begun to record all accidents in a "flying accident book," issued quarterly, which also contained reports analyzing that quarter's accident patterns. Pilots were required to read this book and sign off that they had done so, in the hopes that it would make them more aware of the causes of accidents. These causes were grouped into four categories: inexperience, carelessness, disregard for orders, and bad maintenance.[5] When he arrived at Bomber Command, Ludlow-Hewitt praised the books, but was dissatisfied with their categorizations. For instance, if pilots were proving unable to handle controls in the dark, rather than chalk the failure up to "inexperience," he argued that it would be better to change training procedures to have pilots learn their controls on the ground blindfolded. He also ordered the creation of a post for a "flying accidents inquisitor whose job will be really to get down to the elucidation of the fundamental cause in all these accidents and not rest until the true cause has been uncovered."[6]

For Ludlow-Hewitt, understanding accidents was tantamount to understanding what procedures and training regimens could effectively prevent them. This tying together of heuristics and the revision of training and procedure was also evident in discussions surrounding the establishment of a "Bomber Development Unit" (BDU) at the command, beginning in 1938. The objective of the unit was to perform flying experiments to assist with the integration of new technologies into tactics, and to elucidate "problems connected with the operation of bomber aircraft, to which adequate attention cannot be given in the operational units." The unit would also serve as a point of liaison with the Air Ministry's technical staffs and its experimental establishments. Initial programs of work for the unit suggested twenty-five questions requiring investigation, although this list was soon shortened to fourteen. Ultimately, the BDU that was finally established in November 1940, after Ludlow-Hewitt's departure, was a modest affair, comprising only four planes. And it was shut down in early 1941.[7]

Before Ludlow-Hewitt left, he also attempted to create a more productive flow of information throughout all parts of Bomber Command. In January of that year, he ordered that units coordinate their meetings with those of the bomber groups they were a part of so as to encourage observations and ideas to "ventilate" up from crews to factor into higher-level deliberations. This coordination was also meant to improve the reception of high-level decisions at lower levels.[8] Ludlow-Hewitt's belief that there should be a flow of ideas between higher and lower levels of the command also accorded with his belief in the importance of training regimens as a means of making bomber crews into knowledgeable experts at their jobs. In his memoirs, Arthur Harris suggested that the ultimate cause of Ludlow-Hewitt's transfer from Bomber Command to the post of the RAF's Chief Inspector was his insistence on diverting scarce planes and personnel from operations into specialized training units.[9]

If Ludlow-Hewitt's reforms to the way Bomber Command revised its doctrine were halting, that was perhaps less a function of the military's attitude toward reform than it was a reflection of an attitude that peace was a time for evolution and preparation, and war a time for fighting.

It was only as the war progressed that it became clear that heuristics had to become an integral part of the war-making process. Bomber Command was, as a matter of course, swept up in this transition, particularly once it committed to bombing by night and was found to be essentially incapable of locating targets in the dark. Aside from improving its navigational aids, its tactics, and its training procedures, the command established an Operational Research Section (ORS) in the late summer of 1941, and reinstituted its BDU in early 1942.[10] But even once the war entered a more heuristic phase, and reforms to how the military services adjusted their tactics and procedures were more widely welcomed, there was still a question of just what specific reforms should be adopted.

The decision to implement a reform has mainly to do with perceptions of a current state of affairs. Dissatisfaction with that state could entail a reaction to failure, or a simple belief that things could be done better. These perceptions, in turn, relate strongly to how decision makers make plans and react to their results on a day-to-day basis. In the case of military decision making during the war, officers adjusted plans, and instituted training and procedures, in the belief that their improvements would have an important effect on the outcomes of operations. Few plans could ever be expected to achieve ideal outcomes. Therefore, the question was how to react to deviations from the ideal. Officers could discipline personnel for failing to follow plans, or reward them for following them as best as they could. They could push for better training, invest in new technology, or they could dismiss less-than-ideal mission results as the product of unusual circumstances, or as the ordinary outcome of the chaotic circumstances of war. Importantly, instituting changes in plans, and reforming the way that plans were made, always came at a cost. Developing new plans and procedures, of course, diverted resources from the conduct of operations. Beyond that, making decision making more complex, or changing plans too often, was also apt to lead to breakdowns in the ability of personnel to follow plans.

If military officers had any number of choices concerning whether to change plans and how plans were made, the clear trend by the middle of the war was toward an increased sophistication in both. This sophistication, in turn, led to the creation of an intricate body of military knowledge, which was constantly revised as more was learned, and as circumstances changed. This body of knowledge was never collected into one place, and was rarely visible in issued orders, but it could sometimes be discerned in training materials. In 1942, for instance, the British RAF Coastal Command, responsible for patrolling the seas around Britain, produced a booklet for its air crews called "Submarine and Anti-submarine." The booklet was essentially a summary of the Coastal Command's knowledge of German U-boat technology, tactics, and the best methods of hunting them from the air, with material gathered from sources such as experience, intelligence, and tests conducted on a captured U-boat rechristened HMS *Graph*. Air crews were expected to absorb this body of knowledge, and to act on its recommendations whenever possible. However, their experiences were also critical to this knowledge's revision. The booklet spelled out an elaborate procedure of "interrogation" in which crew members related their experiences attacking a U-boat from the air, first upon landing, and then again once they had had a chance to recover from their mission. Specialists in anti-submarine operations then amalgamated these experiences and compared them with others to derive useful conclusions that would inform preparations for future missions.[11]

Curtis LeMay and the Late Phases of the War

By the time World War II reached its concluding phases, strategic bombing had become an intensively investigated and routinized activity, particularly in the United States Army Air Forces (AAF). AAF Major General Curtis LeMay proved to be a particular master of the heuristic mechanisms of planning. He is usually remembered as the dour, cigar-chomping overseer of the strategic and atomic bombing of Japan, and later as an aggressive advocate of the United States' nuclear arsenal while head of the Strategic Air Command. He was Air Force Chief of Staff from 1961 to 1965, and his 1965 autobiography notoriously suggested that the United States should threaten to bomb North Vietnam "back into the Stone Age." His ongoing advocacy for nuclear assertiveness led him in 1968 to seek a platform as the vice presidential nominee on Alabama Governor George Wallace's segregationist ticket.[12] LeMay's reputation is thus indissolubly linked to his faith in strategic attack, and the threat thereof as a component of a belligerent diplomacy. His career success, however, probably owed more to his keen handling of tactics and procedure than his strategic commitments and adamant demeanor.

While LeMay came off as anything but inquisitive in public, the transcripts of "critiques" he led following missions show that he was endlessly interested in how missions were evaluated and the ways they could be improved both in terms of efficacy and crew safety.[13] He envisioned

Figure 3.1

An illustration of Curtis LeMay. The magnifying glass and the officer with the clipboard connote study, as of a strike photograph, while the evasive expression on the bomber connotes interrogation. From Headquarters, XXI Bomber Command, "Incendiary Bombing Phase Analysis," 1945, in LOC CEL, Box B37, Analysis of Incendiary Attacks.

bombing as a consistent but malleable procedure to be instilled into crews through rigorous training and discipline (see figure 3.1). In a mission critique in November 1944, he explained:

I am working toward a standard procedure to use every time, so that everyone knows what is going to happen. Knowing what is going to happen, the mission will be a success from every standpoint. So let's do these things we decided on down here. Let's have no special occasions. What we want is a standard procedure for bombing those targets—one that will work under all conditions. If we get that type of procedure we will have simpler problems.[14]

Later that month, in another critique, he announced that he was going to "get rough with people who violate our tactical doctrine." While he was willing to allow deviations in highly exceptional cases, he warned that crews "had better have a good reason," and emphasized, "A poor plan well executed is better than poor execution of a good one."[15]

LeMay's rhetoric stressed the managerial virtues of homogenization and control. Yet, he did not intend for his tactical doctrines to be doctrinaire. Rather, for him, discipline made it possible to learn *collectively* (see figure 3.2). In his foreword to a March 1945 manual on tactical doctrine, he wrote:

In its present form, the Tactical Doctrine reflects the overall combat experiences of very heavy bombardment units.... As more experience is obtained and as the tactics

Figure 3.2
Bombing portrayed as a scholarly activity for crewmembers; "precipitation" is being taken away from the puzzle because it was discovered it did not have a major impact on the success of incendiary raids. From Headquarters, XXI Bomber Command, "Incendiary Bombing Phase Analysis," 1945, in LOC CEL, Box B37, Analysis of Incendiary Attacks.

employed by the enemy are modified, new precepts will be involved and must be incorporated. Organization commanders are encouraged to suggest revisions and new techniques which will improve the doctrine. Recommendations will be embodied immediately, following the establishment of the value and need for change.

He concluded, "It must be kept in mind that an excellent plan half-heartedly followed will never put as many bombs on the target as a less perfect plan which is thoroughly understood and is forcefully executed."[16] What was structured, and thus comprehensible, even if not perfectly so, could be compared with experience and amended. If crews acted on their own initiative, it would not be possible to derive any sort of common wisdom from experience, to the detriment of crews and the command alike.

If discipline dictated how crews behaved on missions, mission critiques offered an opportunity to make modifications to procedure and training. Many problems were straightforwardly technical: poor vision caused by bright sunlight reinforced the need for better sunglasses; a gunner lost when the blister he was seated in got blown out raised calls for better seatbelts; dust generated on takeoff hampered the coordination of large sorties; frost forming on windows hindered formation flying and navigation.[17] In many cases problems could be solved through simple agreement of those gathered, but in other cases special action had to be taken. The frost problem required a "special wire to Washington," which ultimately resulted in new heaters being installed. In some cases the causes of problems had to be carefully parsed. For instance, in one case crew members' explanation that their poor bombing accuracy resulted from flaws in their bombs was dismissed, but only after LeMay felt the problem was well understood. He noted that British bombardiers had had similar complaints, but that after "long studies were conducted," it was found that they simply needed more practice with the bombs they were given. The appropriate response, therefore, was to modify training procedures.[18]

By this point in the war, LeMay could support his commitment to obtaining the most nuanced understanding of missions possible by drawing on a wide range of technical specialists, intelligence analysts, statistical analysts, and operations analysts.[19] He depended on the expertise of this staff, but he was also confident enough in his own understanding of missions that he felt qualified to critique expert work. For example, in a criticism of photographic analysts' work, he insisted they were not giving enough credit to crews. He pointed out that, given that strike photographs could not be relied upon to show all bomb hits, and given that all bombs were dropped at once, analysts simply had to assume that all bombs dropped somewhere in the same vicinity as each other. In marking up photographs from bombing runs, he instructed them, "Definitely outline the pattern you can see and draw a dotted line to identify where smoke makes it impossible to see."[20] The intellectual relationships among officers such as LeMay, combat personnel, and technical experts flowed in all directions.

II The Origins of the Sciences of Policy

4 L. B. C. Cunningham and the Mathematical Theory of Combat

Traditionally, the post-World War II sciences of policy have been traced to the development of "operational research" in the British armed services, which, in turn, is traced to prewar work developing radar and integrating it into combat tactics. While it is generally acknowledged that OR activities were not entirely novel, the sheer quality of collaboration between scientific and military personnel on the radar work—and the contributions of outside scientific advisers to that collaboration—are recognized as important steppingstones toward the scientific study of combat itself. However, the novelty and unusual quality of that collaboration have perhaps been overstated at the expense of understanding the radar work as simply an important instance of the collaborations that ordinarily characterize the successful implementation of engineering schemes and technologies. In the case of the military, researchers, engineers, technical specialists, officers, and other personnel already existed in complex networks of liaison with nodes at such places as proving grounds and training schools.[1] To emphasize this point, this history begins with a lesser-known development that was perhaps more radically novel, but took place entirely within these networks: the formulation of a "mathematical theory of combat" by British military statistician L. B. C. Cunningham.

Leslie Bennett Craigie Cunningham spent his career up until World War II as an instructor of ballistics in the British Air Ministry's No. 1 Air Armament School at Eastchurch on the coast of the North Sea east of London. Cunningham had studied mathematics at the University of Edinburgh. Later, while working for the Air Ministry, he earned a PhD under the Edinburgh mathematician E. T. Whittaker for theoretical work on bomb ballistics.[2] In 1937 his career took a significant turn when he set about constructing a "quantitative method" of analyzing the effects of different armament configurations on the anticipated outcomes of air duels. Traditionally, engineers developed specifications of armament performance—caliber, accuracy, rapidity of fire, and so forth—as measured against a passive test target. Armament configurations for aircraft designs were then decided upon based on engineers' and officers' agreements concerning which configurations were practical and desirable. What Cunningham's theory did was probabilistically interrelate armament specifications with mathematical representations of air combat scenarios, which included statements of the vulnerable surface area of opposing aircraft as well as each aircraft's "stated tactical methods either of approach or withdrawal." Doing so, the theory derived expectation values concerning the chances that different armament configurations had for achieving victory in a typical instance of air combat. The underlying idea was that no engineering design had an intrinsic value. Rather, this value always had to be measured in terms of a design's expected impact on the outcomes of combat engagements.[3]

In November 1937 Cunningham sent a short paper detailing his theory to the Royal Aircraft Establishment (RAE) at Farnborough, just southwest of London, to gather more expert opinions. There, it came into the hands of Ben Lockspeiser, a senior researcher.[4] After the paper's mathematics were vetted, Lockspeiser suggested it might prove valuable for laying out an analytical approach to settle not only technical but also tactical questions. For instance, he pointed out, it could help determine if it was better to use explosive or ordinary bullets. Cunningham's theory suggested that the reduced rate of fire of explosive bullets would be compensated for by the increase in the effective area of the target. The theory also might be able to solve whether or not it was advantageous to open fire as soon a target came into range, or to conserve one's fire for shorter ranges.[5] "In terms of actual fighting conditions," Lockspeiser felt that the method was "immediately applicable" to weighing the factors involved in the first problem, but he deferred judgment on the complicated mathematical expression demanded by the second, and was "not prepared to say whether it can be related to practical considerations or not."[6]

Emboldened by the fact that his "fundamental method of attack" had been found valid, Cunningham set out to expand on his work, and in early 1938 he produced a paper entitled "The Theory of Machine Gun Combat: An Amplified Introduction." Building on his previous paper, he improved on his probabilistic tools and then worked out some "simple illustrative examples" of aircraft duels involving various scenarios concerning the strengths, vulnerabilities, and tactics of the combatants, all the while stressing that the paper made "no serious attempt to achieve far-reaching conclusions." These were intricate calculations, but, he argued, "any labour devoted to it which gives an incontrovertible result will be more than repaid by the fact that this result will be quantitative, and may be used to settle finally the optimum choice between alternative proposals regarding equipment, or its tactical employment." He added, "This section [of examples of application] is capable of unlimited development." If his paper was so fortunate as to "survive" the RAE's "serious fundamental criticisms again, then," he felt, "the impetus necessary to ensure [the method's] rapid development will be provided."[7]

This paper, along with a few other papers he wrote on "statistical gunnery and bombing problems," was forwarded to the RAE in May 1938.[8] Some of these were considered valuable, while others were thought to treat their subject too unrealistically.[9] But it was, in any case, enough for Cunningham to build a lasting relationship with the Air Ministry's highest-level researchers. In June he was named an adviser to the RAE on statistical problems by the office of the Air Ministry's Director of Scientific Research. The Ministry, of course, had its own methods of selecting armament. While nobody thought Cunningham's method offered "complete solutions" to their problems, they did feel that "some useful guiding principles may nevertheless emerge."[10]

Because Cunningham's theory could never even approach being a fully realistic representation of combat, actual quantitative solutions to his theoretical formulations were found to be less useful than their ability to clarify the implications of choices that aircraft engineers had to make, sometimes revising their understanding significantly. RAE engineers, in turn, began to send Cunningham data from their trials, to aid him in refining his theoretical representations.[11] By the winter of 1938 Cunningham had grown more confident, and had become "convinced that the method as a whole is capable of giving important results—not merely *ad hoc* solutions of particular problems, but broad generalizations which might short-circuit numerical work to a large extent." He wrote, "There is a full year's work waiting to be tackled once this elementary part has been placed on a satisfactory footing."[12] G. W. H. Stevens, a colleague of Lockspeiser's, told Cunningham not to rush himself as "the theoretical contribution we have so far will help us to sort out some of the fallacies that are existing." Furthermore, the theory was only as good as the test data that were available, and these were "not worthy of application to refined analysis."[13]

By the summer of 1939 Cunningham had worked out the fundamentals of his "Mathematical Theory of Combat," and had also produced another paper detailing combats between a particular number of "fire-sources" on one side versus another number on the other.[14] He later observed that this consideration "led to some quite unexpected conclusions" that he then developed into a "Theory of Multiple Combat," which turned out to be widely applicable to problems such as tank combat.[15] When war began in September 1939, Cunningham found his teaching burden at the Armament School lightened, which gave him more time to devote to practical applications of his theory until he received "pending news of [his] war-time appointment." Finally, in February 1940, he was named the head of a new organization called the Air Warfare Analysis Section (AWAS), which Henry Tizard, the newly appointed scientific adviser to the Chief of the Air Staff, had recommended be set up to investigate how data from combat could best be collected and analyzed.[16]

Cunningham's "Mathematical Theory" became AWAS Note No. 1, and "An Analysis of the Performance of a Fixed-Gun Fighter, Armed with Guns of Different Calibres, in Single Home-Defence Combat with a Twin-Engined Bomber," which Cunningham wrote with members of his staff as an extension of his theories, became AWAS Report No. 1. But, in the first year or so of AWAS's work, Cunningham and his staff had idiosyncratic and wide-ranging responsibilities over matters quantitative. For example, in one report they attempted to divine information about enemy techniques, tactics, and accuracy by collaborating with Ministry of Home Security researchers conducting an analysis of damage caused by German bombers.[17] In another case AWAS staff criticized Bomber Command mission planners for allocating effort in bombing raids based on statistical fallacies, calling their methods "wholly irrational."[18] AWAS also gathered statistics on anti-aircraft activities before handing the responsibility over to the Army Anti-Aircraft Command's new "operational research" group.[19] As time went on, though, AWAS began to develop Cunningham's earlier theoretical work for special applications in order to answer pressing questions. For example, a 1943 paper, "Notes on the Theory of Aerial Gunnery," addressed fighter-bomber duels. In this case, existing theories used rough approximations to address gunnery accuracy, which made them unusable for analyzing gun-sight performance. Cunningham wanted "to place the errors due to these and other approximations between limits in order that rational choice may become possible of the simplest formula or formulae tolerable in a given calculation; instead of letting errors run wild."[20]

The value of Cunningham's paper, and in many ways the value of his entire theory, was predicated on its ability to improve upon prior thinking. Before the war, such improvements involved the use of data from equipment tests to make more informed choices about engineering designs. During the war, actual combat data could be used to test whether the rationales underlying choices addressed the reality to which they allegedly referred. While these theories had their limitations and by no means guaranteed that correct choices would be made, they do seem to have been valued by Cunningham's colleagues. When he died prematurely in 1946 at the age of fifty-one, his obituary in *Nature* noted that he had been in line for a "high appointment" at the Royal Aircraft Establishment.[21]

5 Patrick Blackett and the Anti-Aircraft Command Research Group

Between the world wars, the principles of anti-aircraft gunnery began to be built into fire-control "predictors," which could aim guns automatically based on a "track" of a target manually input into them. These tracks could be input by observers following aircraft visually, but, once radar was developed starting in the mid-1930s, predictors were adapted to accept tracks from radar operators as well. Ostensibly, the use of such mechanisms augured the end of ballistics tables and slide-rule computers in anti-aircraft gunnery.[1] However, when the Battle of Britain began in July 1940, fire-control predictors did not prove up to the task. The trouble was less that the devices were too primitive, and more that the particular calculations built into their mechanisms did not correspond to those needed to shoot down German bombers. Lt.-Gen. Frederick Pile, head of the British Army's Anti-Aircraft (AA) Command, quickly became frustrated with the devices. To address the issue, the Ministry of Supply's new Advisory Council on Scientific Research and Technical Development called a meeting for August, to be chaired by one of its members, A. V. Hill (figure 5.1).[2]

By 1940 Hill had become a very senior figure in the academic scientific community. He had won the 1922 Nobel Prize in Physiology or Medicine and had directed the biophysics laboratory at University College London since 1923. He had been serving as secretary of the Royal Society for biology since 1935. He was also well connected socially and politically. He was the brother-in-law of the eminent economist John Maynard Keynes, and had been elected an "Independent Conservative" Member of Parliament for Cambridge University in February 1940. He had re-entered the world of war research as early as 1935 when he had joined the Committee for the Scientific Survey of Air Defence (CSSAD), which was instrumental in establishing a research and development program at the RAF for radar (or "RDF" as it was then called). When the anti-aircraft conference was convened he had just returned from a five-month visit to North America,

Figure 5.1
Archibald Vivian Hill, 1940. Photograph by Bassano, Ltd. © National Portrait Gallery, London.

where he had worked to establish ties with war researchers in Canada and the United States.[3]

The August 1940 meeting resolved that existing liaison between equipment designers and users required emergency ad hoc augmentation in order to supply fire-control engineers with accurate technical expressions of the field requirements for predictor performance. Pile and Hill recruited the academic physicist Patrick Blackett (figure 5.2) to address the situation. Blackett had made his reputation in the design and use of cloud chambers to study cosmic rays, and would go on to win the 1948 Nobel Prize in Physics for that work. He had been a member of the British government's

Figure 5.2
Patrick Maynard Stuart Blackett, 1942. Photograph by Henry Coster. © National Portrait Gallery, London.

Aeronautical Research Committee since 1934, and had served with Hill as a member of the CSSAD since its inception. More recently, he had become a member of the Ministry of Supply's RDF Applications Committee, as well as of the "MAUD Committee," which had been established in April 1940 to explore the prospects for an atomic weapon development program. In addition to his committee work, from 1939 he had begun devoting his laboratory talents to the design of bomb sights at the Instrument Department of the Royal Aircraft Establishment. His work there included activities such as modifying sights using an artificial horizon manufactured by the Sperry Corporation to keep the sights pointed at the anticipated location of bomb impact even as a plane descended, ascended, turned, and rolled.[4] This experience made him a sensible choice to work on AA Command's Sperry and Vickers predictors.

When Blackett arrived at AA Command, he assembled a team of academic scientists (including Hill's son), which became known as the Anti-Aircraft Command Research Group, or, more informally, "Blackett's Circus." The first several months of their effort were primarily spent making certain that guns were being properly calibrated and adjusted to deal with the terrain and obstructions specific to their locations. The greatest problems, though, were in the mechanics of the predictors themselves. Radar operators tracking the course of enemy bombers were prone to waver back and forth over the true bearing.[5] The predictors, which smoothed the tracks over only a few seconds, took the operators' inputs as the true track of the bomber and then further amplified their wavering through mechanical delays in adjusting the gun's position. The result was guns aiming "in almost every possible direction except that in which the target was flying." Prior to Blackett's arrival, AA Command had already decoupled their predictors from gun-laying radar sets, and introduced old-fashioned plotting methods and range tables to find a temporary solution to the problem.

Some of the work of Blackett's group entailed finding the best way of using this roundabout procedure, but they also decided to have the Sperry predictor "amputated" to accept tracks smoothed over longer time spans. The Vickers predictors could not be so modified. Fortunately, with Vickers predictors, guns were aimed manually to follow the movements of an indicator needle that responded to radar tracks. The object for the gun operators was to keep the needle from wandering by adjusting the gun to mimic the needle's movements. However, savvy gun operators ignored twitches in the needle induced by radar operators and adjusted the gun's aim smoothly. Studies of the method were made, and new rules were issued to gun operators which instructed them to ignore "spurious movements of the needle and to respond to those which genuinely indicated the behaviour of the target."[6]

After Blackett departed the AA Command in March 1941 to become a scientific adviser at Royal Air Force Coastal Command, the group grew rapidly. During its growth, its efforts remained focused on problems of using radar technology in operations. Work was split between helping military personnel use the equipment, and studying how the equipment was used and how it performed under combat conditions—what the group's leaders called "real operational research."[7] In this context, "real" OR referred to the ability of the work to complement actual engineering work, which tailored device designs to the anticipated conditions in which a device was to be used. Ordinary design work, for instance, could take into account questions such as how a gun of finite maneuverability would respond to adjustments in its aim. Without field data, though, there was no good way to answer questions such as how a device could track a target without also responding to signals introduced by human trackers. Understood in this sense, "real" OR also provided an important counterpart to theoretical work done by L. B. C. Cunningham and his Air Warfare Analysis Section.[8]

In the short run, this sort of OR work could help field personnel devise new tracking and plotting methods, training regimens, and user manuals, in order to cope with the innate limitations of their equipment. The work also helped to devise the most meaningful metrics of equipment performance and of the effectiveness of different techniques of using it. However, in the long run, OR could also help engineers determine what the most

Table 5.1

Major radar research facilities in Britain during World War II. Facility names date from 1941, but evolved prior to and throughout the war.

SERVICE (SUPPLY MINISTRY)	PRIMARY RADAR R&D FACILITY (as of 1941)
Royal Air Force (Ministry of Aircraft Production)	Telecommunications Research Establishment (TRE)
Royal Navy (Admiralty)	Admiralty Signal Establishment
Army (Ministry of Supply)	Air Defence Research and Development Establishment (ADRDE)

effective specifications were likely to be for future equipment designs. Because of this long-run value, the group was soon removed from AA Command's authority and placed under the authority of the Ministry of Supply's Air Defence Research and Development Establishment (ADRDE), which was responsible for developing radar devices for use by the Army (see table 5.1). In the summer of 1941 the group was briefly renamed the Petersham Research Group, after its location in southwest London. It then became the ADRDE Operational Research Group that autumn.

6   Scientific Advisers and Operational Research in the British Military Services

At the beginning of World War II, scientific advisers and advisory panels were already commonly used to support the work of military research and development establishments by reviewing their progress, suggesting new ideas, and helping foster contacts with university and industrial scientists and engineers. Patrick Blackett's Anti-Aircraft Command Research Group was unusual in that, before it was transferred to the Air Defence Research and Development Establishment, it reported directly to an officer in the military chain of command. This novel arrangement had been made because the Battle of Britain exposed a need for emergency augmentation of the ordinary liaison channels between equipment developers and the military's field personnel. However, some scientific and military figures felt there was a systematic need for such enhanced liaison. Their advocacy led to a whole series of new advisory posts within the military hierarchy, of which Blackett's was an early example, but not the first.

The first such post was created in the Air Ministry in November 1939 when Henry Tizard (figure 6.1) was named the Scientific Adviser to the RAF's Chief of the Air Staff (CAS). At that point Tizard had been a fixture within state R&D circles for over two decades. Trained as a chemist, he rose during World War I to become the head of the British Army's Experimental Flying Establishment. After the war, he entered the peacetime civil service to become deputy secretary of the state's new Department of Scientific and Industrial Research (DSIR), an organization dedicated to civilian research and development. At DSIR, among other duties, Tizard maintained liaisons with the military's research establishments. He was promoted to the post of DSIR secretary in 1927. He then moved out of the civil service upon being named the rector of Imperial College in 1929.[1] However, he maintained a prominent role in state research throughout the 1930s. In 1933 he became chairman of the government's Aeronautical Research Committee, on which he had long served as member. In late 1934 he also became

Figure 6.1
Sir Henry Thomas Tizard, 1942. Photograph by Henry Coster. © National Portrait
Gallery, London.

the chairman of the Air Ministry's new Committee for the Scientific Survey of Air Defence (CSSAD), which, as we have seen, also included both A. V. Hill and Patrick Blackett among its members.[2]

As chair of the CSSAD, Tizard encouraged intensive liaison between the Air Ministry's new radar R&D effort and the RAF facilities where new radar tracking and fighter interception tactics were developed. In September 1939 an entire group of researchers took up full-time work at the headquarters of RAF Fighter Command at Stanmore, just northwest of London, to assist in the integration of radar into combat tactics. The group soon became known as the Stanmore Research Section. Thus, when Tizard was named scientific adviser to the CAS in November, it was effectively a culmination of his earlier work on radar. What the new post represented was an *official* recognition of the need for scientific personnel to advise on what developments in research were apt to be relevant to the RAF's "tactical requirements." Tizard's numerous contacts throughout academic, industrial, and government research laboratories and establishments ensured he would be privy to a wide range of technical developments. To ensure he was privy to the tactical side of the equation, he was granted a staff of two military officers who would liaise on his behalf with the RAF's three home commands: Bomber, Fighter, and Coastal Command.[3]

The need for the sort of liaison that Tizard's position provided only increased in May 1940 when the Ministry of Aircraft Production (MAP) was separated from the Air Ministry (see table 6.1). But this event coincided with Winston Churchill becoming Prime Minister, which elevated Churchill's friend and adviser, Oxford physicist Frederick Lindemann, to the status of de facto scientific adviser to the military services. Lindemann had already been a member of Tizard's CSSAD in 1935, but had conflicted badly with the other members and had been maneuvered off the committee within a year of his joining. Thus, in June 1940, when Tizard attended an Air Ministry meeting to which he had not been invited and found Lindemann discussing research priorities with the Secretary of State for Air, he felt it necessary to resign as scientific adviser. His survey committee, having long since served out its initial function, was dissolved as well. At the end of the summer Tizard departed on a mission to North America to exchange research advances. When he returned, he was given a position as an independent adviser within MAP. Although he would remain influential within the Air Ministry and RAF, his official mandate to advise military decision makers had disappeared.[4]

Meanwhile, in late 1939, very shortly after Tizard had accepted his advisory position, Robert Watson Watt[5] was named to another new Air

Table 6.1
Nomenclature relating to the British military services, the government's service and supply ministries, and their heads.

CABINET MINISTRY	POLITICAL HEAD	ARMED SERVICE	SERVICE HEAD	SUPPLY MINISTRY
War Office	Secretary of State for War	Army	Chief of the Imperial General Staff	Ministry of Supply*
Air Ministry	Secretary of State for Air	Royal Air Force	Chief of Air Staff	Ministry of Aircraft Production**
Admiralty	First Lord	Royal Navy	First Sea Lord	N/A***

*Split from the War Office in July 1939. **Split from the Air Ministry in May 1940. ***The Admiralty maintained control over its supply bureaucracy.

Ministry post, Scientific Adviser on Telecommunications (SAT). Watson Watt had been a radio engineer with the National Physical Laboratory before being called on by the Air Ministry in 1935 to direct their new radar R&D effort. Since this work had been regularized by 1940, his move to the SAT position was supposed to free him to advise the RAF, and also Dominion and colonial governments, on the adequacy of their radio and radar systems. He could also recommend new research and development projects to the Air Ministry's Director of Production in view of any field requirements he might discover.[6] In September 1940 Watson Watt's position was officially transferred from the Air Ministry to MAP. However, the RAF's Assistant Chief of the Air Staff for Radio, Air Marshal Philip Joubert de la Ferté, insisted that he continue to spend most of his time at the Air Ministry.[7] Watson Watt, who was more than willing, borrowed three scientists from MAP's Telecommunications Research Establishment (TRE) to function as "operational research officers" (OROs).

At that point the term "operational research" had already been intermittently used during the radar development effort to distinguish research related to implementation in field operations from the research housed in the technical establishments. Analogously to Tizard's military assistants, Watson Watt's OROs were supposed to liaise with the three home commands on his behalf. Watson Watt wanted the head of the aforementioned Stanmore Research Section, an industrial researcher named Harold Larnder, to serve as his ORO to Fighter Command. The young academic chemist John Kendrew, who was already a liaison for the TRE to Coastal Command, would become the ORO there. Finally, A. O. Rankine, an established academic physicist, would become the ORO with Bomber Command.[8] In January 1941 Joubert authorized the ORO posts, and put not only Larnder but also the entire Stanmore Research Section under Watson Watt's control.[9]

Meanwhile, shortly following Blackett's September 1940 appointment as scientific adviser at Anti-Aircraft Command, A. V. Hill was named to a new Scientific Advisory Committee, which reported to the Cabinet on matters of civilian science. Though the committee had no real influence over military matters, Hill was still able to use it as a platform to push additional bureaucratic reforms within the military.[10] In January 1941, just as Watson Watt was setting up his OROs, Hill proposed through the committee's political chair that another scientific adviser be posted at RAF Coastal Command. His main concern would be the application of aircraft-to-surface vessel (ASV) radar to the hunting of the German U-boats, which were increasingly threatening transatlantic supply routes, and thus the entire war effort.[11]

Hill had the prominent Cambridge experimental physicist John Cockroft in mind for the new post. Cockroft had just returned from America as part of the 1940 Tizard mission and was presumably free for a new assignment. However, Cockroft's wartime employers at MAP had promised him to the Ministry of Supply where he was to become Chief Superintendent of the Air Defence Research and Development Establishment.[12] Meanwhile, the Air Ministry consulted Watson Watt about the idea. Watson Watt, unaware of Cockroft's unavailability, was aghast that such an "exceptionally good physicist of strong personality, exceptional energy and drive" should be assigned to be an independent adviser to work on a single problem within a single section of a single defense service. He thought that employing Blackett, another skilled and driven physicist, on the particular technical problems afflicting anti-aircraft guns was already "anomalous and embarrassing," but he allowed that it was "partially justified by the absence of any other provision for operational research in the War Office field."[13] Here Watson Watt was evidently identifying Blackett's work with that of his own assistants. To his mind, it was better to take advantage of the breadth of knowledge possessed by elite scientists than to concentrate their attention on particular technical problems, however pressing.

Of course, outside of his concern for the appropriate use of elite physicists, Watson Watt was also acutely aware that, by sending Cockroft to address issues of radar implementation in Coastal Command activities, the Air Ministry would be creating a position very close to the work he had John Kendrew doing there as his ORO.[14] So, Watson Watt crafted an alternative proposal that defined the new advisory post around the *tactical* problem of antisubmarine warfare, and left the *technical* issues of aerial radar to Kendrew. He recommended that Blackett himself be brought in on loan for "a few weeks" to serve on a "roving commission," and to report on the situation, noting, "Blackett is an ex-Navy man."[15] In his teenage years, Blackett had served in the Royal Navy in World War I, which meant he had a chance to include the often-prickly Admiralty in his studies, which further defined his position in a way that did not threaten Watson Watt's RAF-specific bailiwick.[16] This plan immediately began to go awry for Watson Watt. The Admiralty was in no rush to make use of a MAP employee assigned to the RAF, and Blackett's position rapidly came to be regarded as an RAF Coastal Command post. Then, in June, the Commander-in-Chief of Coastal Command was replaced by Watson Watt's main military patron, Air Marshal Joubert, who quickly established a strong rapport with Blackett. Blackett stayed for months rather than weeks, and Kendrew was quickly integrated into a new group of scientists that Blackett began to gather around him.

Meanwhile, in May 1941, Watson Watt made an ambitious move to establish his dominance over scientific advice in the Air Ministry, submitting a plan to the Vice Chief of the Air Staff (VCAS) Wilfrid Freeman to elevate himself to a high Air Ministry position to be called Scientific Adviser to the Air Council (SAAC, "despite the humorists").[17] This position would deal not only with the deployment of radar and telecommunications systems, but with the deployment of all RAF technologies. His control over his operational research officers was key to the plan. He pointed to the success of the Stanmore group—newly renamed Operational Research Section, Fighter Command—as a boon not only to the implementation of radar, but to the planning of operations at Fighter Command as well. The data that the section collected on all facets of the fighter interception problem, it turned out, could also be used to gauge the strengths and weaknesses of both British and German strategies, tactics, and technologies.[18] Watson Watt now proposed to build similar groups of OR scientists in all the other RAF commands, as well to take authority over L. B. C. Cunningham's Air Warfare Analysis Section and the Air Ministry's Assistant Director of Scientific Intelligence, Oxford physicist R. V. Jones. Watson Watt's new scheme received Tizard's blessing and was approved by Freeman. However, after meeting criticism from the Permanent Under Secretary for Air, it stalled.[19]

Watson Watt's increasingly complicated bureaucratic balancing act soon began to collapse. By June 1941 his ORO at Coastal Command, John

Kendrew, had become the "Officer-in-Charge of ORS, Coastal Command," and there was still a nominal separation of duties between his staff and the scientific adviser, Blackett. However, the concept of operational research shifted meaning that summer in a subtle but crucial way. By Watson Watt's definition OR still applied primarily to the study of technology in use in the field, which could have a subsidiary use in command decision making.[20] However, the work of the Coastal Command ORS increasingly turned to the production of studies directly in support of Blackett's advisory work on antisubmarine tactics, rather than Watson Watt's advisory role on telecommunications and radar equipment.

This shift in the OR Section's work does not seem to have been a conscious maneuver for power on Blackett's part. The nontechnological nature of his advisory role had originally been Watson Watt's suggestion, and OR studies' radically new *function* of supporting the adviser to a command officer produced no clear shift in their *content*. OR work still entailed the collection of data pertinent to particular technical concerns. Now, however, some reports began to bear little relation to the technical problems of Coastal Command, including the OR Section's first official report, an intelligence-like digest of enemy aircraft activity detected by radar in the western approaches to Britain in May 1941.[21] Even a year later, one researcher at MAP's Telecommunications Research Establishment had not appreciated the change and complained that many of the studies they received from ORS Coastal Command did not address their radar-oriented concerns.[22]

By July 1941, though, Henry Tizard had already picked up on what was afoot. After Watson Watt's proposal stalled, he had put deeper thought into the question of operational research and had decided that he could not support the scheme after all. He laid out his thinking in a memorandum to Freeman and the Chief of the Air Staff on July 17, which ought to be regarded as a major turning point in the early history of OR.[23] First, Tizard had recognized that Blackett's and Watson Watt's work of integrating field research with military planning was already taking place in a less formalized way throughout the services. Seeking insight into how military officers collected information to inform their decision making, he had asked the Air Ministry to supply him not only with all instances of "the type of operational research that is now being carried out at the Commands and by the various sections of the Air Staff," but also "the operational statistics that are now being collected in any form."

Although his survey remained incomplete, Tizard now felt confident in suggesting that "the object of operational research is to endeavour by scientific study of the results of operations against the enemy to arrive at methods for using existing military equipment in the most economic and effective way, and to guide future technical and operational policy." This definition led him to the important conclusion: "It must therefore cover a wide field and it should be, and no doubt is, regarded as a normal function of a well organized force. If we accept that broad definition it follows that the executive responsibilities should be entrusted to high officers of the Royal Air Force and not to civilians or independent scientific men." Although "scientific study" by definition, OR was effectively identical to longstanding military planning practices. For this reason, Tizard felt it should be directed by military planners and focused on their needs. Nevertheless, he was quick to point out that "the work itself demands a scientific training and outlook," and suggested that personnel ought to be recommended by the administrators of MAP's civil service research corps.

This rethinking of what OR was and who should be responsible for it had consequences for Watson Watt's ambitions. Because OR, in Tizard's reckoning, was of primary use to military planners, and of only secondary use to technical advisers, Watson Watt, as a technical adviser, could not appropriate it for himself. Responding to another OR scheme Watson Watt proposed on September 1, Tizard brushed him off directly, writing to Freeman, "The chief reason why Watson Watt is busying himself so much on the organization of operational research (at which I do not think he is very good) is that he is not being properly fully used in ways which are appropriate to a scientific adviser on telecommunications. I have suggestions to make to you which I think will enable him to pull his weight, which is considerable, in appropriate directions."[24] Tizard never addressed the similar implications his memorandum had for Blackett's position at Coastal Command, possibly because he still thought that Blackett's position was temporary, or possibly because he wanted Blackett to keep it.

In the meantime, the memorandum provoked a remarkable reaction from the military hierarchy. On July 23, 1941, Rankine quit his post as Bomber Command ORO, citing disagreements with Watson Watt.[25] On August 8, the Commander-in-Chief of Bomber Command, Air Marshal Richard Peirse submitted to the Permanent Undersecretary a list of pressing tactical problems:

(i) The method of operating enemy night fighters and searchlights.
(ii) Location of enemy searchlights and A. A. defences.
(iii) Location of enemy night fighter zones.

(iv) Routeing of our own bombers to avoid enemy night fighters and A. A. defences.
(v) Method of approach to the target in order to obtain a good bombing run with the minimum interference from searchlights and flak.

Some of these problems could be handled through traditional avenues such as intelligence gathering, whereas others, such as bomber routing and the methodology of the bombing approach, were simply dependent on assessing plans against mission results. Peirse observed that Bomber Command had already made progress on these problems, but felt "that we are still far from making the best use of the data that is available."[26] Having been shown Tizard's memorandum, he argued that these problems required "a large amount of research and the analysis of the data needs a trained scientific mind in order to draw from it the correct deductions," and asked to be assigned an Operational Research Section. Tizard wasted no time in helping to build up a staff around a small core of technical specialists already at Bomber Command.[27] By the end of the month, he and MAP's Director of Scientific Research, David Pye, had selected Basil Dickins, another civil service scientist, to lead the construction of the ORS. Dickins remained there until the end of the war.[28]

Adding fuel to the OR fire, academic crystallographer J. D. Bernal, then undertaking a survey of the effects of German bombing for the Ministry of Home Security, introduced Frederick Lindemann to Bomber Command's photographic analysts and showed him first-hand the inaccuracy, on the scale of miles, of nighttime bombing raids on Germany.[29] Lindemann ordered his assistant, David Bensusan-Butt, to undertake an examination of Bomber Command's photographic reconnaissance. Butt's report, issued August 18, 1941, was largely indistinguishable from an OR report. It detailed the statistics of bomb plots and likely causes of error, and called for further analysis of the problem, as well as the establishment of a statistical section at the command. Butt noted, though, that his own abilities were limited by his lack of "skill at interpreting photographs and technical knowledge of bomb damage."[30]

The Air Ministry considered the report "fair," but added mitigating context to the statistics, pointing to the inexperience of bomber crews, the fact that crews told to photograph their missions were ones suspected of inadequate performance, and the exceptionally bad weather during the period the study examined. They also pointed out "on the authority of Sir Henry Tizard" that the Germans exhibited a remarkably similar degree of inaccuracy in striking English urban districts.[31] Still, they were quick to add that these caveats were "not put forward in any spirit of self-satisfaction and complacency." The techniques of night bombing had been under study for years, and the recent challenges of the war "served to give new zest and determination to our studies of means to overcome them." They acknowledged Butt's call for "further operational research"—Butt had not actually used the term—and pointed to the establishment of their new ORS, which was news that Lindemann warmly welcomed.[32] The Ministry also promised to develop new means of navigating by radar.[33]

Meanwhile, the RAF's Assistant Chief of the Air Staff for Intelligence informed high ministry officials and Tizard that he had for some time been considering conducting a survey of the German defense system, and that he believed there must be a large amount of "uncorrelated" information already available. He argued that it was important to work out how the various elements of enemy defenses—searchlights, anti-aircraft guns, night fighters, and radiolocations—impacted the effectiveness of the German defense network. He remarked that a "scientific brain" would be best able to "pull together all the various threads and produce a comprehensive picture," and had already discussed the issue with R. V. Jones. Now he was anxious to ally Jones's efforts with the new Bomber Command ORS.[34]

Tizard, of course, greeted the offer with open arms. In September he began concocting a scheme to coordinate the efforts of Jones, the various OR sections, the Air Warfare Analysis Section, and the Ministry of Home Security's bombing effects survey. John Kendrew was dispatched to the Middle East, by way of Malta, to begin doing OR work there. By October the RAF had formalized its institutional arrangements for OR. An Operational Research Centre was created to collect and distribute OR reports. To help plan research projects and the further institutional development of OR, an OR committee was established comprising the heads of the OR sections, high military officers, and other scientists. The OR committee held its first meeting at the end of October 1941.[35]

At some point in the fervor to build and coordinate OR, Watson Watt was included among the other scientific personnel who should attend the new committee's meetings. When he discovered that he "was added, as an afterthought, to the 'reserve team,'" he issued a "mild protest" to the RAF's Deputy Chief of Air Staff (DCAS), and enclosed his plan that Freeman had supported in May and the January order from Joubert establishing his original ORO system. He used these documents to explain his surprise that the Air Ministry had "summarily disposed of a large block of staff still 'operationally under the command' of SAT." The DCAS replied, apologizing, "I am afraid I had no idea that the Operational Research Sections were 'operationally under your command.'" He wrote that he had been working on the matter with Tizard and David Pye, and observed that he could not be responsible for dealing with the "domestic relations of the various authorities" among the scientists. He suggested that if Watson Watt was dissatisfied he should take the matter up with them.[36] Thus Watson Watt's scheme to keep Blackett off his territory and advance his own career backfired spectacularly. His association with OR ended with its rise as an important feature of the military bureaucracy.[37]

7 "Not Yet Enough": The British Rhetoric of Science and Its Coordination

Aficionados of the history of the British state often concentrate their attention on moments of major reform, such as the creation and dissolution of government ministries, or the reorganization of the civil service. By comparison, the reforms chronicled in chapters 5 and 6 were comparatively minor, focusing on the creation of posts and bodies with only advisory powers, albeit reporting to a high level within the military chain of command. These reforms were proposed by a small group of scientists and engineers, most of whom were only temporarily working for the state. They were justified by specific wartime shortcomings in military suppliers' ability to design technologies that met the services' pressing requirements, and in the services' ability to formulate their technical requirements clearly and to develop effective techniques and tactics for using the equipment available to them. The military and the established civil service accepted these justifications, and implemented the proposed reforms with little objection.

Although these developments were minor within the grand scheme of the British state, they quickly took on a profound significance to those who had pressed for them. The reformers certainly understood the specific reasons why the new posts and advisory bodies were appealing to established authorities. But, beyond that, they also understood the changes to be important developments in the more general integration of "science" into the British state and military. By linking their successful proposals to others with a less distinguished history of acceptance as all part of a longer and more onerous struggle for the cause of science, they were able to portray their successes as representing a sudden and laudable change of military opinion toward science. Adopting this narrative, in turn, allowed them to portray new suggestions as clearly progressive steps in further advancing their worthy but neglected cause. This chapter will retread the events detailed in chapters 5 and 6 from this perspective.

The British Tradition of Advocacy for "Science"

The narrative of the struggle of science in British politics and society can be traced back to at least a century before World War II. One important touchstone in this history is the mathematician Charles Babbage's 1830 book *Reflections on the Decline of Science in England, and on Some of Its Causes*. In it Babbage decried British scientific education, the ineffectiveness of the Royal Society as an institution for promoting scientific inquiry, and the relationship between British science and the British state. In fact, he spent 34 of his 212 pages criticizing as inadequate a panel of three "scientific advisers" who had been appointed to serve the Admiralty part-time after the dissolution of its Board of Longitude in 1828.[1] Babbage's polemic was part of a broader movement of scientific reformism, which led to the 1832 establishment of a new learned society, the British Association for the Advancement of Science (BA).[2] Throughout much of the nineteenth century, British technical education was compared unfavorably to state-backed educational institutions in France and the German states. These complaints led to the establishment of a series of technical colleges and institutions, which early in the twentieth century transformed into major educational centers such as the Imperial College of Science and Technology and the "red brick" universities.[3]

The expansion of British scientific and technical education early in the twentieth century coincided with an expansion in the number of state bodies responsible for funding research. The Development Commission was established in 1909, the Medical Research Council in 1919, and the Agricultural Research Council in 1931. At the same time, the state was augmenting its own research capacity. The National Physical Laboratory (NPL) was established in 1900. World War I led to the establishment of the Department of Scientific and Industrial Research (DSIR) in 1916, which oversaw its own laboratories, including the NPL, as well as a series of industry-supported "research associations." Some government ministries, such as the new Ministry of Agriculture and Fisheries, also began to conduct their own research. The military services, meanwhile, expanded their already substantial in-house research and engineering activities. The advent of radio electronics and aeronautics demanded new facilities. New and expanded facilities, in turn, spurred new organization: in the interwar years each service ministry appointed its own civilian director of scientific research (DSR) to help prioritize and oversee work.[4]

Amid this rapid institutional development, a new generation of scientist advocates also arose. The British Science Guild was founded in 1905, the

Association of Scientific Workers (AScW) was founded in 1918, the BA established a Division for Social and International Relations of Science in 1938, and the Parliamentary and Scientific Committee lobbying group was founded in 1939. Publicly minded scientists took their opinions to the pages of journals such as *Nature*, participated in radio programs, and wrote books and articles for the general public. Private clubs such as the venerable Athenaeum on Pall Mall in London, and the "Tots and Quots" dining club, convened by urbane young zoologist Solly Zuckerman, brought elite scientists, political leaders, and other intellectuals together in unofficial settings to discuss the issues of the day.[5]

Most of this advocacy was limited to generic calls for new academic research and the expanded application of research to society's problems, and much of it was directed at the scientific community rather than outward. Interest in the "social relations of science" ranged across the political spectrum, but it particularly appealed to scientists on the Marxist left, who often expressed a "frustration of science." This frustration related, first, to the academic scientific community's purported disinterestedness and self-isolation from social and political issues, and, second, to the social and political ends to which most scientific labor was in fact being put.[6] J. D. Bernal was an important figure in this line of thought. His widely read 1939 book *The Social Function of Science* surveyed the distribution of scientific effort among various parts of the British economy, universities, and state. He complained that, far from being concerned about the domination of most technical research by militaristic and capitalist interests, the bourgeois academic elite treated science as an "amusing pastime." He sniffed, "It has all the qualities which make millions of people addicts of the crossword puzzle or the detective story."[7]

When the war began in September 1939, scientist activists' traditional focus on social welfare easily translated into suggestions that scientists could make vital contributions to the wartime economy and to the conduct of the war itself. Zuckerman's Tots and Quots group, which included Bernal, made just this argument in its book *Science in War*, which was rushed to publication in the summer of 1940.[8] But, once activists became ad hoc members of the scientific civil service, they ceased to argue that science was being employed insufficiently. Instead, they argued that scientists were being employed *improperly*. In October 1939, for example, the twenty-six-year-old physicist Bernard Lovell joined the radar development effort, eager to make his contribution. However, he was frantically upset by what he regarded as the poor working conditions and management in the Air Ministry's radar research facility. He confided to his peacetime supervisor Patrick Blackett, who was prominent among the socialist scientists, "You may remember from my book"—*Science and Civilization*—"how keenly I felt the disillusionment and frustration in science. But when even frustration is being frustrated by inefficiency, it is very hard to sit under!"[9]

Bureaucratic Reformers

When academic scientists began to join war research full-time and en masse, Henry Tizard and A. V. Hill were already attempting to reform the government science bureaucracy from within. Neither, though, was an activist reformer. Both established insiders, they were motivated by their practical concerns that academics would not be easily integrated into military research, that problems in the coordination of research and development projects would lead to wasted resources and the needless duplication of effort, and that equipment developers would not develop designs in accord with the needs of equipment users.

Although both Tizard and Hill had experiences with research and development reaching back to World War I, their key point of reference in advocating bureaucratic reform was their experience in pre–World War II radar development.[10] The Air Ministry had established its ad hoc Committee for the Scientific Survey of Air Defence in late 1934 in response to widespread concern that there was no feasible defense against enemy bombers. Chaired by Tizard, the committee comprised Hill and Blackett, as well as Air Ministry DSR Harry Wimperis, and his assistant A. P. Rowe serving as secretary. As academics, Hill and Blackett were both impressed by Tizard's work overseeing the program, and particularly by his ability to marshal the expertise of researchers, officers, and pilots in guiding the development effort and integrating new equipment into effective daytime interception tactics.

Hill and Blackett also had the opportunity to contrast Tizard's leadership with the contributions of the abrasive Frederick Lindemann, who joined the committee in the summer of 1935. Lindemann's friend Winston Churchill was a well-known figure in British politics. At that time, though, Churchill was consigned to the Conservative Party's backbench, and had started a campaign to remake his reputation by aggressively demanding that Britain confront the Nazi threat that had arisen in Germany. Lindemann shared Churchill's views, and together they argued that the Air Ministry should be goaded into a large and diverse research and development effort. Lindemann was particularly disappointed that Tizard's CSSAD was created within the Air Ministry, rather than above it. When Churchill maneuvered Lindemann onto the committee, Tizard warned him against antagonizing the Air Ministry's research staff. Undeterred, Lindemann pushed new projects contrary to those researchers' opinions of their viability, and complained privately to Churchill about the CSSAD's and Air Ministry's lethargy. In the summer of 1936 Hill and Blackett quit the CSSAD in protest, seeing Lindemann's behavior as a threat to radar development. They rejoined the committee once the Air Ministry reconstituted it without him.[11]

For Hill and Blackett, the conflict with Lindemann was not simply a clash of personalities—it was an object lesson. Failing to take advantage of available expertise, and offering opinions at odds with it, was, of course, bad administrative practice. But, beyond that, it could be construed as negligence of the scientific duty to gather relevant evidence and to confront pertinent arguments. This was, furthermore, a lesson that could be applied not only to the imperious Lindemann, but also to career civil servants who did not cooperate as readily as those in the Air Ministry did. In late 1938 Tizard suggested that, with war approaching, a committee should be established to help coordinate research programs across Britain's three military services. This suggestion was rebuffed by the Admiralty's Deputy DSR, who denied the need for it. Over lunch, Tizard fumed to the Cabinet Secretary, a top-level civil servant, that the Admiralty scientists' attitude was not one that would be "adopted by really first-class scientists, who would be perfectly willing to try and learn from anybody."[12]

Tizard let the matter drop, but, at the beginning of 1939, Hill, apparently unaware of Tizard's overture, pushed, using his position as secretary of the Royal Society, for another high-level coordinating committee. He, too, met resistance, not only from the Admiralty, but also from the secretaries of the Medical and Agricultural Research Councils, and from academic physicist E. V. Appleton, who had replaced Lindemann on the CSSAD and had recently been named head of the DSIR. Their objection was that the proposed committee superseded their existing posts, and that additional bureaucracy would introduce an undue burden. But one of Hill's allies in the Cabinet Secretary's office was certain that objections reflected the "*amour propre*" of the leaders of the research establishments and their desire to keep their "skeletons in the cupboard" well hid. He expressed his amusement that Appleton, in particular, appeared to be "*plus royaliste que le roi*"—more royalist than the king—so soon after entering government service.[13] Hill's proposal collapsed toward the end of 1939, just as Tizard was appointed scientific adviser to the RAF's Chief of the Air Staff.

When Tizard resigned his advisory post in June 1940 in view of Linde-mann's ascendancy under Churchill, he framed his resignation as a neces-sary step if the state's advisory machinery were to function properly. He wrote how he had worked "in close personal touch with the Air Staff," and how he had made it his "business to study the operational needs and dif-ficulties of the Commands." But he continued that it was not appropriate for conflicting sources of advice to exist. To avoid "confusion" and the possibility that technical advice would not be "properly co-ordinated," he was compelled to step down.[14] Tizard would later regain some influence. At the time, though, his allies regarded the dissolution of his post as a grievous blow to the responsible integration of science into the war effort. In August 1940, when Hill arranged to install Patrick Blackett as an adviser to Anti-Aircraft Command, he reflected, "You would be able to do for the AA Command what Tizard ought to have been given a chance to do for the RAF."[15] Around the same time a private memorandum began circulat-ing among disgruntled scientists and their allies in the military and civil service detailing Lindemann's follies as an adviser.[16]

Once Blackett was successfully installed at Anti-Aircraft Command, Hill used the existence of the position to argue once again for his floundering Royal Society proposal.[17] He had already threatened, following the resigna-tion of Tizard, to make his grievances public using his new position as a Member of Parliament.[18] However, this time his demands were met. In September 1940 the government established a Scientific Advisory Commit-tee, which reported to the Cabinet and was chaired by the politician and former Cabinet Secretary Lord Hankey. This new committee had no author-ity to advise on military research and development, and it was not espe-cially influential in civil research.[19] However, coming as it did on the heels of *Science in War*, it was seen by the government as a good investment, "if only to keep the scientific people quiet," as Alan Barlow, the Undersecre-tary of the Treasury, privately observed to Hankey.[20]

The committee's creation, however, only marked a point where out-sider scientific advocacy and the insider bureaucratic reformism began to cohere into a single cause. Hill, notably, was by no means diverted from reform by his new committee assignment. As he wrote to Tizard in March 1942, he was not content to put his trust in the "rather casual cross link-ages by 'co-ordination' or 'liaison'" that already existed.[21] Throughout 1941 and much of 1942, Hill attempted to embarrass the government into implementing more coherent bureaucratic structures. In a January 1941 speech, he inveighed against "the danger that science will be planned by administrators in offices instead of by young men with their sleeves rolled up, in laboratories or workshops."[22] But in July 1942 he blasted the Army's tank designs in *The Times*, complaining, "There is no central technical staff or individual to advise the Cabinet or the Chiefs of Staff, directly on the scientific and engineering aspects either of operations or production, or to ensure, on their behalf, that design and development are efficient and far-seeing. All such functions are left to departments."[23] These complaints were not necessarily paradoxical. Hill's goal appears to have been to try to emulate an academic model, where scientists used journals and other means to survey ongoing research so as to choose projects appropriate for their skills. An academic system was, of course, impossible in a compartmentalized and secretive wartime bureaucracy. His hope was that it could be approximated through a formalized bureaucratic apparatus capable of surveying the knowledge and needs in its various parts and conveying them to the people who could make the best use of the information.[24]

When the RAF instituted its system of operational research sections in the summer and fall of 1941, the development fit neatly within this discursive framework, and extended its essential principles from research and development to military planning. Blackett, who was initially only a rider on this wave, became a full-fledged bureaucratic reformer at this time. In a widely circulated October 1941 memorandum, entitled "Scientists at the Operational Level," he explained how OR groups could help interpret "the operational facts of life to the technical establishments," and, conversely, "the technical possibilities to the operational staffs." He therefore concluded, "The head of the Operational Research Section should be directly responsible to the Commander-in-Chief [of a command] and may with advantage be appointed as his scientific adviser." But operational research was not simply a new mechanism of liaison used to better inform scientific advisers' advice and military decision making. Members of operational research sections went a step beyond that by actually researching the conduct and results of military operations, which provided a stronger evidentiary basis for advice given. As Blackett wrote, "The atmosphere required is that of a first-class pure scientific research institution, and the calibre of the personnel should match this." This imperative, of course, went hand in hand with proper liaison. He immediately added, "All members of an Operational Research Section should spend part of their time at operational stations in close touch with the personnel actually on the job."[25]

The rise of operational research galvanized scientists' public activism. Just as the RAF was formalizing its new system of OR sections in September

1941, the BA's Division of Social and International Relations of Science held a conference on the lofty subject of "Science and World Order." Hill and J. D. Bernal both spoke in a session on the topic of "Science and Government." In his speech, entitled "The Function of the Scientist in Government Policy and Administration," Bernal used the war experience to that date to argue for planning scientific research and for basing policy planning on the rigorous collection and analysis of information. "With such an integrated body of information, research, development, execution, and control," he observed, "we have the backbone of a scientific administration, one which is scientific through and through and not merely by the addition of a few eminent scientists." He deferred to the discussion of the "importance of operational research" in Hill's speech, but argued that OR-type research would play an essential role in ensuring effective, scientific planning, in peacetime as much as in war.[26]

In Hill's speech, "The Use and Misuse of Science in Government," he built up the rudiments of a historical narrative conveying the uneven progress of the proper integration of science into government. He recalled the work of the CSSAD on radar as "the most important scientific development of the war." He publicly idealized Tizard's skills as a coordinator of research, development, and military needs. More than a year after Tizard's resignation as scientific adviser to the Chief of the Air Staff, Hill approvingly observed, "Sir Henry Tizard is at last in a position to give his unique knowledge and experience unhindered to the scientific needs of air warfare." Hill also discussed the need for more independent scientific advisers, more advisory teams, and for generally more contact between "scientists" and "users" of technology. He expressed his satisfaction with recent developments, but, as was his wont, he urged that it "would be wiser ... to stabilize the use of external scientific advice by some constitutional organization."[27]

Tizard, usually content to work through inside channels, also began to make a broader appeal in this period. Speaking to the Parliamentary and Scientific Committee lobbying group in February 1942, he began by reflecting on a conversation in which he had been involved:

In a group of friends I made some probably quite unnecessary critical remark; and one of them said, "Oh you scientists, you get so absurdly exasperated at any inefficiency. You seem to think this war is like a game of golf between two scratch players, and it exasperates you to see the ball in the rough.... [I]t's much more like a game of golf between two 18 handicap players. The ball is usually in the rough, and every now and then one of the players gets a good niblick shot out on to the green, scuffles down the putt and that's one hole up."

While acknowledging his meddlesomeness, Tizard did not think the integration of science into the war effort could rest content with the standards of amateur golf. He allowed that no one could "deny that the influence of science is now greater than it has ever been, and that the present Government and Parliament attach a value to the help and guidance of scientists that no previous Parliaments have ever done." Nevertheless, while praising the concrete accomplishments of Britain's scientists—what he called the "tactics" of science—he felt that "strategy," i.e., coordination, was still wanting. Scientists' expertise had to be put to still better use. It could not be done by "scientists alone, working in the void, however eminent." Rather: "The safest way of reaching the right decision is to have scientists working side by side and in the closest collaboration with those who have the administrative and executive responsibility." In these environments, "the first thing that the scientist learns when he has the benefit and privilege of such collaboration is that he has a lot to learn." He recognized the progress made at the RAF, and even noted that the cause had "spread to other Departments. But," he observed, "not yet enough."[28]

8 Operations Research and Field Liaison, 1941–1945

Great Britain

The coherence and widespread acceptance of operational research sections by top Royal Air Force officers encouraged bureaucratic reformers, led by Henry Tizard, to leave behind ad hoc suggestions for reform and to concentrate their energies on the new system's expansion. On October 31, 1941, the first meeting of the RAF's new OR committee was held to clarify OR's role with respect to Ministry of Aircraft Production researchers, the Air Warfare Analysis Section, Robert Watson Watt's advisory position, RAF intelligence, the RAF's signals and armaments staff, and the RAF's "other statistical organisations." The meeting also addressed how the OR model could be expanded to the RAF's field commands.[1]

The reformers, however, were already looking beyond the RAF, and made a successful overture to the Admiralty. Patrick Blackett moved from RAF Coastal Command to the Admiralty in December 1941 under the title of Chief Adviser on Operational Research (CAOR). He was joined by a handful of junior scientists, and part-time by the eminent but ailing Cambridge physicist Ralph Fowler. However, Blackett's role was not exactly what he expected. Before he arrived, the Admiralty's Director of Establishments (D of E) pointed out that it would be necessary to define OR firmly. "It might, for example, be taken to mean research into the conduct of operations," he warned with no hint of irony. Instead, the Admiralty defined OR as "the scientific investigation, in the light of results of actual operations, of the performance of existing weapons, and of the lines upon which they should develop."[2]

In the Ministry of Supply, the Air Defence Research and Development Establishment's OR Group began to diversify its work under the leadership of the commissioned South African physicist Basil Schonland, who was named its superintendent in August 1941. In the first half of 1942,

Table 8.1
Operational Research Sections in the Royal Air Force

September 1939	**ORS Fighter Command**
September 1939	Stanmore Research Section established
January 1941	Harry Larnder named Operational Research Officer for SAT
June 1941	Renamed ORS Fighter Command
1943–1944	Temporarily renamed ORS Air Defence of Great Britain
March 1941	**ORS Coastal Command**
January 1941	John Kendrew named Operational Research Officer for SAT
March 1941	Patrick Blackett named Scientific Adviser to C-in-C
May 1941	Renamed ORS Coastal Command
September 1941	**ORS Bomber Command**
January 1941	A. O. Rankine named Operational Research Officer for SAT
August 1941	Rankine resigns
September 1941	ORS Bomber Command established
September 1941	**Operational Research Center**
January 1944	Renamed RAF Deputy Directorate of Science
February 1942	**ORS Middle East**
July 1942	**ORS Army Co-operation Command**
June 1943	Renamed ORS Tactical Air Force
February 1944	Absorbed into ORS Allied Expeditionary Air Force
October 1944	ORS 2nd Tactical Air Force becomes independent
October 1942	**ORS Air Headquarters India**
November 1943	Renamed ORS Air Command South-East Asia
April 1943	**ORS Northwest African Air Forces**
December 1943	Renamed ORS Mediterranean Allied Air Forces
November 1943	**Scientific Adviser to the Air Ministry appointed**
December 1943	**ORS Allied Expeditionary Air Force**
October 1944	Disbanded
March 1944	**ORS Flying Training Command**
September 1944	**ORS Bombing Analysis Unit**
October 1944	**ORS No. 38 Group**
	(previously part of Allied Expeditionary Air Force)

Sources: Air Ministry 1963; Waddington 1973; Kirby 2003b; and Wakelam 2009.
Abbreviations: SAT, Scientific Adviser on Telecommunications (Robert Watson Watt); C-in-C, Commander-in-Chief.

Table 8.2
Operational Research in the Ministry of Supply

September 1940	**Anti-Aircraft Command Research Group**
November 1940	Anti-Aircraft Wireless School branches off
April 1941	Authority transferred to Anti-Aircraft Wireless School
July 1941	Renamed Petersham Research Group
August 1941	Transferred to ADRDE in Ministry of Supply
September 1941	Renamed ADRDE (Operational Research Group)
1940–1941	**Army Operational Research Sections 1 and 2**
	AORS 1 (A. A. Radar and Operations), originally ADRDE (ORG)
	AORS 2 (Coastal Artillery and Radar, Misc. Radar)
Spring–Summer 1942	**Army Operational Research Sections 3 to 5**
	AORS 3 (Signals)
	AORS 4 (Tanks, Anti-Tank, Field Artillery)
	AORS 5 (Airborne Forces)
January–February 1943	**Army Operational Research Sections 6 to 10**
	AORS 6 (Infantry)
	AORS 7 (Lethality of Weapons)
	AORS 8 (Mines, Obstacles, Optics, and Misc.)
	AORS 9 (Time and Motion Study)
	AORS 10 (Bombardment)
February 1943	**Army Operational Research Group**
	AORG oversees work of the AORSs

Sources: Austin 2001; Kirby 2003b; and typescript histories in National Archives of the UK: Public Record Office, WO 291/1286–1301, especially "Operational Research in the British Army, 1939–1945," October 1947, in WO 291/1301. Abbreviation: ADRDE, Air Defence Research and Development Establishment.

ADRDE(ORG) expanded to include five sections devoted to different problems, but still mainly involving the integration of technologies into operations. By the time it became independent of the ADRDE in February 1943 and was renamed the Army Operational Research Group (AORG), Schonland's organization comprised ten sections. Meanwhile, the Director Royal Artillery—who had previously been a staff officer at Anti-Aircraft Command—extolled OR to the War Office, which led to the establishment of a Scientific Adviser to the Army Council (SA/AC) in May 1942. As with Blackett's CAOR post—and in spite of Tizard's recommendation that the War Office should invest in OR and leave technical advice to the Ministry of Supply—the SA/AC was tasked primarily with monitoring weapons effectiveness and suggesting priorities for future development. The director

Table 8.3
Operational Research in the British Army

September 1940	**Anti-Aircraft Command Research Group**
	See table 8.2 for further history
April 1942	**SD(6), a small group in the Middle East**
July 1943	Disbanded
May 1942	**Scientific Adviser to the Army Council appointed**
May 1943	**No. 1 Operational Research Section (Italy)**
July 1943	**No. 10 Operational Research Section (11 Army Group)**
July 1943	Established as No. 1 ORS
February 1944	Renamed No. 10 ORS to avoid confusion
August 1943	**No. 2 Operational Research Section (21 Army Group)**
February 1944	**Operational Research Section (Australia)**
October 1944	**No. 11 Operational Research Section**
Date unknown	**Artillery Operational Research Section**
Date unknown	**Biological Research Team**

Sources: Austin 2001; Kirby 2003b; and typescript histories in National Archives of the UK: Public Record Office, WO 291/1286–1301, especially "Operational Research in the British Army, 1939–1945," October 1947, in WO 291/1301. Abbreviation: ADRDE, Air Defence Research and Development Establishment.

Table 8.4
Operational Research in the Admiralty

December 1941	**Chief Adviser on Operational Research**
May 1944	Renamed Naval Operational Research Directorate

Source: Kirby 2003b and additional archival sources cited in this chapter.

Table 8.5
Operational Research in the Canadian and Australian Military Services

1942	**Operational Research Section, Royal Canadian Air Force**
1943	**Operational Research Directorate, Royal Canadian Navy**
1944	**Operational Research Group, Canadian Army**
1944	**Operational Research Section, Royal Australian Air Force**

Sources: Morton 1956; Mellor 1958; and Air Ministry 1963.

Table 8.6
Operations Research in the U.S. Navy

January 1942	**Mine Warfare Operational Research Group (BuORD)**
March 1942	MWORG formalized
April 1942	**Anti-Submarine Warfare Operations Research Group (ASWORG)**
	Also known as Group M until incorporation into OSRD
October 1942	**Air Operations Research Group (AirORG, Air Intelligence Group)**
November 1943	**Submarine Operations Research Group (SORG)**
September 1944	**Anti-Aircraft Operations Research Group (AAORG)**
July 1945	Renamed Special Defenses Operations Research Group (SpecORG)
October 1944	**Operations Research Group**
	Initially encompasses ASWORG, SORG, and AAORG
October 1944	AirORG joins
October 1944	Operations Research Center created
October 1944	Amphibious Operations Research Group (PhibORG) created

Sources: Thiesmeyer and Burchard 1947; Tidman 1984; and Rau 1999; as well as Memorandum No. 2, "Operations Analysis in the U.S. Army and Navy," in U.S. National Archives and Record Administration, Record Group 218, "Joint New Weapons Committee, Subject File, May 1942–1945," Box 57; and Operations Research Group, "Summary Report," December 1, 1945, in U.S. National Archives and Record Administration, Record Group 227, "Office of Field Service: Manuscript Histories and Project Summaries, 1943–1946," Box 28. Abbreviation: BuOrd, U.S. Navy Bureau of Ordnance.

of the National Physical Laboratory, Charles G. Darwin, was appointed SA/AC on top of his other duties. An academic physicist, Charles Ellis, served as his deputy until it became clear that Darwin could not balance his obligations to the War Office and the NPL. In early 1943 Ellis was promoted to take his place.[3]

In June 1942 Tizard began holding "informal" meetings in his office, attended by the "independent advisers" from across the three services. Meeting attendees were largely likeminded scientists, and regularly included Watson Watt, Blackett, Fowler, Schonland, Darwin, and Ellis. John Cockroft attended as chief superintendent of the ADRDE. J. D. Bernal attended as a scientific adviser, first to the Research and Experiments Division of the Ministry of Home Security, then later to the Chief of Combined Operations Lord Louis Mountbatten. Scientific administrators from the Ministry of Aircraft Production also attended every meeting, but none did from any

Table 8.7
Operations Analysis in the U.S. Army Air Forces

March 1942	**Operational Analysis Group, Directorate of Air Defense**
	Operations Analysis Sections
September 1942	OAS AAF School of Applied Tactics
October 1942	OAS Eighth Air Force
December 1942	OAS AAF Headquarters
May 1943	OAS Eleventh Air Force
May 1943	OAS IX Bomber Command (disbanded October 1943)
June 1943	OAS VIII Fighter Command (disbanded October 1944)
July 1943	OAS Mediterranean Allied Air Forces
September 1943	OAS Second Air Force
October 1943	OAS Fifth Air Force (merged into FEAF in July 1944)
October 1943	OAS Thirteenth Air Force (merged into FEAF in July 1944)
November 1943	OAS Fifteenth Air Force
December 1943	OAS Ninth Air Force
December 1943	OAS XX Bomber Command
January 1944	OAS AAF India-Burma Theater
March 1944	OAS Fourth Air Force
May 1944	OAS Twentieth Air Force
July 1944	OAS Third Air Force
July 1944	OAS Fourteenth Air Force
July 1944	OAS Far East Air Forces (FEAF)
August 1944	OAS AAF Pacific Ocean Area
October 1944	OAS XXI Bomber Command
January 1945	OAS AAF Weather Wing
May 1945	OAS Continental Air Forces
July 1945	OAS 301st Fighter Wing, Very Long Range

Sources: United States Army Air Forces 1948, and Shrader 2006.

Table 8.8
Other American Operations Research Groups

March 1942	**Operational Research Group, U.S. Army, Directorate of Planning**
March 1944	**OSRD OFS, Research Section, Southwest Pacific Ocean Area**
June 1944	**OSRD OFS, Operational Research Section, Central Pacific Area**

Sources: Thiesmeyer and Burchard 1947; Shrader 2006; and Memorandum No. 2, "Operations Analysis in the U.S. Army and Navy," in U.S. National Archives and Record Administration, Record Group 218, "Joint New Weapons Committee, Subject File, May 1942–1945," Box 57. Abbreviations: OSRD, Office of Scientific Research and Development; OFS, Office of Field Service. The OFS sections were more field branches of research and development activities, but the U.S. Army preferred to designate the Central Pacific Group, based in Hawaii, as an "operational research section."

of the other services' research establishments, save for Cockroft.[4] These meetings dealt with the coordination of research and development work with the needs of the war effort, but also focused heavily on developing the network of OR groups in the British military, and diverting scientific resources into it. There was some disagreement as to whether the military services had become universally open to receiving advice on operational planning. At the seventh meeting, in September 1942, Blackett suggested that an "O.R.S. should attempt to infiltrate tactfully ... into the more general study of tactics," using technical problems as a "foothold." He believed that such infiltration was an important "measure of success" for an OR section.[5]

Blackett had evidently had some success in subverting his own group's confinement to issues relating to weapons. This success continued in October 1942 when the Admiralty appointed Charles Goodeve to a new Assistant Controller for Research and Development post. Goodeve was a Canadian chemist who had been commissioned into the Royal Navy, and had been working in the Admiralty's research corps for several years. His post made Blackett's attachment to weapons redundant, and in October 1943 Blackett requested that his group report to the Vice Chief of Naval Staff (rather than the assistant chief for weapons) in light of Goodeve's appointment and the fact that "the work of C.A.O.R. is developing more in the direction of operations than of weapons."[6] This request was granted, and Blackett and his view of OR thrived in the Admiralty thereafter. In January 1944 the D of E reported that "Operational Research has come to stay," and that there was "general agreement that the time has come to give the group enhanced status." That spring Blackett's office became the Naval Operational Research Directorate (NORD)—a major promotion within the Admiralty bureaucracy.[7]

Meanwhile, in 1943 the British Army began to set up a new series of OR sections to work with the army groups in Italy, Asia, and, following D-Day, northwest Europe. Unlike Schonland's groups, these new groups reported to SA/AC, and SA/AC duly became the head of the Army's new OR organization as well as a link between the War Office and the Ministry of Supply.[8] Finally, in the summer of 1943 the scientific advisory and OR reform movement came full circle as Tizard contemplated retirement from his advisory post in the Ministry of Aircraft Production. He recommended that the Air Ministry establish a "full time operational research adviser" as a counterpart to Blackett at the Admiralty and Ellis at the War Office, and as an Air Ministry counterpart to MAP's Director of Scientific Research. This suggestion resulted in the creation of a new Scientific Adviser to the

Air Ministry (SAAM) position, which was given to the Imperial College physicist George Thomson.[9] Although Thomson returned to academic work as soon as victory against Germany seemed assured, the existence of his post brought coherence across all three services to Britain's system of scientific advisers and OR groups.[10]

The United States

The RAF's OR model also proved influential in the United States, but in that country there was no push for bureaucratic coherence behind it, nor was any coherence achieved while the war lasted. Within months of America's entry into the war in December 1941, various military officers who were aware of the British OR model separately created four different groups. In January 1942 the Research Division of the United States Navy's Bureau of Ordnance established what by summer was called the Mine Warfare Operational Research Group (MWORG). In March, the Army Air Forces' (AAF's) new Directorate of Air Defense and the Army Signal Corps's Directorate of Planning both set up OR teams led by academic engineers. In April, the Navy's new Antisubmarine Warfare Unit established an organization alternately called Group M and the Anti-Submarine Warfare Operations Research Group (ASWORG), which was led and built up by Massachusetts Institute of Technology physicist Philip Morse.[11]

Among these four groups Morse's was unusual, both in that it was immediately devoted to tactical as well as technological problems, and in that it was established on a contract with the National Defense Research Committee (NDRC). The NDRC was a subsidiary organization of the Office of Scientific Research and Development (OSRD), and the director of that organization, Vannevar Bush, had already decided that it was inappropriate for it to sponsor activities beyond the R&D contracts it was explicitly set up to administer. Since OSRD contracts, rather than civil service posts, were the major path by which America's top academic scientists were integrated into the war effort, Bush's reluctance to get involved with OR left the activity with no obvious way to grow.[12] In spite of Bush's preferences, ASWORG remained under OSRD contract and became the core of a burgeoning Naval OR organization. However, Bush had more influence on the adoption of OR in the AAF. When they set up the first of what ultimately became seventeen "operations analysis sections" in the autumn of 1942, it comprised both military and civilian members working directly within the Air Forces hierarchy, and initially employed a large number of lawyers. The head of the AAF's first section was John Harlan, who would later be named a justice of the United States Supreme Court.[13]

Meanwhile, an entirely separate model for integrating civilian technical experts into war planning was built up by the MIT electrical engineer Edward Bowles. Since 1940 Bowles had served as the secretary of the NDRC's early Microwave Committee, dedicated to the development of high-powered radar devices. However, the committee's main offshoot, the Radiation Laboratory (Rad Lab) at MIT, was set up on the initiative of three physicists: Microwave Committee chairman Alfred Loomis, MIT president Karl Compton, and Rochester professor Lee DuBridge, who became the lab's director. The engineer Bowles was left as the odd man out. But, within days of leaving the radar effort, he was approached by the War Department to work directly for the Secretary of War, Henry Stimson.[14] He took up his new post in April 1942.

Stimson gave Bowles the authority to examine all aspects of radar development, procurement, training, planning, and operations in order to advise him as to what actions should be taken to facilitate the implementation of radar technologies. However, the first problem to which Bowles turned was the U-boats that had begun prowling off the east coast of the United States after the declaration of war. Where ASWORG studied the tactical aspects of this problem, from his position Bowles was able to recommend broader organizational changes. In this case, his suggestions led the Air Forces to establish what became known as the Sea Search Attack Development Unit (SADU) at Langley Field in Virginia. Similar to RAF Bomber Command's Bombing Development Unit, SADU was responsible for experimenting with the tactics and technologies of submarine hunting.[15]

Meanwhile, Bowles's first official task was to evaluate and improve the radar defenses of the Panama Canal Zone, which Robert Watson Watt had criticized while on a visit there. Bowles flew to Panama accompanied by Bell Laboratories researcher Ralph Bown to assess the situation, and pushed for the improvement of early warning and airborne radar technologies in the region. Then, in the summer of 1942, Bowles recruited Julius Stratton, an MIT physicist who was working for the Rad Lab. Stratton, in turn, selected Harold Beverage, the vice president for R&D at RCA Communications, to assist him in a study of communications problems along North Atlantic ferry routes. Not long thereafter, Bowles brought in the geophysicist David Griggs to work on problems associated with airborne radar. Toward the end of 1942, he recruited Rad Lab physicist Louis Ridenour as well.[16]

As 1942 wore on, Bowles came to understand his own work to be to conduct preliminary surveys to make informed assessments of the adequacy of equipment and its implementation in overseas commands. Planners in Washington were typically only exposed to immediate, short-term troubles, such as shortages of supplies, and Bowles felt they could not easily conceive of where more fundamental improvements might be made. Once Bowles had used his basic assessment of the situation to envisage such possibilities, he arranged for appropriate experts to be sent to the commands. In 1943 Bowles was appointed to the additional post of Communications Consultant to the Commanding General of the Army Air Forces, Henry Arnold, and used the position to expand his consultants' reach still further. At the top level, he assigned Hartley Rowe, the chief engineer of the United Fruit Company, a global produce giant, to be a field adviser to the Supreme Commander of the Allied Forces, General Dwight Eisenhower. Bowles also began recruiting teams of eminent scientists and engineers to form what he called "Advisory Specialist Groups." Though he had initially worried that he would not have enough work in his post to occupy even himself, by the end of the war Bowles's office had deployed over seventy "expert consultants" to the field.[17]

Like the British bureaucratic reformers, Bowles emphasized what he referred to on one occasion as the "fusion of the technical and the operational." And he was aware of British OR work and its resemblance to his office's work.[18] At the same time, he was suspicious of the OR model as it developed in the AAF, preferring his own method of conducting a preliminary survey before dispatching specialist consultants. In April 1944 Warren Weaver, the head of the NDRC's Applied Mathematics Panel, wrote in his work diary following a meeting he had had with Bowles: "It is his general idea that it is very unfortunate and unprofitable to ship out to a Command a basket full of scientists and announce to the Command, 'Here these experts are: They will study problems for you.' He never said so, but he more or less implied that this was a description of the way Operational Research Groups have been sent out by the AAF." Weaver feared that Bowles's consultants could endanger the operations analysts' work, and was not encouraged when Bowles suggested that they "would get along all right if they restricted themselves to their proper field of activity," defined as the "post mortem analysis of operational data."[19]

In the end, though, Bowles's office experienced greater conflict with "field laboratory" researchers than with the operations analysts (see figure 8.1). In late 1943 the MIT Radiation Laboratory set up a "British Branch" (BBRL) under Rad Lab physicist Lauriston Marshall, primarily to support

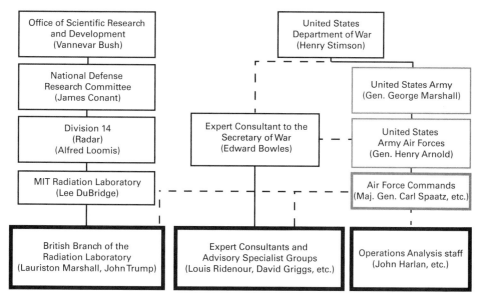

Figure 8.1
Organizational relations between the British Branch of the Radiation Laboratory, Edward Bowles's Advisory Specialist Group, and Operations Analysis Sections. Gray boxes represent military personnel; thick boxes represent field-based staff.

its "crash" development program. Ordinarily, Rad Lab prototypes were turned over to industrial suppliers who adjusted their designs to conform to military standards and to set up large-scale production. However, pressure from the field led the laboratory to begin manufacturing a limited number of devices to be shipped directly to combat theaters. Since these devices had not been properly tested under controlled conditions, Rad Lab personnel had to go to the field to glean performance data from their use in operations. Moreover, oftentimes only Rad Lab personnel had the knowledge of the devices necessary to install the equipment properly, and to develop techniques, standards, and manuals for its use. Setting up the BBRL simply systematized this work. Following the liberation of Paris in the summer of 1944, BBRL personnel went on to establish an "Advanced Service Base" there as well.[20]

During the campaign in Europe, the BBRL and the Advanced Service Base ended up sharing responsibilities with Bowles's first Advisory Specialist Group. For his part, Bowles felt his staff should be preferentially consulted. In September 1944 he confided to Louis Ridenour, his group's senior radar expert, "I fully expect the BBRL people to be unable to go beyond the bounds of technical comprehension of the operational problems."[21] Ultimately, though, it proved easier to make the BBRL's director a scientific adviser than for one of Bowles's consultants to speak for the Rad Lab. The tension was finally relieved in November 1944 when the BBRL's new director, physicist John Trump, took Ridenour's place in Bowles's Advisory Specialist Group, linking the two organizations. Ridenour proceeded with Bowles to the Pacific Theater to survey the need for expert advice there (figure 8.2), before ultimately returning to his former base at the Rad Lab in Massachusetts.[22]

With Germany's defenses crumbling, operations in the Pacific were a growing concern. And, when field scientific work began to take shape

Figure 8.2
Edward Bowles, front, and Louis Ridenour arrive in Australia. © The MIT Museum.

there, it was primarily with the OR model in mind. By late 1943 OR groups had expanded markedly throughout Navy and the Army Air Forces, and were increasingly responsible for staff being spirited away from OSRD research facilities. For this reason, among others, Vannevar Bush relented on his reluctance to become involved with the subject, and approved the creation of a new OSRD Office of Field Service (OFS), to be headed by MIT president Karl Compton. The anomalous status of Philip Morse's OR Group in the Navy was relieved by shifting it into the new office, and it would remain the OFS's largest project. However, the OFS's newer ventures were shaped more in the mold of the BBRL. Although the name of a new "Operational Research Section" in Hawaii was clearly inspired by OR, it was led by the first BBRL director, Lauriston Marshall, and its personnel were overwhelmingly technical specialists. Of forty-seven members listed in the ORS's final report, only five were "operations analysts," and the duties of four of these were listed as "work simplification," while the other was listed as "statistics." A similar group was established in Australia, but was left behind when the island hopping campaign began. Toward the end of the war a full-scale Pacific Branch of the OSRD was established under Karl Compton, with Yale physicist Alan Waterman taking over OFS, but the war ended before it could begin its work.[23]

Coda

On April 20, 1945, in a shattered Germany, John Trump interviewed two of the heads of the German company Telefunken's radar development program. He recorded in his diary that "this interview proved to be quite thrilling because we had long been deeply interested in the organization and thinking of the German radar people." In addition to the historical details of the development effort, he found out that industrialists did not work closely with physicists in university and private laboratories, and that the industrial laboratories had received their orders from technical personnel at the Luftwaffe. Furthermore, just "as there was a gap between German scientists and industrialists, there was also an even greater gap between both of these and the military. It was virtually impossible for a scientist or engineer to accompany radar equipment into combat areas to observe its performance or to assist in training." Trump also recorded his doubts that the Allies would learn much from the Germans beyond the occasional technical detail.[24]

As the head of the British Branch of the Radiation Laboratory, Trump was clearly satisfied that the subtle virtues needed to coordinate the technical war effort had paid dividends that were not realized by the enemy. This sentiment was widely shared—certainly by bureaucratic reformers like Henry Tizard who had relentlessly tried to build these virtues into organization charts and committee assignments, but also by those like Trump who were simply aware of how difficult it was to choose development priorities and technology designs wisely, and to ensure that equipment was used effectively in combat. In the end, the scale and lethality of the Allied war machine ensured a final victory, but those subtler virtues still seemed crucial to the aversion of defeat in the critical early moments of the war, and to preventing unnecessary casualties as victory approached. This sensitivity would go on to inform the transition of the Allied military services to peacetime and the Cold War.

III A Wartime Intellectual Economy

9 The Legitimacy of Wartime Operations Research

For British scientist activists and bureaucratic reformers such as Henry Tizard, A. V. Hill, Patrick Blackett, and J. D. Bernal, the extraordinary success of scientific advisory posts and operational research suggested that their efforts, stretching back to 1938, to promote the coordination of "science" with the war effort had at last succeeded. There is no doubt some merit to their view. As instances of successful collaboration multiplied, military officers gained an increasingly better sense of how people outside their ordinary spheres of collaboration could contribute to their work. But it is certainly also the case that the activists and reformers themselves gained a better sense of what kinds of reforms would constitute legitimately productive contributions, and began to tailor their proposals accordingly.

In accord with their general emphasis on "coordination," reformers understood that, to be productive, their contributions necessarily had to mesh with an existing intellectual economy populated by officers and personnel in combat roles, intelligence officers, engineers, technical specialists, and others. As their reforms took root, the reformers began to focus their intellectual energies less on how to extend their system, and more on making sure the people who worked within it abided by the principles that ensured its continued acceptance and success, and thus its legitimacy. Accordingly, they began to articulate more carefully how their activities fit together with military activities within the overarching intellectual economy of the war. In March 1943, for instance, Tizard wrote to Solly Zuckerman, who was stationed in North Africa as a scientific adviser to the British Chief of Combined Operations, Lord Louis Mountbatten. Tizard chastised him for some of the advice he was giving, writing, "You might be exposing yourself to the criticism which is now levelled with some justice against scientific men, namely that you are getting away from science and making recommendations on strategy and tactics." Tizard's objection was not of course that scientists had no business in such matters. Rather, he was concerned that Zuckerman was not supporting his advice with proper evidence. He urged Zuckerman to "put a little more science" into his work.[1]

Later, in May 1943, Zuckerman reported back to Tizard on the progress of the RAF's Middle East OR Section and a new section in the Northwest African Air Forces. In spite of uneven success, some military officers saw the virtue in the work. One senior commander in the region had asked Zuckerman "to pick up [an] operational research officer who could do his 'thinking' for him." This request led Zuckerman to reflect on the proper qualities of an OR scientist. A "narrow specialist would hardly fill the bill"—it was important to be able to discern and adapt to whatever issues might bear upon operational success. OR scientists also had to be able to get along with military personnel. He explained, "An operational research man will not get the information he wants unless he has the confidence of the flying man, and he will not be accepted unless he behaves reasonably like an ordinary R.A.F. officer." Indeed, "An operational research officer with a poor personality not only finds himself handicapped in carrying out his job, but also handicaps his colleagues in so far as he gets them associated with the idea of eccentricity." Eccentricity essentially equated with irrelevance, which threatened the idea of OR itself. Zuckerman reported how he was told by "a man holding a good position … that operational research officers were nothing more than a nuisance. His view was that they only collected information to pass back to other operational research officers and that they did nothing to help the men who have to see to to-days and to-morrows battles."[2]

OR scientists' ability to get along well with military personnel was not, however, simply a question of ensuring their institutional acceptance and access to data. There were also deeper intellectual issues at play. In his letter to Tizard, Zuckerman related the tale of a scientist who had been "an extremely bad choice for an overseas job." In "a rather unfortunate occurrence," a station commander outside of Cairo had tried to have the scientist removed from his station because, Zuckerman wrote, the commander "did not regard him as completely rational!" This notion that rationality was in the eye of the military officer and not the scientist spoke to an important point that Blackett had picked up on as early as his 1941 "Scientists at the Operational Level" memorandum. Discussing how OR sections went about their work, Blackett wrote, "Operational staffs provide the scientists with the operational outlook and data. The scientists apply scientific methods of analysis to this data, and are thus able to give useful advice."[3] What he probably meant by the term "operational outlook" was a sort of mental model that military planners had about what makes military operations work. It was the intellectual substrate on which all OR work had to be based, and the standard against which military officers regarded all suggestions for reform as more or less rational.

Military officers' operational outlook derived from an economy of information encompassing a bewildering array of intelligence reports, technical reports, personnel reports, exercise reports, mission reports, meeting minutes, statistical digests, bulletins, instructions, orders, doctrines, training syllabi, maps, charts, and communiqués. Officers somehow had to absorb and synthesize the contents of this economy, judge their reliability, and use them to formulate effective plans of action. They were generally aware that it was impossible to do so in anything like a systematic way, particularly given the exigencies native to war. Moreover, because military culture vested authority in rank and chains of command, officers were rarely called upon to articulate their reasoning thoroughly, as figures in science, engineering, or law might. In fact, because military success hinged on discipline and order, requiring that actions be explicitly justified could have been regarded as throwing commanders' authority open to doubt and objection. Prerogative and reliance on experience and intuition were unquestionably important elements of military culture.

Yet, at the same time, successful military planning did require sound reasoning, and articulating the rationales underlying plans could help improve those plans and inspire confidence in them. Therefore, military officers were apt to see value in improved means of drawing better conclusions from the economy of knowledge available to them. This, no doubt, was something like what the officer in the Middle East had in mind when he asked Zuckerman to "do his 'thinking' for him." In fact, because military officers rarely expressed their reasoning fully or all at once, an important part of OR work was attending the meetings where military officers discussed their plans and problems mostly openly and flexibly. It was also important for OR staff to gain access to all the different kinds of information flowing through the economy of military knowledge to gain a clearer sense of what informed the reasoning of military officers. Beyond any recommendations OR scientists might make, simply articulating officers' thinking and the various arguments for and against proposed courses of action was a valuable activity because it could clarify officers' own thinking and help integrate it with that of others.

This intermediary sort of function is well captured in figure 9.1, which shows how RAF Coastal Command coordinated its "Planned Flying and

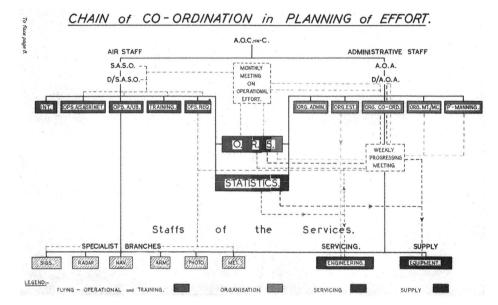

Figure 9.1
Organization chart for the coordination of the Royal Air Force Coastal Command's Planned Flying and Planned Maintenance scheme. The original is in color. Note, though, the position and color-coding of the Operational Research Section ("O.R.S.") straddling the various aspects of planning. From "Planned Flying and Planned Maintenance in Coastal Command, Part V: The Approach to a System," August 22, 1945, TNA: PRO AIR 15/154, facing p. 8. Reproduced with permission from the National Archives of the UK.

Planned Maintenance" scheme, which was designed by its OR staff. The RAF's ordinary hierarchical organization filtered the work of specialist staff through high-level officers. This flow of information was supplemented by weekly and monthly meetings to ensure that knowledge about strategic goals, operational plans, personnel assignments, supplies, and equipment maintenance schedules cohered into working plans. The command's statistical staff aided this process by providing figures on activities throughout the command. The OR staff, which was separate from the statistical staff, supplemented the cursory work of meetings by undertaking more long-term investigations.[4]

Once OR scientists assembled pictures concerning the rationales underlying existing plans and policies, officers could then determine to what degree it mapped onto their own outlook. If they felt it failed to encompass their own reasoning, they could offer criticisms. However, they might also find that it presented information or arguments that they had not previously considered, either tacitly or explicitly. If so, they could reconsider their thinking, and determine whether the novel aspects of the analysis caused them to draw different conclusions about their plans and policies. Of course, OR scientists might themselves identify points where officers' logic seemed suspect, or where up-to-date information suggested the need for a change of policy. But, at all times, it remained the prerogative of military officers to accept or reject any criticisms or suggestions that OR scientists might make. In the final account, responsibility for decision making rested with them.

The importance of constructive relations with the military as a prerequisite to legitimate work was also recognized in America, where Philip Morse articulated it particularly forcefully. New members of his Navy OR Group were given a handbook upon joining, which they were instructed to reread on a monthly basis. It detailed the importance of maintaining good relations with the military, and of respecting the experience and knowledge of established military personnel.[5] The group's final report, written by Morse with his deputy George Kimball, reflected on the issue as one that actually determined whether or not OR could claim the name of "science." It declared:

Operations research is fruitful *only* when it studies *actual operations*, and that a partnership between administrator and scientist, which is fundamental in the process, requires an administrator *with authority* for the scientist to work with. Operations research done separately from an administrator in charge of operations becomes an empty exercise. To be valuable it must be toughened by the repeated impact of hard operational facts and pressing day-by-day demands, and its scale of values must be repeatedly tested in the acid of use. Otherwise it may be philosophy, but it is hardly science.[6]

More prosaically, decades later, Morse remained sufficiently impressed with his ability to work within the naval bureaucracy that he boasted in his autobiography about how he had been able to acquire an air conditioner from the bureaucracy for Admiral Ernest King, the Chief of Naval Operations, where King's own staff had failed.[7]

10 The Methods of Operations Research

1. Analysis of Past Operations
    1.1 Obtain the Plans for Husky, Torch, Avalanche, Dieppe.
            "   "   account of what happened.
    Compare these, using some few suitable variables as functions of time.
            No. of men ashore
            " " *vehicles shore* [sic].
            Area of bridgehead occupied.
    Compare actual enemy reaction with that allowed for in the plans.
Attempt to deduce degree of rigidity of a plan which is useful; also how much deliberate flexibility should be planned for.
Patrick Blackett, personal notes, 1943[1]

If the legitimacy of wartime operations research always hinged on its ability to mesh with military thought, the work itself was not characterized by any particular methodology. Leaders in OR tended merely to describe their work as following the scientific method. Accordingly, OR staff were drawn from a variety of fields, albeit mostly from physics and mathematics. Many worked in universities, some worked in industry or in government research laboratories, some were schoolteachers, and many were graduate or even undergraduate students. A significant number were well-established scientists, and several would go on to win Nobel Prizes.[2] Some were not scientists or engineers at all. Some groups became very large, but there were also some comprising only one or two members. Table 10.1 shows the backgrounds for members of the United States Navy's OR Group, which offers some sense of what sorts of personnel the group leaders recruited. However, it should be noted that different groups pursued different recruiting strategies.

As the war progressed, an emphasis on generic scientific expertise gave way to specialization with particular kinds of military problems. In RAF

Table 10.1
Membership of the U.S. Navy Operations Research Group and its predecessor groups.

Membership by Top Degree (Total 86 over course of war)	
PhD	39
Master's	13
Bachelor's	31
Unspecified or "Abitur" (1)	3
Membership by Field (Total 86 over course of war)	
Mathematics	32
Physics	28
Chemistry	6
Engineering	5
Biology	4
Economics	2
Geology	2
Astronomy	2
Library Science	2
Other (zoology, architecture, unspecified)	3
Membership by Peacetime Employment (Total 59 in Oct. 1945)	
University (Professor or Instructor)	20
University (Research Assistant or Graduate Student)	10
Insurance (Actuarial)	11
Government, Military, or Nonprofit Research Facility	5
Industrial Research	3
Other (Non-Technical*)	7
Unspecified	3

Sources: Operations Research Group, "Summary Report," December 1, 1945, in NARA RG 227, "Office of Field Service: Manuscript Histories and Project Summaries, 1943–1946, Box 28; and "Members of the Operations Research Group (as of 16 Oct. 1945)," NARA RG 38, New Development Section—Subject File, 1942–46, Box 9, "Operations Research Group—Correspondence" folder.

Coastal Command, William Merton, a graduate student, was put in charge of studies relating to bomb sights and camouflage. The biologist Cecil Gordon worked on personnel assignments and aircraft serviceability, and designed the mathematical logic underlying the Planned Flying and Planned Maintenance scheme. Other team members worked on such issues as aerial attacks on U-boats, photographic reconnaissance, navigation, and the implementation of aircraft-to-surface vessel (ASV) radar.[3] Meanwhile, the RAF Bomber Command OR Section was divided into teams associated with the efficacy of night bombing, the "cost" of night bombing (i.e., aircraft losses), and problems related to day bombing. These teams, in turn, divvied up their work according to problems within each category. One set of problems in night bombing efficacy, for example, included "research on the problem of visual identification of the target," "study of general navigational problems (including the use of non-radio aids)," "study of the use of decoys and camouflage by the enemy," "determination of the effect of experience on target location," and "study of the effect of routeing on target location."[4]

One approach to studying these sorts of problems was simply to evaluate how well plans with well-considered rationales worked. Report No. 19 of the Operations Analysis Section at Curtis LeMay's XX Bomber Command, for instance, was a study of how an incendiary raid on the dock and storage area of Japanese-occupied Hankow, China, played out. Plans for this mission had worked out a specific order in which "components" of the dock area were to be bombed, and with what kinds of weaponry and in what quantities. The architecture of the attack was designed by measuring the economic functions of the target, its flammability, the location of fire fighting facilities, and the size and density of the buildings against how many bombers could be mustered against the target, how many would actually reach it, and how accurately they could bomb it. The report recounted how this plan went wrong from the start. Due to weather, the raid began forty-five minutes early but only one bomber group received the message. The smoke resulting from its attack, which would have blown over areas that had already been bombed had the aiming points been hit in order, hampered the targeting of the later formations. Only 30 percent of the total bombs dropped hit any of the components of the target, let alone the assigned one, and one formation ended up hitting the "Chinese Quarter," which was to have been avoided. The report analyzed patterns of devastation (see figure 10.1) and the spread of fire through the areas that were bombed, sorting out the factors—such as density of bomb pattern, the likely presence of firewalls in buildings in targeted areas, and

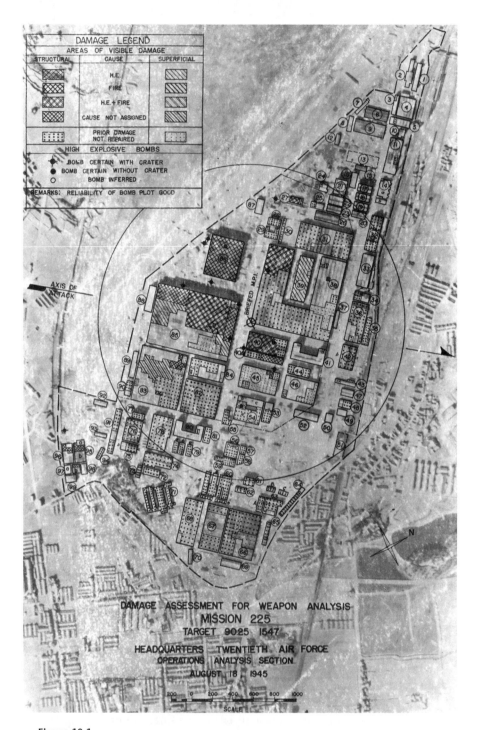

Figure 10.1
Analysis of weapons effects from a bombing attack on Kawasaki Aircraft Company, Akashi plant. From Weapons Analysis Report, No. WA-3, August 26, 1945, LOC CEL, Box B37, Official Document Summaries, Weapons Analysis 1945, (OAS) XXI Bomber Command.

the distribution of firefighting facilities—that contributed to the observed effects. Areas hit properly were almost entirely devastated. In the Chinese Quarter, however, the fires were contained, which was "best explained by the tremendous number of individual firefighters living in the area and by the fact that the type and tonnage of bomb superimposed on this area was neither the type nor the quantity of the bomb designed to destroy such an area." The report made a number of "pertinent observations" about the conduct of the mission, such as how important knowing wind speed and velocity were and how aiming using an offset point of reference across the river could have improved accuracy; and, of course, it admonished how disruptive even simple last-minute changes could be.[5]

In other cases military plans were not nearly so intricately worked out. In such cases, OR reports might help establish an analytical framework. A good example is a report issued by the Ministry of Supply's Army Operational Research Group (AORG) in January 1945 at the request of the British Army's Director Infantry, which was simply entitled "Street Fighting." The report had few pretensions, its second paragraph beginning: "To enter a house with minimum casualties inflicted on the attackers, it is frequently necessary to blow a hole in the wall"—the enemy often had the entrances to buildings covered by gunfire. The report listed various accepted methods of blowing holes in walls, and evaluated the effectiveness of each in light of the fact that a hole in the wall had to be at least two feet in diameter to permit entrance to troops wearing equipment. To aid in this task, the AORG made use of weapons tests conducted on specially built nine-inch-thick brick walls, a house used at the London District School of Tactics, a derelict barn with twenty-inch-thick stone walls, and a derelict farm house with twenty-inch brick walls.

The "standard method" of "holing" a wall—setting a pole charge against it—was effective but dangerous in that the soldier setting the charge was highly vulnerable. Another method, firing an anti-tank PIAT bomb against the wall, had been recommended "from various theatres of the war." OR staff found it required between two and twelve such shots to blow a sufficiently large hole in the wall, and speculated that perhaps walls encountered in the field were weaker, or that the effect on the morale of those in the house was "considerable." Other methods were also tested. The report also considered other problems, such as killing defenders from the outside, killing defenders once inside, the various means available of setting fire to a house, the uses of smoke, problems of weapons accuracy, and the use of wire netting over windows as a defensive measure. None of this work was especially novel, and at fourteen pages the study was certainly not an exhaustive guide to street fighting. The object was simply to map out certain aspects of the rationale behind certain street-fighting tactics, to place what would have been a largely anecdotal body of knowledge within an explicit experimental framework, and to set the stage for further discussion. For instance, the study suggested that, given a choice, it might be more useful to fire a PIAT bomb through an open window of a house to kill those within than it was to use it for blowing holes in the walls.[6]

In some cases, OR analyzed highly disorganized information about missions into a discussion of the quality of that information and what provisional conclusions might be drawn from it. In April 1945 the U.S. Navy's AirORG released a sprawling "analytical report" on the activities of carrier-based aircraft during the Battle of Leyte Gulf in the Philippines, which had taken place from October 24 to 26, 1944. At over fifty pages, many of them taken up by statistical tables, the report dissected not one mission, as the report on the Hankow raid did, but many different missions taking place at once. The first part of the report compiled accuracy rates against various classes of naval targets, noting that claims of hits were based only on the reports of pilots who doubtless had other things on their minds than accurately recording their actions. The report noted that, on the basis of the limited photographic evidence available, pilots certainly overstated their accuracy and that "there was gross misidentification of the vessels attacked."

Nevertheless, it was possible to reconstruct some provisional facts about the battle. The study combined reports of the types of tactics and weapons used against various classes of target, observations of damage caused, and known or suspected technical specifications and vulnerabilities of Japanese ships to ascertain how likely it was that the kinds of hits claimed would produce the damage that was actually observed. The report offered caveat-laden conclusions about how effectively the aerial effort was coordinated versus how effectively it was said to be coordinated. It also offered some tentative suggestions concerning how weapons selection, target selection, and strike tactics might be modified to produce a more effective attack in the future.[7]

While the Leyte Gulf study focused on the overall results of a large and complicated operation, another common function of OR was to clarify particular questions of interest. The British Army's No. 10 ORS traveled on a rotating circuit through its jurisdiction in Southeast Asia. Before their return to one of their stops in 1944, they sent out a request for suggestions for research into "facts that will be of use when conducting further operations," and offered a list of "useful facts" from the previous season's research. These included what Japanese weapons caused the most casualties, what percentage of casualties were likely to be prevented by wearing helmets, and the inadvisability of leaving behind small arms ammunition to reduce loads. This solicitation resulted in the compilation of a list of thirty-four suggestions for investigation, which ranged from "the value of issues of dehydrated meat to Indian troops," paying attention to such pertinent points as "what types and classes refuse to eat it" and whether the refusal to eat was "religious or prejudice," to a study of the efficiency of the "present system for evacuation of repairable and unserviceable stores."[8]

Some such problems were more intensively technology oriented. All three RAF home commands released a large number of OR reports relating to the equipment used for navigation, enemy aircraft detection, radar signal detection, and so forth. At RAF Coastal Command, for instance, ASV radar equipment changed continuously throughout the war, so it was important to keep current with what kinds of results could be expected from the equipment to know how effective it was, and whether it was being used properly. One study charted the differences between theoretically ideal long range ASV radar efficiency and actual results, measured by the range at which vessels were located. These discrepancies were charted against variables such as aircraft altitude. The report offered some tentative reasons for the discrepancies. For instance, a higher than theoretical search efficiency at lower altitudes was not a technical artifact. Vessels that would have been seen by radar at short ranges actually tended to be detected visually instead, leaving only those spotted at long range to contribute to the average figure. A severe decrease in efficacy at higher altitudes presented more problems, and ultimately there was not enough information to make more than educated guesses. Even the act of translating data into a curve introduced a host of caveats. The expectation values given by the curve could not represent the results to be expected on any given mission. When making actual plans it would be necessary to consider variations in the efficiency of individual radar sets, the production of interfering return signals by large ocean waves, and other factors such as the experience of the radar operator. Nevertheless, the curve was more useful as a benchmark for developing expectations than the theoretical values, and therein lay its worth.[9]

Developing expectation values could prove useful in analyzing other components of operations. A separate Coastal Command study analyzing disappearing radar contacts noted that the range distribution on first pickup of those contacts that later disappeared (and were known not to have other clear causes) appeared to adhere to a cube-root law, which was consistent with observed radar efficiency. But the range at which the signals disappeared adhered to the more ideal square-root law, which would have been consistent with expectations concerning U-boats equipped with radar detectors that submerged once the signal achieved a definite strength. Thus, the data appeared to confirm that U-boats were detecting radar signals. Since radar only gave a 20 percent increase in efficiency over visual detection, the study suggested that radar should not be used in daylight. Although OR studies were not essential to arriving at such conclusions, they ultimately made the decision-making process more robust.[10]

Of course, OR was only a part of the effort the military put into making decision making more robust during the war, the bulk of which involved the basic gathering and parsing of statistics. Although much reviled by those scientists pushing the development of OR, Frederick Lindemann was essentially responsible for the establishment and administration of statistical groups at the Admiralty, where Churchill was First Lord before he became Prime Minister, and then in the Cabinet and in the Prime Minister's office.[11] But the most pressing demands for statistics came from within the military services themselves. Most notably, the U.S. Army Air Forces established a behemoth Office of Statistical Control with remote statistical control units to handle the burden of managing logistical planning and processing the mountain of reports produced by operations.[12] OR groups sometimes undertook these tasks as well. For example, the OR Group at the British Army's Anti-Aircraft Command took over the compilation of results from "ZZ" forms detailing anti-aircraft activity from the Air Warfare Analysis Section.[13] Similarly, the RAF Coastal Command ORS, the Admiralty's OR staff, and the U.S. Navy's Anti-Submarine Warfare Operations Research Group, all produced regular reports detailing patrol activity by region, convoy activity, encounters with U-boats, and estimated U-boat activity, which was generally determined by intercepting radio transmissions.[14]

While OR scientists, especially early in the war, often pressed for the implementation of routine statistical work where it was needed, they tried to keep their own work from being absorbed by this effort. After the war, a few instances when OR groups were able to overturn a gross statistical fallacy were held up as more illustrative of OR work.[15] Most often, though, OR groups used statistics simply to guide their investigations into particular topics. When commanders thought they understood an aspect of a mission, they could gather statistics to see if their assumptions were borne out. Alternatively, if OR personnel discovered an unsought statistical correlation, they could try to find a suitable qualitative explanation. While the need for statistics sometimes called for more intensive surveillance of operations, useful fictions could also be used to produce necessary numbers. For instance, in studying attacks on U-boats, the Anti-Submarine Warfare OR Group was faced with reports of U-boats that were "probably sunk," "probably seriously damaged, may have sunk," and "seriously damaged and had to return to port." To cope with this uncertainty, OR scientists tabulated fractions of kills: a U-boat probably sunk, for instance, was equated with four-fifths of a sunken U-boat. The defensibility of the assumptions could then later be corroborated by obtaining more conclusive reports from espionage.[16]

OR scientists also had to be keenly aware of the relationship between data and the means by which they were produced. The way questions on forms were asked and answered, for example, could produce inconsistent or inaccurate information. In one case, researchers attempting to understand how often disappearing radar contacts represented submerging U-boats suspected a dearth of data on contacts tracked for less than a mile was due to radar operators not bothering to record brief contacts.[17] Similarly, Patrick Blackett related after the war how anti-aircraft kill rate discrepancies along the coast and inland had been resolved by discovering that inland results were confirmed through reports of wreckage whereas kills over water were obviously not. He portrayed the issue as a particularly clear application of scientific method, but one must take care in identifying science with scientists. In September 1940 some RAF Fighter Command personnel saw that combat statistics were skewed by the uncertainties surrounding kills made over water. A year later the RAF Fighter Command ORS suspected the disparity was a direct effect of geography.[18]

It was common for statistics and their interpretation to evolve as a program of research proceeded. Following the Blitz, the German Luftwaffe continued to mount raids on Britain. Part of RAF Fighter Command's response was to develop "intrusion" tactics, where small groups of fighters were sent to France to intercept bombers on their return to base. One report, issued in November 1941, noted that the value of intrusion versus traditional defensive interceptions over Britain had "become the subject of some controversy." Over a six-month period there had been 5,000 sorties over Britain, whereas intrusion entailed only a couple of hundred sorties. Comparisons were not *prima facie* obvious, but parsing the statistics already collected produced a revealing comparison: it took eleven intruder sorties versus thirty-two defensive sorties to down a German plane in the first ten months of 1941.[19]

A subsequent report, issued in April 1942, delved deeper into the issue, noting, "A simple examination of success achieved, measured in enemy aircraft casualties, is not enough to determine the overall worth of an operation. There are sure to be less obvious or even intangible factors to be taken into account." But even basic statistics revealed a changing situation. After raids tapered off in the latter half of 1941, twin-engine fighters using defense tactics had become three times more effective against German planes than intruders, and four times more so if intruder losses were taken into account. While intrusion patrols using single-engine craft were still extremely effective, the reduced size of German effort had dented these planes' success rates. A major source of declining success was German bombers' increased employment of bases outside of fighters' range. Intrusion was also becoming more dangerous. The number of sorties per loss of single-engine craft decreased from 58 to 9.25, while the defensive figure rose from 104 to 130. Still, the report remained optimistic about adapting intrusion tactics to new uses, such as catching bombers on takeoff, and offered several characteristics of a successful intrusion sortie, privileging aircraft endurance, maneuverability, and armament over high speed.[20]

There is little that can be said synoptically about wartime OR, except that OR scientists were expected to adapt their work to whatever issues and uncertainties bore upon successful mission planning. Their work had to take seriously the experience and intuition of military officers and combat personnel both to suggest problems and to help think through them. At the same time, scientists had to use evidence from a variety of sources to test and surpass military thinking. Sometimes OR scientists could grasp onto concrete realities, but oftentimes they struggled alongside military personnel around them to capture glints of insight piercing the fog of war. Sometimes OR work had a decisive impact on tactics, and sometimes it was ignored, while, most often, it formed one part of a larger economy of knowledge that gradually shaped military thinking.[21] OR could not, moreover, alter the basic fact that while the military sometimes changed its plans by anticipating problems and sometimes by reacting to them, sometimes changes only came following long and bitter loss, if at all.[22] The best OR could accomplish was to allow better decisions to be made more often.

11   Meta-Calculation and the Mathematics of War

Search Theory

Mathematical theorization was not an integral part of wartime work in operations research. In a 1943 memorandum detailing the methodology of OR, Patrick Blackett termed theorization the "a priori method" and observed that its uses in OR were limited compared to the statistical analysis of empirical data.[1] However, as we have seen, L. B. C. Cunningham made extensive use of theorization in order to anticipate the effects that choices in armament designs would have on combat outcomes. When the RAF regularized its system of OR sections in the fall of 1941, Cunningham's Air Warfare Analysis Section had already been established for a year and a half. At the first meeting of the RAF's Operational Research Committee that October, AWAS was clearly regarded as different from the new OR Sections. Blackett suggested, and it was agreed, that AWAS could "act as a pool of specialist mathematicians for high grade mathematical analysis, which was outside the scope of the staff normally available at O.R.S's."[2]

This division between mathematical analysis and OR was not quite so clear in America. In fact, the first project pursued by the U.S. Navy's Anti-Submarine Warfare Operations Research Group was a theoretical analysis of submarine-hunting operations. The problem was how best to locate a U-boat that had submerged following an unsuccessful attack, in the knowledge that its battery charge and air supply were limited. It was not imagined that a theoretical treatment could easily produce answers. Rather, as the first report on the subject, issued in May 1942, related, the idea was to use theory "to clarify the minds of the Operational Research Group on this problem." It went on:

The only way a theoretical subject of this sort can be attacked is to set down a series of preliminary hypotheses, which can then serve as a basis for discussion and modification. It is to be hoped that if the comments are read by the various Naval A/S/W [Antisubmarine Warfare] Units they will be considered in this light. Suggestions, particularly contradictory opinions, by members of these A/S/W Units, will be particularly valuable. Only in this manner can the O/R/G modify its hypotheses to come closer to actual cases.[3]

The decision to use theory as a means of articulating existing military knowledge might have owed to ASWORG director Philip Morse's experience as a theoretical physicist concentrating on the highly practical problems of acoustics.[4]

Both Cunningham's theory of combat and ASWORG's search theory were concerned with what we might call the problem of meta-calculation, that is, some implicit calculation bearing upon the validity of a calculation explicitly formulated as a mathematical statement. One kind of meta-calculation was a calculation that heuristically determined whether or not the explicit calculation would yield a satisfactory answer to the problem it was intended to solve, and, if it did not, in what ways it was inadequate. In both Cunningham's theory of combat and ASWORG's search theory, the formulation of theoretical expressions was intended as a way of revealing their own inadequacies compared to as-yet unarticulated expressions. Beyond the simple validity of a theory, though, meta-calculation could also entail some implicit calculation that dictated how an explicit calculation should be used to arrive at a genuinely good decision. In such cases, a theoretician might attempt to express the implicit calculation and integrate it into the explicit formulation. Cunningham's theory, for instance, made explicit previously implicit calculations concerning the anticipated effects that competing engineering designs would have on the outcome of combat. Importantly, in all cases, the fact that a decision ultimately had to be made was central to the meta-calculative epistemology of wartime theorization. The objective was decidedly not to develop a theory that perfectly corresponded to reality, but one that was sufficiently well developed as to improve on the decision that would otherwise have been made. In fact, most wartime theoreticians were apt to look askance at needless refinement when there were so many problems that needed to be addressed.

Over the remainder of the war, Morse's group found the subject of search theory capable of accepting extensive development. After only a year of work on it, search theory had become so sophisticated that mathematical theorization dissolved into ersatz experience. On August 10, 1943, ASWORG's research director, Bell Laboratories physicist William Shockley, gave a presentation entitled "Simulated Hunts for U-boats" at one of the regular Navy conferences held on antisubmarine warfare. The specific subject was search patterns to be used in "exhaustion" tactics, where a U-boat was hounded from the air until its air supply and batteries were depleted, forcing it to stay on the surface. The trouble was that the longer a U-boat was submerged, the more aircraft were needed to guarantee a sighting when it resurfaced. A hunt required a minimum of fifteen aircraft, and, if the position estimate of an initial U-boat sighting was off by as little as twenty-five miles, many more were required. Yet only as many as three aircraft were typically available for such a hunt, yielding only a particular probability of contact. The problem, then, was to design a search pattern that would maximize this probability.

When two members of ASWORG visited Britain, they had the opportunity to question three Royal Navy submarine captains, two of whom had commanded the captured U-boat HMS *Graph*. The OR scientists carried out a "highly idealized exhaustion hunt plan" with the commanders on paper, and each commander explained what his tactics would be upon being sighted by an aircraft at a given time and place. Even though the idealized search was supposed to be sufficiently thorough to guarantee a contact, Shockley related that the regularity of the search plan "permitted the U-boat to surface and dive on schedule between aircraft transits." This experience drove home the reality that "a U-boat hunt is not a geometrical problem, but a sort of game, with opposed intelligences on either side."[5] In due course, the applicability of mathematician John von Neumann and economist Oskar Morgenstern's as-yet-unpublished theory of games would be recognized.[6] ASWORG's response, however, was to move from theorization to simulation.

Back in Washington ASWORG staff developed a paper-based game, which took into account the fact that U-boats tend to spot aircraft and submerge without notice about twice as often as aircraft manage to spot U-boats:

The U-boat commander plots on tracing paper his position at the end of his first 30 minutes of travel and at the end of the first hour. Similarly the A/C [aircraft] squadron commander plots the position of each A/C at the end of each 10 minute interval, during the first hour. These two charts are turned over to a referee, who compares them to see whether or not a sighting is possible. If one is possible he decides whether or not it occurred by doing the equivalent of flipping a coin. If, during the hour, either commander made a sighting, he is so notified by the referee and permitted to alter his plans if he so desires. This procedure is repeated for each hour of the hunt.

These kinds of simulations actually took an amount of time comparable to a real hunt, and it was decided to carry out the investigation "theoretically along mathematical lines, based on the results of the few hunts actually run."[7] ASWORG designed an electromechanical device, constructed by the Special Devices Section of the Bureau of Aeronautics, which eliminated the referee and calculated the sighting probabilities automatically. Philip Morse later recalled, "For several weeks the men in our group played dozens of games on this device. Within these few weeks we learned more about the complicated problems of submarine search than 6 months of analytical work had taught us."[8]

One early result of these simulation exercises was the development of a tactic called the "continuous gambit," in which aircraft would patrol just outside the area where a U-boat could possibly be. Not sighting an aircraft through a periscope, the U-boat would be tempted to surface and run with greater speed, but quite possibly into the patrol area. In his presentation, Shockley pointed out that the continuous gambit had already been implemented by the Navy's Fourth Fleet. He reckoned, though, that the method would permit other effective tactics to be developed, and that it could be adapted to special situations, such as those involving restricted waters, shallow bottoms, and navigational errors.[9]

Warren Weaver and the Applied Mathematics Panel

During World War II there were few people with a keener appreciation than Warren Weaver of the potential and limitations of theorization as applied to war problems. A native of Wisconsin, Weaver earned a master's degree in civil engineering in 1917 before deciding to become a mathematician. He joined the Army during World War I and spent much of his time designing equipment to aid Army aviators. Afterward, he took a professorship at the University of Wisconsin and coauthored a book on electromagnetic field theory with the physicist Max Mason. In 1935 Weaver succeeded Mason as the director of the Division of Natural Sciences at the Rockefeller Foundation. Because there was little federal funding for university work at that time, Weaver's position made him a powerful figure in the world of American science, as he essentially was in a position to choose what lines of research were likely to yield the greatest future benefit. Most notably, he played a crucial role in supporting the growth of molecular biology.[10]

In the summer of 1940 Weaver was named the head of Division D-2 (later Division 7) of the new National Defense Research Committee, making him responsible for supporting research and development projects relating to fire-control devices. Where in his peacetime position Weaver tried to develop research programs with prospects for long-term benefits, in his wartime position he focused on projects likely to be of more immediate value. The wartime career of the brilliant but eccentric mathematician Norbert Wiener as a Division D-2 contractor reflected these priorities. During the war, Wiener's work on anti-aircraft gunnery revolved around an attempt to develop a fundamental theory of information feedback. After the war, this work formed the basis of the field of cybernetics. In 1942, though, Weaver wondered whether Wiener's work was a "useful miracle or a useless miracle." Whatever intellectual virtues it had, Weaver felt it was not apt to lead to improved fire-control devices in the near future, and he did not renew Wiener's contract.[11]

In a wartime context, useful work on fire control related to addressing the immediate inadequacies of existing devices, while respecting various ad hoc constraints, such as the complexity and bulk of equipment that would be tolerated, or the typical behaviors of enemy targets. General theories were of little use in such circumstances, while inventive, easily implemented stop-gap measures were regarded as quite valuable. Much of this stop-gap work was bibliographical rather than theoretical, involving the cobbling together of piecemeal data sets from equipment tests and combat reports.[12] British experience prior to American entry into the war was of particular value, and when results were filtered through technically qualified individuals—such as those in Basil Schonland's Army Operational Research Group—they often proved especially useful.[13]

A major difficulty was determining how cobbled-together theoretical expressions should bear upon engineers' decision making. To solve this problem, Weaver was impressed by the potential of L. B. C. Cunningham's work. At the end of the war, Weaver recalled receiving copies of Cunningham's first AWAS papers, "Mathematical Theory of Air Combat" and "An Analysis of the Performance of a Fixed-Gun Fighter, Armed with Guns of Different Calibres, in Single Home-Defence Combat with a Twin-Engined Bomber." He "showed and praised" them to various military officers until they asked him to "try to digest and simplify" them, and to explain their contents "in terms not so formidably mathematical."[14] Thus, in the summer of 1942, Weaver approached Columbia University statistician and economist Harold Hotelling to recruit a team that could help him with the task. Hotelling referred Weaver to his former students, Allen Wallis from Stanford University and Jacob Wolfowitz, who was teaching in New York Public Schools. Wallis and Wolfowitz formed the core of a new Statistical Research Group (SRG) at Columbia University, and set to work on Cunningham's theory.[15]

The SRG's first application of theory was to determine whether eight 0.50-inch or four 20mm guns constituted a preferable armament configuration for a fighter making a tail attack on a bomber. The first task was to set down what kinds of information "(or of estimates, or of guesses)" were needed "before the answer would be forthcoming." Weaver recalled, "These questions (nature of combat, bomber and fighter armament, value and variations of accuracies, ammunition, vulnerabilities, etc.), were then discussed at very considerable length with experienced officers. As a result estimates were arrived at which everyone agreed were almost certain to bracket the true values, although in many instances the true values were admittedly unknown." The questions could then be explored mathematically to ascertain whether or not the uncertainties in certain factors bore upon the final decision between the two configurations. For four out of five assumptions of vulnerability, the analysis favored the 0.50-inch gun configuration; for the fifth assumption, the 20mm guns had a 20-percent advantage in securing a favorable outcome of the duel. Although the theoretical treatment was unable to arrive at a definitive conclusion, it performed an important meta-calculative function by identifying which uncertainties mattered most, and what tests could be performed to resolve them. This was a valued virtue when time, equipment, and personnel for testing were scarce.[16]

In the waning months of 1942, Weaver became the head of a new NDRC organization called the Applied Mathematics Panel (AMP), based in New York City. The SRG at nearby Columbia was preserved and made into a subsidiary contractor. The idea behind AMP was to provide mathematical assistance for the various NDRC divisions, an ambit that soon expanded to include the military services as well. Thus, much of AMP's work involved fairly straightforward mathematics.[17] However, Weaver and the SRG also remained deeply interested in the deeper meta-calculative issues surrounding Cunningham's work. These issues even pervaded the SRG's lunchtime "mathematical recreations." One problem they considered was: "Given twelve coins, all of the same weight except for one, using only a two-pan balance, find the odd coin and determine whether it is heavier or lighter making only three trials." Using an unlimited number of trials, it is, of course, not difficult to find the odd coin. However, accomplishing the task using *only* three trials requires an extremely efficient trial sequence that must be modified mid-sequence based upon possible information gathered during a trial, which can only be validated in a subsequent trial. The SRG not only was able to determine a solution to the specific problem, but also to grasp and formulate the general meta-calculative principles informing the design of efficient trial sequences for any number of coins.[18]

This problem of designing efficient and effective test procedures was the core of the SRG's major intellectual triumph: the development of sequential analysis in early 1943 as a means of improving the Navy's quality-control procedures.[19] The central problem in industrial quality control is to determine whether or not manufactured products contain flaws making them unsuitable for use. The industrial production of consumable goods generally takes place in batches called lots, which are consistent in quality to the extent that lots are either accepted or rejected in their entirety, but are inconsistent to the extent that some items within lots may pass a quality trial while others do not. Statistical quality control selects a random sample from a production lot for trial, and the results of this test determine whether or not the lot is to be discarded. Taking larger samples ensures they are more representative of the lot as a whole, but additional trials cost more time and resources, and, in cases such as munitions, they also involve destroying the product. So, wartime inspectors tried to keep samples as small as possible. For similar reasons, if a number of product failures occurs sufficient to reject the lot before the sample is fully tested, inspectors would truncate the test sequence.

However, SRG statisticians soon learned that the accepted testing procedure of predetermining a sample size was not the most efficient one. Experienced inspectors often truncated test sequences before the criteria for accepting or rejecting the lot were satisfied if they regarded the sequence as having overwhelmingly established the lot's acceptability or unacceptability. Intuitively, this policy had some appeal, but it was not clear just how many trials would have to succeed or fail before the inspector was justified in stopping the test sequence. The Navy posed this problem to Allen Wallis and his SRG colleague, the young economist Milton Friedman. The two of them saw in the Navy inspectors' meta-calculative intuition the potential for rigorous "super-colossal" tests that were more powerful than the "most powerful" tests articulated by statisticians to that point. Unable to work the problem out for themselves, they presented it to their senior colleague, Columbia University statistician Abraham Wald. Though initially skeptical that such test sequences were even possible, Wald soon found that, by incorporating information obtained during the testing process into decisions made about whether to truncate the sequence, newly powerful tests could indeed be formulated. In fact, the insight seemed to suggest the possibility of opening up entirely new vistas in statistical thought, though these could not be explored until after the war's conclusion.[20]

The Place of Theory

Although most work done under AMP auspices did not concern deep meta-calculative problems, Warren Weaver and many of the mathematicians and statisticians who worked under him fostered a strong meta-calculative attitude toward the allocation of their own efforts. While they never tried to formulate that attitude in mathematized terms, at times their commitment to it approached the level of philosophy. As with equipment trials and quality-control tests, calculative work required time and resources that had to be expended wisely by the limited number of qualified mathematicians available to do it. In order to decide on how to allocate their effort, it was important to understand how theoretical work related, both institutionally and intellectually, to the decision-making problems faced by those actually running the war. Operations research would prove to be a crucial link to those problems.

By late 1942 Weaver's curiosity about OR was already piqued from reports he received from Britain in relation to his work on fire-control devices. Then, in December, when he began to set up AMP, he discovered at nearly every turn how deeply the subject had already influenced events in America. His initial plan was to have a small core of mathematical specialists working directly for AMP, and then another set of people who would liaise with particular NDRC divisions. When he inquired with the head of Division 6, dedicated to underwater technologies, to find a good representative, he was informed that the best person would be Philip Morse, who had been directing ASWORG for almost eight months. At that moment, though, Morse was in England with his deputy, William Shockley, surveying OR work there.[21] A week later Weaver sought out Princeton statistician Samuel Wilks for his core group only to find out that he was working part-time developing analytical methods for Morse. Wilks agreed to join AMP full-time while maintaining a connection with ASWORG.[22] Five days later, consulting with the head of Division 2, dedicated to impacts and explosions, Weaver was informed that the physicist H. P. Robertson (known as "Bob") was their "chief theoretical man," but that he, too, was "in England doing operations research for the AAF."[23]

In early 1943 Weaver decided that there was a close relationship between the sort of mathematical analysis involved in fire-control computations and the analysis done in some OR groups.[24] Although he was aware that

OSRD director Vannevar Bush felt that OSRD and NDRC should remain aloof from OR, Weaver wrote to him urging that thought be put into the relationship between OR and NDRC work. He explained the importance of OR in "getting information back concerning operational results," how his mathematical work appeared "to fall within some broad definition of Operational Research," and how the Navy had already expressed interest in adopting AMP personnel into a new OR group. He observed that the subject of OR had "suffered distinctly from a lack of clear definition of words." The term itself, he claimed, was "being applied to activities of very different characters," and that "certain observations and conclusions concerning Operational Research" were "confused and vague."[25]

Weaver drew upon the antisubmarine problem to illustrate his point: "I suggest that the subject would be clarified if we agreed … to speak on the one hand of 'Operational Research—Submarine Warfare', and on the other hand of 'Submarine Warfare Analysis.'" Weaver felt that OR, proper, was empirical research and analysis geared toward helping the military to make better decisions. Weaver saw the work of the RAF's OR Sections as undertaking this sort of work, and he agreed with Bush that that work ought to be separate from the OSRD and NDRC because it more or less had to be done under military auspices. However, "warfare analysis" entailed working out mathematical theories that linked equipment design with the design of tactics. NDRC Division 7 and AWAS did this sort of work, and Weaver felt it could take place away from combat theaters and should be done "under fundamental scientific (OSRD) auspices." Some work, such as that done by the British anti-aircraft OR group and ASWORG, seemed to fall under both categories. But, Weaver argued, however the two kinds of work were organized, they needed to be put in much stronger contact with each other. He was particularly concerned that the quality of the mathematical theorization that AMP undertook depended on the perspective of combat realities that OR provided.[26]

Weaver soon reached out to ASWORG, and AMP became heavily involved in its more theoretical work. In March 1943 Morse joined a meeting of AMP's executive committee, and informed its members of various theoretical problems arising in the work of NDRC Division 6 and ASWORG.[27] Weaver saw similarities between torpedo-shooting problems and anti-aircraft gunnery problems, and was especially excited about the search problem. It was, he wrote to the head of Division 6, "entirely new to us," and he felt it was "of sufficient importance and novelty to justify consideration by more than one group."[28] The Applied Mathematics Group at Columbia University (AMG-C), an AMP contractor separate from the

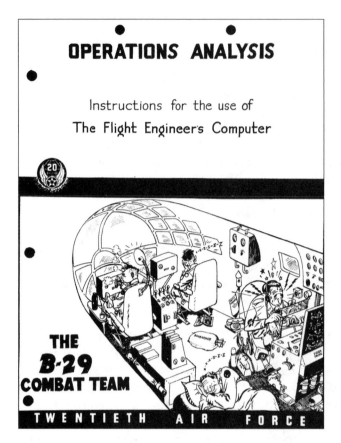

Figure 11.1
From "Instructions for the Use of the Flight Engineer's Computer," Operations Analysis Section, Twentieth Air Force, Report No. 27, in LOC CEL, Box B38. The computer is the slide rule held by the harried flight engineer.

SRG, subsequently took up search theory, which soon absorbed Columbia mathematician Bernard Koopman. In the summer of 1943, Koopman, who was not a member of the group, passed on to it a paper he had written entitled "A Quantitative Aspect of Combat," detailing the attrition rates that groups of different sizes and fighting strengths could be expected to inflict on each other.[29] The paper generated interest at AMP, which distributed it for comment. Morse and others informed Weaver that the paper actually replicated work that British engineer Frederick Lanchester had done during World War I. But, even though the work was not novel, it earned Koopman a place in AMG-C beginning in 1944.[30] Almost immediately, though, he was transferred to Morse's group, where he took on a central role in the ongoing development of search theory.[31]

Most work within the Navy ORG took place in combat theaters, where OR scientists could gain access to the experience of combat personnel. However, theoretical work such as Koopman's, like the work of AMP, was based in the United States. This arrangement reflected the view that, while valuable, the ultimate impact of theory on military decision making was oblique, and needed to be filtered through intermediaries, both to ensure that theorization tackled real problems, and that theoretical results were translated into forms that field staff could use. As the war entered its last phases, theoretical and empirical work became more tightly intertwined, and less easily distinguished. For example, field-based OR scientists began to develop ad hoc theories that could be translated into action through the use of nomograms and slide rules, such as the flight engineer's computer (figure 11.1), which was used to help manage fuel usage on B-29 bombers.[32] Similarly, theoreticians began to liaise more consistently with OR and military staff, and even sometimes took to the field themselves.[33]

12 John Williams, Edwin Paxson, and the Stop-Gap Mathematical Life

Talked a little with [operations analysts James] Clarkson and [Jack] Youden regarding [Berkeley statistician Jerzy] Neyman's embattled paper. Clarkson said he thinks it is the best paper on bombing that he has ever seen—apparently it isn't possible to like or dislike JN's papers just a little. I remarked that we had felt the recommendations were overstrong and that the tacit assumptions were more-than-passably unrealistic. He replied that yes these things were true, but that nobody would take the results of a paper written here and rush out to apply them in the field, and he is sure JN is such a reasonable fellow that you'd only have to mention these objections in order to get him promptly to make the necessary changes.

Work Diary of John D. Williams, February 13, 1945[1]

It has never been clear to me why mathematicians adopt a trade union attitude toward attempts to apply mathematics to such matters as biophysics and psychology. The fields are sensibly virgin, complex, and demand an unrelenting comparison with experiment. And experiment is mathematical prophylaxis.

Edwin Paxson to Oswald Veblen, February 15, 1945[2]

Among the people who worked for the Applied Mathematics Panel, John Williams and Edwin Paxson were especially enthusiastic about the potential benefits of applying mathematical analysis to the problems of combat. Although both recognized that there were pragmatic limits to what mathematics could accomplish, neither was afraid of pushing analysis to those limits. They understood that their work was not apt to find immediate application, and that, therefore, they had a freedom, which military leaders did not have, to treat the problems of the war in an intellectually adventurous spirit. Nevertheless, they did believe that the conceptual clarity of mathematics, if not its precise numerical outputs, could have real practical value once formulations were sufficiently developed and responsibly translated into the language of the choices that people in combat theaters faced. The key to doing so was harnessing the hectic flow of information running to and from those theaters.

John Williams did not graduate from high school. He later recalled that he chose instead to devote his later youth to "such matters as racing cars, pocket billiards, and philosophy," which included a dose of "the physical sciences and mathematics" in recognition of an "old fashioned notion of natural philosophy." His lack of diploma did not prevent him from entering the University of Arizona, where he trained in astronomy. However, he switched to mathematics as a graduate student at Princeton, working under theoretical physicist Bob Robertson. After World War II began in Europe, Robertson introduced Williams to a senior ordnance officer at the Aberdeen Proving Ground in Maryland, who gave him a number of problems to work on. Then, "to get a little diversification," he "went touring." He recalled that he "showed up at the front door" of various military agencies "with no credentials except my mouth and talked my way in, talked the problems out of them, and talked my way out." These problems mainly involved evaluating the design of weapons and their effects. For instance, "the Field Artillery Board wanted to know the optimum way to use fragmentation shells in pattern fire against personnel protected in various ways; in trenches or standing or prone, etc."[3] Williams reckoned that he was working for "three or four government agencies" before he met Warren Weaver, who thought it would be wise to "'legitimatize' these bootleg activities." Weaver arranged a contract for Williams to work on war problems at Princeton. Williams then became the liaison between the Applied Mathematics Panel and Division 5 of the NDRC, which was dedicated to the design of guided bombs and missiles. He was also a consultant to the Pentagon-based staff of the Twentieth Air Force. There were occasions in the course of these duties when he visited military proving grounds and flew in bombers on practice flights. However, he supposed that he spent more time arguing with such people as the eminent statistician Jerzy Neyman about mathematical problems. It was, in any case, a hectic life, with "too much time in bad airplanes, in bad hotels, producing bad studies too quickly."[4]

What Williams meant by a "bad study" was an analysis of a problem that one knew to be inadequate, but that was nevertheless necessary if one hoped to make any progress at all. This spirit is well captured in a discussion that Williams had with Allen Wallis, Jacob Wolfowitz, and Abraham Wald in November 1943 on problems concerning torpedo evasion faced by the Anti-Submarine Warfare Operations Research Group. Williams recorded in his travel diary that he suggested

that one should attack this problem, in the first instance at least, by "quick and dirty" methods; rather than, say, to square off promptly on a six month's computing program. To this end it was recommended that they begin with a first rough approximation; however owing to the things which they thought should be considered in a first approximation (e.g. errors in torpedo speed, maneuvering, etc.), we gradually worked backwards until I recommended that they start with the 0th approximation and finally that they start with the -1th approximation.[5]

As with ASWORG's early development of search theory, the object was to find a foothold for understanding the problem. By stating a theory, it became easier to discern how the theory could best be improved, leading to more nuanced formulations until, ultimately, responsible recommendations could be made.

And, as with search theory, the best way to improve formulations of a problem was to draw on the experience of military personnel. In his wartime travels, Williams spent time talking with officers and technical experts at military experimental establishments, attempting to ascertain what equipment and tactics were being used, what the rationales behind their policies and decisions were, and what kind of data was available for determining the effectiveness of equipment and tactics. Oftentimes, he found, the most useful gatekeepers to military knowledge were OR scientists, who were more able to translate back and forth between mathematics and the realities of combat. Once the bombing campaign against Japan began to escalate in 1945, Williams became especially well acquainted with the Army Air Forces' operations analysts.

AMP work was, on the whole, more mathematically sophisticated than OR, and AMP mathematicians were, as a rule, more mathematically proficient. In an October 1943 meeting with operations analysts on their way to working with the Thirteenth Air Force, Williams and some of his AMP colleagues gave them a swift overview of AMP work on areas such as the "train bombing problem, principally as applied to a single airplane, including some discussion of the bombardier's calculator. There was also some discussion of the maneuvering target problem; a very brief discussion of guided missiles; still briefer mention of incendiaries; balloon barrages were converted into mine fields and discussed as such." The group also discussed tracer fire, the "adaption of a lead-computing tail gun sight to the chin turret position" and "problems of aerial free-gunnery." Williams recorded in his travel diary that after giving them all of this information, "the visiting firemen were turned loose for the night (at their request), to cope with the intellectual indigestion as best they could."[6] Of course, Williams and other AMP mathematicians relied heavily on OR scientists to provide them with data, although they were occasionally able to work their own connections to holders of data on behalf of the OR groups. In early 1945 Jack Youden, a noted statistician and senior operations analyst, called Williams "his first class moocher" after Williams secured a promise of data from Lt.-Col. Robert McNamara, a member of Army Air Forces' Statistical Control Office (and, much later, U.S. Secretary of Defense). Two weeks later, McNamara indeed "came through very nicely" in providing Williams with a "nice fat bunch of Consolidated Statistical Summaries from the 21st [XXI Bomber Command]."[7]

Data, of course, did not quite equate with knowledge of combat reality. Data were merely a means of testing knowledge of certain aspects of that reality. Knowledge of reality itself, to the extent it could be known, came from understanding how it impacted combat planning. OR scientists generally boasted the most comprehensive view of combat reality, because they had the best access to the thinking of military planners and were usually privy to more of the various kinds of information flowing through the economy of military knowledge. Neither AMP nor statistical analysts such as McNamara had remits to assemble such multi-perspectival pictures. These limitations became clear during a discussion of bridge bombing that Williams and Samuel Wilks had with the mathematician George Dantzig, who was working in the Combat Analysis Branch of the Statistical Control Office. Dantzig explained to Williams and Wilks that conducting a study "from the operational side" would take time, "since it would have to be pursued along slightly informal lines." A formal study "would never get to first base, because collecting the data would necessitate cutting across half a dozen departmental lines such as Photo Interpretation, Intelligence, etc, and ... it would just naturally get bogged down somewhere."[8] Dantzig's branch was only tasked with determining bombing accuracies in practice and combat, and was not in a position to undertake such wide-ranging inquiries. People like Williams had to develop a nuanced understanding of the different roles of different elements of the military's heuristic apparatus played in order to gain the understanding needed to develop valid theoretical formulations.

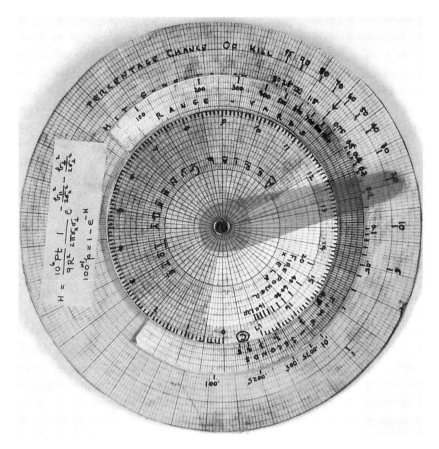

Figure 12.1
A slide wheel crafted by Edwin Paxson to calculate the influence of factors bearing upon the chance of scoring a kill in an aerial battle. NARA RG 227, Applied Mathematics Panel: General Records, 1942–1946, Box 8, "Paxson, E. W. Correspondence" folder.

Edwin Paxson had a rather different background from Williams. In 1934 he received his bachelor's degree in mathematics from the California Institute of Technology. Three years later he obtained his PhD from there as well. From 1937 until 1942 he worked as a professor of mathematics at Wayne University in Detroit. After Pearl Harbor, though, he developed an expertise in the mathematics of aerial gunnery while designing training films for gunners for the Detroit-based Jam Handy Organization, a film production company. These films used stop-motion models to demonstrate what kinds of combat situations gunners on bombers were likely to witness, and instructed them where they should aim in each case.[9] In January 1944, at the instigation of Edward Bowles, the OSRD Office of Field Service obtained Paxson and sent him to the proving ground at Eglin Field in Florida to become a scientific adviser to its commander, General Grandison Gardner. The idea, Paxson thought, was to take on a position similar to Bowles's other consultants. Upon arriving, though, he was surprised to find that Gardner had been informed by General Henry Arnold that the Germans had "a sort of braintrust who 'sit in a room and think.'" Arnold wanted to try something similar at Eglin. According to Paxson, the "Technical Planning Committee" that resulted comprised himself, an industrial physicist, "a man with considerable ordnance experience from the Sperry Company," and "a small-town lawyer from New Jersey, who had for several years been an attic enthusiast about radio." Once this experiment had been tried long enough to report back to Arnold that it did not work, Paxson took on a role much closer to the one he had been expecting.[10]

In his position Paxson soon came into contact with AMP, and had a number of ideas about what role it ought to be playing. At one meeting Paxson chided Weaver and other members of the Panel for not doing any work on a certain kind of sight. Weaver, taken aback, replied that the Panel had nothing to do with the sight. According to Weaver's notes on the meeting, Paxson, in turn, pointed out that "certain tactical problems have to be solved quickly and on the spot in the various theatres of war. These are presumably to be handled by the Operational Research Group. Then there are certain basic long-range developments ... which the NDRC is particularly well equipped to handle." However, there were also "a lot of stop-gap problems." These could be addressed with "methods or pieces of equipment which are admittedly not perfect, but which are nevertheless very much better than nothing." Included in this category was the sight in question: this intermediary category of technology was something on which AMP should be working as a matter of course.

Paxson had chosen the correct words. Weaver replied that AMP's leadership had "been preaching exactly this same philosophy on every possible occasion during the last twelve or eighteen months." Returning to New York, he felt that the Panel should do everything in its power to help Paxson.[11] By July 1944 Weaver decided that he wanted Paxson to work for AMP as the Panel's resident expert on aerial gunnery (see figure 12.1), and, after securing the permission of Gardner and Bowles, Paxson moved the next month.[12] To replace him, Eglin Field was to receive an operations analysis section.[13] Paxson's new duties at AMP included correlating "the large number of existing activities in this field," collecting and maintaining "a complete library of the subject," and preparing and distributing "frequent bulletins which could be distributed by AMP to all interested agencies."[14] However, his most important job was to assist on a new project the Panel had landed to determine the best tactical use of the B-29 bomber. The B-29 problem, however, was much larger than Paxson and AMP's work on it, and should be seen from a broader perspective.

13   The B-29 Problem

The Boeing B-29 "Superfortress" bomber entered service in 1944. A marvel of aeronautical engineering, its ability to fly exceptionally high, fast, far, and heavily loaded made it the primary weapon used by the United States against mainland Japan. However, like many other technologies of the war, it had been rushed into service. It was initially plagued by technical malfunctions, but, beyond that, very little thought had been put into the doctrines governing its use.[1] In the end Curtis LeMay decided the B-29 would be best employed dropping incendiary bombs on cities and other targets by night at high speed from low altitude. But there were many other doctrines under consideration, including stripping out all except the rear defensive armaments to fly as high and as fast as possible. Moreover, there were pressing technical questions that had to be answered, such as how the plane's state-of-the-art General Electric defensive fire-control system could be most effectively used. Officers in the Army Air Forces thought extensively about these questions, but also consulted outside engineering and scientific experts.

Among those consulted was Edward Bowles, who assigned William Shockley to the problem. In January 1944 Bowles had enticed Shockley to leave the Navy's OR group for his office. He later recalled that Shockley had become frustrated by the OR Group's inability to obtain operational data from the Navy. And, for his part, Bowles felt Shockley's OR experience made him well suited for tackling the problems of technological implementation his office handled.[2] Accordingly, Shockley's first task was "to assist in the exploitation of the operational potentialities of the B-29 bombardment aircraft, particularly by integrating into the [Very Heavy Bombing] program all those special devices, instrumentalities and techniques which will appreciably enhance its effectiveness."[3] After surveying the B-29 problem, Shockley soon identified H2X navigation, radar that produced a crude image of surrounding terrain on the radar's scope as a crucial area requiring further scrutiny. Over the course of 1944, Shockley dissected the problem and made recommendations focusing largely on the training of navigators in its use. Shockley's advocacy for training resonated well with the meticulous LeMay, and a new emphasis was placed on it forthwith.

Following LeMay's changes, Shockley made studies comparing bombing runs using well-trained and poorly trained personnel, which showed that even a small amount of effort taken from running operations would yield eight- or even ten-fold improvements in bombing accuracy.[4] Bowles approved of these "economic" studies showing the payoff of decisions such as augmenting training. So, upon finishing his work on radar navigation training, Shockley devoted himself to determining to what extent B-29 bombing Japanese cities was itself worthwhile by making estimates of economic damage done and comparing it to the costs of bomber losses and of running the bombardment program. The result was a study report entitled, "A Quantitative Appraisal of Some Phases of the B-29 Program," which estimated that damage caused exceeded losses by a factor of sixty—put another way, an investment of 0.5 percent of the American war effort would do 25 to 50 percent damage to the annual Japanese war effort.[5]

The strategic evaluation of the B-29 bombing program also seems to have put Shockley in the mood for playing with statistics, leading him to propose still more ambitious studies, even as he returned to the work at Bell Laboratories that would soon lead to the invention of the transistor.[6] In a report issued in February 1945 entitled "Discussion of a Proposed Program on the Quantitative Aspects of Modern Warfare," he drew inspiration from British studies done on ship production allocation by Patrick Blackett's Admiralty OR group, and on the effects of bombing on German industry by the Ministry of Home Security. Making some quick calculations, Shockley now questioned the B-29 campaign by comparing casualty rates from different kinds of combat, pointing out that the Air Forces were losing one man on bombing for every four man-years of "Jap labor" destroyed. This estimate compared to a ratio of seven Japanese losses to one American loss due to infantry combat, and ten-to-one losses—rising over time to thirty-to-one losses—in fighter combat. He felt that more study was needed, but this suggestion was never taken up.[7]

Meanwhile, in the summer of 1944, Bowles gathered together a mixed military and civilian team to see what means of attacking the Japanese mainland would be feasible from bases then available, or expected to be soon. The idea was to assess "the instrumentalities available to do the job, including aircraft, radar, television [H2X radar] and other means available for guiding aircraft to the targets, to study the mission to be performed, to formulate a specific technical plan for accomplishing it, to coordinate preparations for the execution of it, and to give the necessary technical support to the ultimate operations themselves." As a first step, Bowles signed on Arthur Raymond and Frank Collbohm of the Douglas Aircraft Company to assess the problem from the standpoint of the capabilities, limitations, and possible modifications to existing aircraft. After bringing on Edward Wells, the chief engineer at Boeing, attention focused on the B-29. By October the group had concluded that it was imperative that advantage be taken of Japanese defensive weaknesses by refitting B-29s to fly as high, as fast, and as heavily loaded with bombs as possible. They urged that the bombers be stripped of all their defensive armament, except for rear-facing guns, allowing it to escape Japanese fighter attacks. By the war's end the suggestion had been implemented in the 315th Bombard-ment Wing with an attendant improvement in accuracy, though the impetus behind the idea had, by that time, been overridden by LeMay's doctrine of low-altitude incendiary bombing.[8]

Warren Weaver's Applied Mathematics Panel was also asked to address the B-29 problem, which it did through an unusually large project known by its contract number, AC-92. Mindful that it was not always easy to judge which components of a complex problem would prove the most important to solve, AMP took the contract as an opportunity to demonstrate the power of mathematical analysis.[9] Unfortunately, there was very little theory, test data, manuals, or training standards on the B-29's defensive armament, nor were there any controlled tests concerning the efficacy of the different defensive formations in which the planes might fly when under attack.[10] Thus, there was essentially no existing intellectual frame-work on which to structure an analysis. To address the problem at all adequately, Weaver and his colleagues felt that a wide array of studies had to be conducted. These included a large number of experimental test flights to be undertaken by the Second Air Force at the testing range near Alamogordo, New Mexico. Another set of experiments, done with the Mount Wilson Observatory in California, entailed affixing special lights representing guns to formations of model bombers, and using pho-tocells to measure the resulting light patterns to estimate the spatial dis-tribution of the formation's defensive fire (figure 13.1). In addition, the Panel felt it was necessary to conduct tests on the central fire-control system of the bomber, gun camera tests at Eglin Field, component studies in collaboration with General Electric, psychological studies in collabora-tion with the NDRC's Applied Psychology Panel on subjects such as

Figure 13.1
Model experiments undertaken at Mount Wilson Observatory in Pasadena, Califor-
nia, to test the defensive gunfire distribution of different bomber formations. From
Report No. 7, "Preliminary Report on the Methods of Study of a Squadron of 11
Airplanes," 12/14/1944, NARA RG 227, Applied Mathematics Panel: Studies and
Notes 1943–1946, Box 18, Fire Control Research Group Bulletins.

gunners' perceptions and reaction times, analytical studies of special problems such as aircraft vulnerability and flak, and, to tie all the other work
together, "broad analytical studies, such as those which relate to bombing
effectiveness, general 'economic' theory of bombing, etc" to be done under
contract with Princeton University.[11]

Throughout the course of the AC-92 contract, Weaver "steadily and
stubbornly" insisted that all aspects of the problem needed to be worked
out simultaneously in order to interrelate component test designs properly
and to understand the significance of results achieved. However, the AAF
leadership did not agree and only took a real interest in the test flights in
New Mexico. They eventually came around to support the Mount Wilson
model experiments as well, but were colder toward the others, and, according to Weaver's end-of-the-war report, were "totally (and in most cases
emphatically) uninterested" in the integrative Princeton studies.[12]

It is doubtful that, even if the AAF had been more amenable, a study of
the kind Weaver imagined would have been practicable. A sense of the
theoretical and empirical disarray in which work on AC-92 took place can
be gained from an internal working paper on the question of whether the
B-29 should be modified, which John Williams and fellow mathematicians
Andy Clark and Jimmie Savage produced in the autumn of 1944. Williams

warned Weaver in a cover letter that he and his team had been working on the report until 7 a.m. that morning. He wrote, "We have some fears that we may have exceeded our directive. If so, we apologize and confess that the urge to do so was too strong to be resisted in our weakened condition. The data from Muroc never arrived, so there was nothing to impede us." In the working paper itself, Williams recounted, "We have been asked to estimate the effectiveness of the B-29 at very high altitude, say 35,000 feet, compared to its effectiveness at a more moderate altitude, say 25,000 feet." He was forced to admit, "The assignment embarrasses us: Little time is available in which to contemplate the subject: few data exist; not all of these are at hand; some which are, are [sic] of questionable pertinency [sic]; and the theory is fragmentary. Hence, while numbers will sometimes be written, the reader should not mistake the discussion for a completely quantitative one; it becomes essentially qualitative before we finish." Going through a long list of potential factors bearing on the success of operations, Williams and his team concluded that, in the end, there was no overwhelming reason to suppose that a stripped, "stuffed," high-flying B-29 would be any more valuable than a conventional B-29, and so the report recommended that the program continue as planned until some evidence arose to the contrary. In light of the notion that it "is dangerous in warfare to reject any weapon without a trial," they decided it would be worthwhile to outfit a group or a wing with revised armament after some battle experience had actually been obtained.[13]

Ultimately, AC-92 came to an end as soon as the work in New Mexico and at Pasadena had produced its major conclusions. In his final report as director of AMP, Weaver did not hide his consternation that the study was not developed in the way he envisioned. He was particularly disappointed that the integrative Princeton work, which he thought of as the "crowning jewel" of the project, "was a total flop." He was bitterly convinced that the AAF simply refused to give the project a proper chance. While he allowed that some members of the Twentieth Air Force in the Pacific had been interested in principle, he remarked that they thought "such studies should be carried out within their own organization, say by their own Operations Analysis personnel." Most AAF officers, he felt, had never really meant it when giving the NDRC such a broad directive.[14] It seems likely that the officers responsible for the contract did not understand how Weaver would interpret a vaguely worded request to study the B-29's tactical applications.

Even still, a substantial amount of AMP, NDRC, and AAF effort went into the AC-92 contract. Weaver estimated that some 350 military and civilian personnel had been involved, and it dominated his own schedule for much of the latter half of 1944.[15] Despite the effort put into the project, all of the studies were done under the pressure of looming deadlines and with limited data availability. Weaver and his mathematicians were continually torn between the need to at least improve on almost wholly arbitrary decision-making methods, and the concern that the inadequacies of their own methods would prevent their own conclusions from realizing any real improvement over arbitrary choices. This point should not be taken to mean that AC-92 did not yield results that made decisions more informed, but rather that the project failed to yield the definitive conclusions that Weaver had hoped to produce.[16]

14   Raisins in the Oatmeal

Although the AC-92 project fell well short of Warren Weaver's expectations, he nevertheless believed that he had caught in it a glimpse of the future of military engineering and tactical planning. Much as he felt that the value of any individual equipment test could not be ascertained except by placing it in the context of other tests as well as an operational framework, he had come to believe that no decision in warfare could be made without considering it in the context of all the other decisions that had to be made, and in view of its contribution to the overall war effort. He further believed that mathematics had a crucial role to play in clarifying just how all of these decisions were thought to relate to each other. Once AC-92 wound down, Weaver's next opportunity to reflect on this theme came in February 1945, when Edward Bowles approached him about the prospect of establishing a "gunnery research unit" at the Eglin Field proving ground.

In a memorandum replying to Bowles, Weaver drew directly on L. B. C. Cunningham's work to outline a central role for mathematical "air warfare analysis" within the cyclical process of the design, production, implementation, and evaluation of weapons (figure 14.1). He pointed out that "liaison" continually took place between all steps of this cycle. The military's needs had to be integrated into designs; specified design features informed the criteria by which the performance of finished prototypes would be tested; knowledge about expected performance bore upon the design of training regimens; actual performance in combat bore upon the military's formulation of its needs. Weaver believed that if the implications of work in each part of the cycle on its other parts could be expressed as clearly and precisely as possible, decisions in all parts could be made more confidently, and, it was hoped, efficaciously. In Weaver's scheme, OR played a crucial role by interpreting combat experience for air warfare analysts. The role of the proposed gunnery research unit was to develop

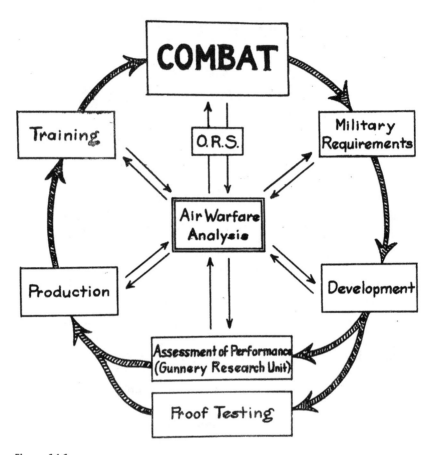

Figure 14.1
Warren Weaver's diagram of the place of operations research, air warfare analysis, and a new gunnery research unit in the ordinary equipment development cycle. Warren Weaver, "Comments on a General Theory of Air Warfare," AMP Note No. 27, January 1946, Appendix B, in LOC ELB, Box 43, Folder 5.

special equipment tests that would take into account assumptions about how weapons were to be used, so that a better sense of their worth could be ascertained before designs went into production.[1]

Weaver's memorandum for Bowles paved the way for—and became an appendix in—his chapter of AMP's final report. This document was less a summary of AMP's work than a rambling and often bizarre manifesto for the potential for mathematics to serve as a guide not only to equipment development but military decision making in general. It was self-consciously speculative, and, tonally, it was intended to pique its readers' interest. Weaver entitled an early draft "Raisins in the Oatmeal." As he explained to colleagues, who might think its presentation too informal, "I am assuming that the chance of it being read by persons in a position to affect future policies is low indeed: and that there may be a somewhat better chance if a few raisins are put in with the oatmeal."[2] Its final title, "Comments on a General Theory of Air Warfare," more explicitly signaled the debt that Weaver's thinking continued to owe to Cunningham's theory of combat.[3]

The report itself reflected on the original development of AMP's interest in Cunningham's work, as well as on the fate of the AC-92 project. However, the bulk of the report was taken up by Weaver's ruminations on the intellectual problem of how military officers could make decisions that benefited the war effort as a whole—this was a calculus that aimed to maximize a mysterious quantity that Weaver called "military worth" or "M.W."[4] Weaver illustrated the problem using the scenario of a colonel charged with selecting a new radar-guided bomb sight for development. The colonel's choice meant weighing such factors as the weight and performance of different bomb sights, how complicated they were to use, and the time it would take to develop them. Weaver observed,

He cannot possibly himself know all the necessary things, but somehow he should certainly bring to bear on these questions a wide and precise knowledge of the probabilities of bombing accuracies; the logistics of the theatres in which these sights are to be used; the nature of the enemy targets; all the vast field of terminal ballistics; the importance of the time factor (which means war plans, among other things); the psychology and physiology of operation of bombsights, selection, and training of bombardiers; accessibility of qualified personnel; the basic cost of accomplishing the same objectives otherwise; the present and potential future effectiveness of the enemy's fighter attack against our bombers; etc., etc.[5]

Clearly the choice was monstrously complex, and, even if all the necessary data were available, the colonel could never be expected to make a perfectly informed decision.

Weaver imagined a fantastic solution to the colonel's problem: the "Tactical-Strategic Computer (TSC)." One could enter into the TSC certain "basic variables" that described factors that could not be altered, such as how the enemy and nature might be expected to impact a mission. Some values would be set precisely, others could accept a range of uncertainty, and others still could accept probabilistic distributions for time-varying factors such as weather. Then, once those parameters were set, the user could begin "to twiddle the decision variable dials." The decision variables encompassed all the different components of a military operation, such as choices of equipment and tactics. Certain decisions, such as choice of weapon, determined others, such as the range at which one could open fire. These variables would be suggested by military experts, "for example, by the Air Forces Board, or by Eglin, or by some general with an astronomical number of stars." How the process of entering the variables worked drew heavily on Weaver's experience with mechanical computers such as bomb sights. He wrote, "It is perfectly possible that, after setting in a certain number of decision variables, the machine is found, when one tries to set in another decision variable, to be locked. This means that the decisions already made in fact determine the remaining decisions." When one had put in enough variables so that only one set of feasible decisions was possible for all remaining choices, one would push "a special button," whereupon, Weaver wrote, "all the remaining decision variable dials move automatically into their necessary positions, thus showing what further decisions we are perforce committed to." When all were set, the "the Military Worth dial lights up and displays the numerical value of M.W."[6]

Weaver did not suppose that even an imaginary TSC could fully automate the decision-making process. In fact, he noted that it was only once the computer took over that "the really interesting things begin to happen." One could "play with specific ones of these variable dials," and one could watch as the computer mechanically worked through all the combinations of basic variables for which ranges rather than specific values were set, and actually see how the M.W. value fluctuated. In places where that value fluctuated widely, one could identify which uncertainties in one's knowledge mattered most, and then conduct additional research to reduce those uncertainties. Weaver further suggested that the machine could be set to shift "all the decision dials through cycles of accessible values, the resulting values of M.W. being recorded so that the maximum can be located and the corresponding set of optimum values of the decision variables D_n determined."[7] In other words, the computer could determine, within the limits of what was known about the relevant variables, what the best plan would be.

Years later John Williams joked, "The best design I ever heard for such a machine was one that contained Warren's home telephone number!"[8] The remark contained a grain of truth. Weaver's vision was grounded less in any heady optimism concerning the possibilities of computing—if anything, the war had made him highly conscious of its shortcomings—and more in his appreciation of how good decisions integrated diverse perspectives. As the natural sciences director of the Rockefeller Foundation, Weaver had a good view of how new scientific research programs developed at the confluence of different streams of existing work, and how those programs then influenced a diverse array of scientific and technological fields. Similarly, in his war work, he could see how a vast array of decisions incorporated experienced military officers' expertise, technical expertise, and data gleaned from diverse sources. He also understood how the decisions that were, in fact, made often failed to make use of all the perspectives that they could. In this respect, Weaver's concern was quite similar to that of British bureaucratic reformers like A. V. Hill and Henry Tizard, albeit expressed in very different terms.

Weaver's own interest in the language of mathematics was rooted in his conviction that, while mathematical formulations had clearer meaning than the vague terms found in other forms of reasoning, they were not different from them in any essential way. Since, in his view, every decision "*has* to be made either by analysis, or magic, or blind guess," some form of reasoning, whether "excellent, good, mediocre, or disastrous," must always be in operation whenever a decision is made. So, for him, the more intuitive forms of judgment to which military figures sometimes appealed were nothing more than "disorganized and feebly intuitive shadows of a real analysis." But he was careful not claim to offer any sort of radical alternative. He allowed that his ambitious vision, "with its Rube-Goldberg Machine, runs the risk of giving certain wrong impressions." He insisted that he was "simply arguing for facing the complexity and the facts, and pushing analysis to its usable limit."[9] He felt that arbitrary facets of a decision should be scrutinized, and, where possible, amended, often by taking meta-calculative considerations into explicit account. This could mean placing decisions concerning the design of a weapon in the context of its intended use, as Cunningham's warfare analysis techniques did. It could mean reconsidering everyday decisions concerning "micro-problems" in the context of more strategic "macro-problems." It could mean taking enemy counter-strategies into account, which was an issue formalized by

John von Neumann and Oskar Morgenstern in their recently unveiled "theory of games." No analysis would ever be transcendentally correct. The point was that, whether deeply developed or "quick and dirty," some analysis of decision problems was better than no analysis.[10]

How these lessons could best be applied to the future remained an open question. Edwin Paxson reflected on this point in his response to Weaver's "Raisins in the Oatmeal" draft. He observed that Weaver's approach seemed to suggest a framework for action that was most appropriate to the drawn-out pace of the war that had just ended, where decisions would be made in terms of their piecemeal contribution to an unfolding war plan. "If," he argued, "one is carried along by the current jeremiad, then World War III will be at a greatly accelerated tempo." This possibility had important consequences. He continued, "I would recommend that most considerable emphasis be placed on carrying out these ideas *between wars*. TSC's should be developed in peace to explore the consequences of all conceivable Pearl Harbors." However, the task was complicated by the vast uncertainties of the future. Thus: "TSC's should also be developed flexible enough to adjust to the special nature of a Pearl Harbor of unforeseen nature."[11]

While actual Tactical-Strategic Computers remained well beyond the pale of reality, the need for analysis persisted as the Soviet Union morphed from ally into dire threat, and the United States and Britain assumed a posture of perpetual readiness. This raised an important question: what did it mean to be ready? All of the issues examined in the first half of this book remained pertinent. Whether technologies were being developed in accord with the needs of the military services, and whether those needs were being articulated in light of what technology might provide and make necessary, were the critical problems driving debates as to what constituted appropriate military policies. The experience of World War II would be regarded as carrying important lessons, certainly for facing these specific problems, but also for building a new postwar world.

IV   The Lessons of the War

15   Operations Research without Operations

The idea of assembling teams of scientists to study military operations had been an ad hoc response to the problems involved in introducing novel equipment into military tactics and coping with the swiftly changing conditions of war. When the cessation of combat removed these teams' reason for being, they began to disband. By 1946 all of the major names in academic science who had been employed in wartime OR work had returned to their peacetime work. However, some wartime OR staff remained dedicated to studying military problems, and military planners agreed that independent analysis would continue to have significant value in peace. By the 1950s military OR would develop into a much larger activity than it had been during the war. The initial institutional organization of these groups in Britain and America is summarized in boxes 15.1 and 15.2.[1]

The Nature of Postwar OR Studies

Postwar military OR continued to address questions concerning the appropriateness of tactics and techniques in view of what was known or expected about how future missions would unfold. Without the benefit of actual combat operations to research, OR staff approached their work using a variety of methods. In the immediate postwar years, a significant portion of their work involved distilling and analyzing the experiences of the war, and extrapolating from whatever information might be pertinent to future combat. One example of such work was a 1947 report produced by the operations analysts at United States Air Force headquarters, assessing the impact of various factors on the bombing accuracy of the Eighth Air Force. Using the "working assumption" that the factors combined linearly, the study's authors produced a slide rule calculating what degree of accuracy would have been expected under varying

Box 15.1

Organization of operations research in the American military services, 1945–1962

UNITED STATES AIR FORCE

The large network of "operations analysis sections" developed by the United States Army Air Forces was completely dismantled after the surrender of Japan, but an Operations Analysis Division (OAD) was retained at headquarters as a civil service organization. When the Air Force became an independent service in 1947, a series of Operations Analysis Offices (OAOs) were attached to individual commands, and operated largely independently of each other and the OAD. Until 1957 the OAD was overseen by LeRoy Brothers, a member of the wartime operations analysis effort. He was replaced by Carroll Zimmerman, another member of the wartime operations analysis staff, who in the interim served under Curtis LeMay as the head of the OAO at Strategic Air Command. In 1946 the Air Force also began to contract with Project RAND of Douglas Aircraft, which became the independent RAND Corporation in 1948. RAND concentrated on longer-range studies, with a heavy emphasis on the development and procurement of new equipment. It will be discussed in chapter 16.

UNITED STATES NAVY

The wartime Operations Research Group was renamed the Operations Evaluation Group (OEG), which was led by Jacinto Steinhardt, a member of the wartime group, until 1962. The OEG was run under a contract administered by the Massachusetts Institute of Technology. In 1955 a separate Naval Warfare Analysis Group (NAVWAG) was established to supplement work of the Navy's own Long Range Objectives Group. In 1960 the OEG established an Applied Sciences Division to study R&D planning. In 1962 all three groups were combined into a new Center for Naval Analyses, administered through a contract with the Franklin Institute.

UNITED STATES ARMY

The Army established an Operations Research Office (ORO, briefly called General Research Office) in 1948, which, following the OEG model, was run under a contract administered by Johns Hopkins University. Ellis Johnson (who had worked in the Navy's wartime Mine Warfare Operational Research Group during the war) led the ORO until the office's 1961 dissolution, when it was replaced by the Research Analysis Corporation (RAC). The ORO was

Box 15.1

(continued)

> supplemented by a number of other organizations: the Chemical Corps Operations Research Group (CCORG) and Human Resources Research Office (HumRRO) in 1951, the Combat Operations Research Group (CORG) and Signal Corps Evaluation and Analysis Group (SCEAG) in 1953, and the Special Operations Research Office (SORO) in 1956
>
> UNITED STATES DEPARTMENT OF DEFENSE
>
> The 1947 separation of the Air Force from the Army coincided with the establishment of a National Military Establishment, which was renamed the Department of Defense in 1949. In circumstances to be described in chapter 16, the department established the Weapons Systems Evaluation Group (WSEG), a civil service organization that was supplemented in 1956 by the Institute for Defense Analyses (IDA), modeled on the RAND Corporation.

conditions of weather, enemy defenses, and so forth.[2] A 1951 study by the RAF Coastal Command Research Branch reexamined instances of actual combat between U-boats and bombers between 1943 and 1945. Even though shooting back had proved a poor tactic for U-boats, it was suspected that Soviet submarines might attempt to improve its efficacy.[3] A 1954 U.S. Navy Operations Evaluation Group study of captured U-boat logs estimated that the interception and decryption of Allied signals offering timely intelligence on convoy locations had allowed U-boats to double their contact rate with convoys.[4] The Admiralty's Department of Operational Research did not finish its last study of captured U-boat logs until 1966.[5]

As time progressed, however, and anticipated conditions of combat changed dramatically, real wartime experience became increasingly suspect as a guide for future action. Speculation concerning actual combat became more necessary. Most data for postwar studies came from elaborately staged equipment tests and tactical exercises. While staged operations could never predict with certainty how future combat would unfold, in many ways the data gleaned from them were more reliable than real combat data because it was easier to observe what transpired.[6] Some OR work was simply the analysis of such tests. For example, in December 1946, the RAF Fighter Command Research Branch issued a report entitled "Training in Air to Air Firing with the G.M.II. Fixed Ring Gun Sight and the Gyro Gun Sight, Film

Box 15.2

Organization of operations research in the British military services, 1945–1962

ROYAL AIR FORCE

The RAF retained the post of Scientific Adviser to the Air Ministry as well as its OR sections, which it renamed Research Branches. These were intended to oversee not only "operational" but also "administrative" research.

ROYAL NAVY

The Naval Operational Research Directorate was renamed the Department of Operational Research (DOR). It was integrated into the new Royal Naval Scientific Service, but reported first to the Deputy Chief of Naval Staff, and later to the Vice Chief of Naval Staff.

ARMY

From 1945 to 1948, the Army continued to be served by two OR organizations: the Operational Research Group (Weapons & Equipment) in the Ministry of Supply, and the Military Operational Research Unit (MORU), which remained under the War Office's Scientific Adviser to the Army Council (SA/AC). In 1948 these organizations were united into the Army Operational Research Group (AORG) under SA/AC. In 1959, when research, development, and supply functions reverted to the War Office, the SA/AC position was abolished. In 1960 Army OR was placed under a Director of Army Operational Science and Research (DAOSR), who reported to the War Office's Chief Scientist. In 1962 AORG was renamed the Army Operational Research Establishment (AORE).

Assessment as the Method of Recording the Results from the Training Exercises."[7]

However, most postwar OR studies focused in some way on problems of choosing between prospective tactics. For example, the Army Operations Research Office's 1955 study, "An Operational Evaluation of the LACROSSE Guided Missile, the 155-mm Gun, and the 8-in. How[itzer]," assessed the tactical suitability of the weapons in question to the tasks for which they might be used.[8] Some OR studies took a broader view of operations by investigating the services' ability to deal with potential strategic threats. A 1950 study done by the Admiralty Department of Operational Research examined the ability of the United Kingdom to counteract a Soviet mining campaign in coastal waters. It estimated how many mines the Soviets could lay based on assumed capabilities, and how many minelayers the British military would have to be prepared to destroy to avoid being defeated by such a strategy.[9] These broad studies could identify specific problems requiring further analysis. Thus, a follow-up DOR study examined at what depths different kinds of available minesweepers could safely operate. Knowing how minesweepers were to be assigned was important for establishing what the country's requirements for its minesweeping force would be in the event of a mine war.[10]

Military OR groups also studied problems concerning the array of unconventional weaponry that could transform the character of future wars. The USAF's Operations Analysis Division had a team devoted specifically to atomic warfare, which produced such studies as "A Method for Estimating the Radius of Blast Damage from High Explosive or Atomic Bombs," and "The Effect of Thermal Radiation from the Super Bomb on the Bombing Aircraft."[11] Similarly, Army ORO Project ATTACK dealt specifically with atomic warfare, while Project COBRA studied the potential impact of biological, chemical, and radiological weapons on operations. Study titles included "Vulnerability of the Infantry Rifle Company to the Effects of Atomic Weapons," "The Feasibility of Chemical Warfare in the Defense of a Perimeter in the Naktong Valley Basin," and "Redstone Missiles as Atomic Warhead Carriers."[12]

Army OR, in both the United States and Britain, tended to be more broadly defined than in the other services, encompassing essentially any aspect of military planning that was poorly understood (figure 15.1). This tradition went back to the war—recall the British Army OR group in Southeast Asia being asked to investigate the dietary habits of Indian troops. In peacetime, physiological and psychological studies were handled by dedicated specialists. Under the leadership of physicist Ellis Johnson (figure 15.2), the ORO established programs dedicated to assessing and improving infantry performance (Project DOUGHBOY), psychological warfare (Project POWOW), and partisan and guerilla warfare (Project PARABEL). This type of work was quickly supplemented by the creation of two new social scientific research organizations serving the U.S. Army, the Human Resources Research Office (HumRRO), and the Special Operations Research Office (SORO).[13]

Actual combat studies resurfaced with the Korean War in the early 1950s, and later with the Vietnam War in the 1960s and 1970s. These wars did not feature the technological novelties of World War II, or its intense measure-countermeasure battles. Nevertheless, operations such as naval

Figure 15.1
Illustrations showing an assortment of ORO study topics. From Johns Hopkins University Operations Research Office, "A Survey of ORO Accomplishments," July 1961, in JHU 03.001, Series 1, Subseries 4, Box 28, "ORO Transfer" folder.

bombardment, aerial interdiction of roads and railroads, and ground combat still had to be assessed and improved upon. For example, one USAF operations analysis study evaluated the effectiveness of the MiG-15 against American aircraft.[14] The Korean War also provided an opportunity for Ellis Johnson and the ORO to pursue its broad vision of OR. While the conflict lasted, field teams produced studies on such issues as the effects of combat on infantry performance, the effects of propaganda on soldiers and civilian populations, and issues involved in administering occupied territories (Project LEGATE).[15] The ORO's most famous study—and one of the few to reach the public literature—was "The Utilization of Negro Manpower" (Study CLEAR), an extensive sociological analysis of racially integrated units in Korea. Project staff, many of whom were consultants and subcontractors, interviewed soldiers to gauge their attitudes and morale, and collected official assessments of combat effectiveness. They found that integration improved the morale and performance of black troops, and did not adversely affect overall morale or effectiveness. Their report

Figure 15.2
Ellis Johnson, director of the ORO. From David B. Parker, "Our Greatest Secret Weapon," *This Week Magazine*, August 5, 1951. Photograph by Joe Cavello.

recommended that the U.S. Army should continue to implement desegregation, and argued that there were no substantial barriers to doing so.[16]

The military OR organizations participated only tentatively in the mathematical innovation that characterized the professional field of operations research that developed in the 1950s. Partially, this methodological divergence resulted from a discrepancy in subject matter. Nonmilitary operations researchers often studied logistical problems, which, in the military, occupied an entirely different branch of the services' hierarchies from operational planning. In fact, as we will see, some of the most important developments in OR mathematics, such as linear programming, came from the military, but not from military OR. Of course, the division between professional and military OR was not absolute, and it grew more permeable as military OR groups began to recruit personnel from university-based OR departments and training programs. Even in the early postwar years, military OR groups undertook occasional logistical studies, and some mathematicians employed by them produced memoranda on mathematical techniques. The ORO even dedicated a small group, Project OPSEARCH, to mathematical methodology.[17] Military OR also made major contributions to the development of the techniques of military gaming and digital computer simulation, which soon worked their way into the toolkits of operations researchers working in industry and the academy.[18]

The Legitimacy of Postwar Military OR

As with wartime OR work, postwar OR did not command any peculiar authority simply because it was scientists who conducted it. Rather, its legitimacy continued to be based on its ability to address and augment the usual thinking of military planners. If anything, OR's credibility was linked to the fact that it was explicitly deprived of authority. Although OR groups often made recommendations, no decision could ever flow directly from a report produced by a civilian organization. For this reason, OR personnel were presumed to be more at liberty to maintain an intellectual detachment toward their work than disputants within the military who had a direct stake in the outcomes of decisions. The value the military saw in OR certainly derived from its perception of the need for such objective opinions. But the military perhaps placed more value on OR's ability to clarify the arguments in military debates than on its ability to resolve them. In fact, because no decision rested upon an OR result, operations researchers were freer to identify areas where they deemed their own conclusions to be tentative. OR risked losing credibility if researchers boasted certainty on claims that were obviously speculative or that were later shown to be false.[19]

Document distribution procedures were probably the most important tool for regulating the relationship between OR work and military planning and authority. Only official policies were distributed widely. OR groups, by contrast, distributed their work only to those offices anticipated to have some immediate need for it. To even reach that level of distribution, though, OR documents had to be vetted by the particular offices with which the OR staff worked in preparing them. If a document produced in consultation with an office failed to anticipate criticisms, it could embarrass that office as well as the OR group. Documents were also categorized to clarify their intellectual status. One common categorization divided documents into reports, technical memoranda, and working papers, indicating, in decreasing order, the strength and finality of their conclusions. The documents themselves were filled with qualifying language, and assumptions narrowing the validity of the claims were explicitly stated, and sometimes underlined. OR documents were never endorsed by any military office, and were usually marked with a disclaimer stating so. Working papers were not even given the official approval of the head of the OR group, and were distributed mainly to garner feedback.[20] The comments OR groups received from military officers were often well informed, offering line-specific critiques, singling out where the officers concurred, where they disagreed and why, where limited sampling threw results into question, where additional study was most needed, and so forth. These officers could also make recommendations for further distribution. Ultimately, these discussions about OR documents merged with other discussions taking place within the military hierarchy, and might, working indirectly through them, result in changes in policy.[21]

The intellectual relations between authoritative plans and OR studies were not always clear-cut. In a 1951 letter to the U.S. Army's Chief of Psychological Warfare, Ellis Johnson explained the status of OR as scientific work in response to a query as to whether ORO reports found to be in error should be recalled. When military doctrines and orders were updated, previous versions typically were recalled or destroyed, but Johnson did not feel this process should apply to OR reports and memoranda because OR was a "pioneering enterprise." He pointed out, "The very greatest scientists have published papers which have … proved erroneous. Such papers are not withdrawn, they simply become known to those concerned with the subject as wrong or negligible." He recognized that the "Army audience is not exactly the same as a scientific audience, and many of those who read our papers may not be regular readers of scientific work on the same subject, and there may not be the usual safeguards that exist in professional scientific work by which students are guided to ignore past papers which are either wrong or irrelevant." Even still, he felt that OR inquiries should retain their histories, and that it was the responsibility of military planners to deal with OR on scientific grounds. Of course, he allowed that the Army had "a perfect right to control its distribution." Cognizant of the authoritative status of military orders and doctrine, he pointed out that the Army could recall distributed papers when it was deemed they "might lead to instances of confusion."[22]

The great existential danger to an OR group was the possibility that its work would cease to engage military officers' interests and problems. This could happen, for example, if groups failed to keep their studies timely. Because studies took time to design, execute, and prepare for distribution, there was a real possibility that, by the time they appeared, circumstances would have changed, data would have become obsolete, or an irrevocable decision would have already been made. Studies could also be neglected if they failed to take advantage of the knowledge of all of the relevant military experts and decision makers. If studies did not reach the appropriate individuals, or take account of their perspective, OR was more likely to languish. Of course, the worst problem was if reports were consistently judged to contain unrepresentative, misrepresented, or outright false information. The Army ORO set up a "murder board" in 1955 to subject studies to an additional layer of withering criticism before distribution, and made repeated surveys to determine how studies were used and by whom, and how they were regarded.[23]

Because OR had to engage military officers' interests, it could never claim complete intellectual independence. OR scientists were, of course, not in a position to question the missions undertaken by the services for whom they worked, nor did they have great latitude to pursue their own interests. At the ORO Ellis Johnson continually pressed the bounds of what constituted OR, sometimes straying well beyond the bounds of Army concern. For a military group, the ORO made an unusual number of contributions to the theoretical OR literature, and Johnson pressed to move into such areas as emergency aid and economic development.[24] Under his leadership, the ORO also started a high school summer program (figure 15.3), where students worked on such problems as assessing the likely effects of an atomic attack on Washington, D.C. That particular project was vetted by the ORO, published in the scholarly OR literature, and led to attention-attracting newspaper coverage. A headline in the *Washington Post* blared, "A-Attack to Kill 80%, 11 Boy Scientists Predict."[25] These initiatives, and Johnson's propensity to speak to the press about nonclassified issues, won him few new friends in the U.S. Army. The leaking of controversial ORO studies to the press by people within the Army, the Secretary of the Army's subsequent requests that the ORO staff undertake less controversial studies and refrain from speaking publicly, and Johnson's incensed reaction finally led to the ORO's closure in 1961. The ORO was quickly reconstituted without Johnson as a new nonprofit organization called the Research Analysis Corporation, or RAC. The joke at the time was that the initials stood for "Relax and Cooperate."[26]

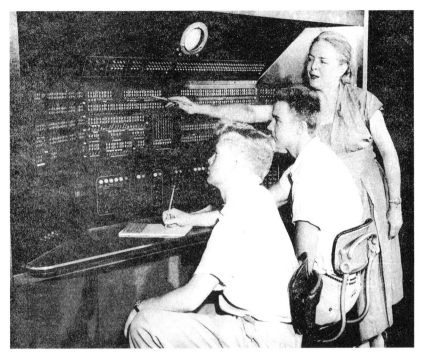

Figure 15.3
An ORO staff member, probably Jean Taylor, instructs students from the office's summer high school program on the use of an ERA 1103 computer. From Jean G. Taylor and John Balloch, "Report on ORO Summer Training Program," November 2, 1956, in JHU 02.001, Series 1, #47.2, Box 34.

Although Johnson was a singular figure in the military OR community, the tensions the ORO increasingly encountered were not unique to it. The director of the Navy's Operations Evaluation Group, Jacinto Steinhardt, was certainly a more demure figure. But in 1959 the Navy's desire to use OEG studies to further its agenda in Congress and the public sphere led him to suggest, to no avail, that entirely separate organizations be set up for research done for "political" purposes, in order to insulate research done for "practical" purposes. He reckoned that the correlation between OEG recommendations and implemented Navy policy, which was internally a sign of the OEG's success, could be used in external disputes to implicate the OEG as biased toward Navy interests.[27] A couple of years later, American military OR groups became even further drawn into interservice politics when President John Kennedy's new Secretary of

Defense, Robert McNamara, implemented a new Planning, Programming, and Budgeting System (PPBS) to help organize and justify his department's appropriation requests to Congress. OR groups were unusually skilled in helping the services assert their point of view within this new funding rubric.[28] By this time, however, the entire terrain of contract analysis had changed, and it will be necessary to backtrack in order to appreciate the nature of these changes.

16 The Challenge of Rational Procurement

Inasmuch as the end of World War II presented problems for planning future tactics, the great strategic dilemmas of the postwar era revolved around procurement. Maintaining war-ready forces and investing in entirely new classes of military technologies demanded an ongoing reflection on the future nature of warfare, all within view of rapidly ballooning, but nevertheless finite budgets. Atomic weaponry was only the most obvious addition to new lines of research and development that included rocketry, jet engines, automatic guidance control, and radar. All of these new technologies had to be considered on top of, and in conjunction with, ordinary developments in armaments, armor, transportation, communication, and medical technology.

The military services struggled continuously to cope with this postwar reality. In Britain, where budgets were much more constrained than in the United States, the government sufficed by rearranging and expanding its traditional civil service bureaucracy. At the end of the war, the Ministry of Aircraft Production was folded into a single enormous Ministry of Supply, which oversaw procurement for both the Royal Air Force and the Army, but still not the Admiralty. In 1954 the UK Atomic Energy Authority became independent of this ministry, as did the Army's supply function in 1959, whereupon the Ministry of Supply became the Ministry of Aviation. In 1964 this new ministry was in turn folded into a new bureaucratic experiment, the Ministry of Technology, which was itself reorganized out of existence by the end of the decade.[1]

While individual services retained a great deal of control over their procurement policies following the war, these policies were meant to be coordinated through a Defence Research Policy Committee (DRPC), which was founded in 1947. The postwar Labour government coaxed Henry Tizard back into government service to chair both this committee and a parallel Advisory Council on Scientific Policy (ACSP), and he served on

them until 1952. The DRPC comprised high-level military officers responsible for procurement, military R&D budget controllers, chief scientists of
military R&D organizations, and the scientific advisers to the services (who
oversaw the services' OR staffs). Though the committee was vested with
no executive authority, the services used it on a voluntary basis to coordinate their R&D and procurement policies until the Ministry of Defence
was strengthened and reorganized in 1963. At that time the DRPC was
replaced by a Defence Research Committee, a Weapons Development Committee, and an Operational Requirements Committee. Solly Zuckerman,
the final chairman of the DRPC, was retained in what had been his ex
officio role as Chief Scientific Adviser to the Ministry of Defence. The next
year, when Harold Wilson became Prime Minister, he was elevated to Chief
Scientific Adviser to the entire government.[2]

The American services likewise remained, for all practical purposes,
independent of each other until the 1960s. During this postwar period,
they competed vigorously for control of new, lavishly funded weapons
programs, creating vast institutional infrastructures to sponsor and guide
their individual procurement and R&D efforts. In addition to new development and testing facilities, the services founded powerful new policy-
building and project-funding bodies, such as the Office of Naval Research,
established in 1946.[3] New high-level advisory committees proliferated,
including the Defense Research Board, and aerodynamicist Theodor von
Kármán's Scientific Advisory Board in the Air Force.[4] Of the numerous
organizations dedicated to undertaking military R&D and formulating
R&D policy in America, here we will mostly concern ourselves only with
the two most germane to the history of the sciences of policy: the RAND
Corporation and the Weapons Systems Evaluation Group (WSEG).
However, neither organization should be thought of as isolated from, or
especially unique within, the rest of the R&D advisory and policymaking
infrastructure.

The origins of the RAND Corporation can be traced to the advocacy of
Edward Bowles. Unlike many of his university colleagues, Bowles did not
leave his wartime post at the end of the war. Ultimately, this move would
not turn out well for him. His main bureaucratic patrons, Henry Stimson
and General Henry Arnold, both retired within a year of the war's end.
Then, in 1947, shortly after deciding to cut his ties with MIT, Bowles was
dismissed from his position when the War Department was absorbed into
the new National Military Establishment (renamed the Department of
Defense in 1949).[5] However, before that happened, he did have at least
some impact on the shape of the postwar R&D bureaucracy. Following his

recommendation, the Army Air Forces created a new Deputy Chief of Staff for R&D post, which was filled by Curtis LeMay. When the Air Force broke from the Army in 1947, LeMay moved on to command the Air Force's European operations, while his post was dissolved and replaced with one an echelon below it.[6] Meanwhile, though, Bowles also initiated the process that established the RAND Corporation, which would prove a more enduring achievement.

At the end of the war, Bowles remained impressed with the studies he had initiated on the possible uses of the B-29 bomber, which were conducted by William Shockley and representatives of Douglas Aircraft and Boeing. He therefore advocated the creation of new venues where military planners and industrial contractors could work closely together in proposing future R&D projects, and in studying the "economics of warfare" so as to estimate the relative payoffs of competing R&D choices.[7] In October 1945 Bowles and Arnold met with Donald Douglas of Douglas Aircraft, as well as with Douglas employees Frank Collbohm and Arthur Raymond, who had both worked on the B-29 study. Together, they agreed to establish a new contract organization that would lay the theoretical groundwork for the future development of missiles and the tactics for using them.[8]

Air Force Project RAND was set up in early 1946 in Santa Monica, California, as an independent wing of Douglas Aircraft, with Collbohm serving as director. However, the purview of the project had expanded from missiles to encompass all potential means of non-land-based intercontinental warfare.[9] Although at first very pleased with Project RAND, Bowles soon began to sour on it. The open-endedness of RAND's contract allowed it to expand beyond the point he felt was amenable to practical work, and he feared that the project would lose its focus on concrete R&D problems. In his mind it should have been followed by the establishment of a series of RAND-type projects under contract with other organizations.[10] Bowles's fears were further stoked by the 1948 decision to sever RAND from Douglas Aircraft and to form the independent nonprofit RAND Corporation. By this time, though, he was helpless to influence the matter further.[11]

It would take RAND several years before it carved out a strong identity for itself, a process to be detailed in later chapters. For the moment, our concern is only with RAND's status as a contract organization with a permanent civilian staff mandated to study the feasibility and comparative desirability of potential R&D projects. RAND was different from most military contractors in that its staff neither conducted R&D, nor consulted on others' R&D projects, at least at first.[12] For this reason, and because it evaluated potential R&D projects in view of their likely operational uses, RAND was often grouped with the military's OR groups. Since the Air Force's own operations analysts had little concern with R&D or procurement, their concerns were usually defined as short term, where RAND's were referred to as "long range" or "blue sky."[13]

In the 1960s, RAND would extend its influence from the Air Force to the Defense Department as the latter organization began to exert tighter control over procurement spending. In the meantime, though, procurement presented high-level defense administrators with severe problems. Prior to World War II, America's military services set their R&D policies more or less independently. During the war, it became advantageous to set up a Joint New Weapons and Equipment Committee, which was chaired by Vannevar Bush. In 1947 relations between the now-three services were formalized with the creation of the National Military Establishment, which included a Joint Chiefs of Staff organization rooted in ad hoc wartime arrangements. Bush's committee was also reconstituted under the new Establishment as a more expansive Research and Development Board (RDB), which Bush continued to chair for its first year. The RDB was intended to play a more intensive role in the prioritization of projects than the wartime committee, and to approach the task it employed over two thousand technical consultants. This small army immediately became buried beneath eighteen thousand potential R&D programs, which had to be individually evaluated before it was felt the RDB could give responsible advice.[14]

The RDB's task was further complicated by the fact that R&D projects had to be evaluated in view of their anticipated value in future combat operations, rather than by some intrinsic merit. The implications of future operational contexts for development, design, and procurement were supposed to be considered by a small group of part-time scientific advisers to the RDB's "policy council," which was responsible for evaluating the board's allocation of its consultants' efforts.[15] This group was chaired by the Columbia University physicist I. I. Rabi, who had been an assistant director of the MIT Rad Lab during the war. The other members were the wealthy independent physicist Alfred Loomis, the wealthy independent entomologist Caryl Haskins, and William Shockley, who had returned to Bell Laboratories following his war work.

At the advisory group's fifth and final meeting in October 1947, held in Loomis's New York City apartment, the group decided to recommend that they be succeeded by a larger organization, which they described in a memorandum to Bush. Prepared from Shockley's notes, the memorandum outlined the need to pursue what it called the "conceptual war" in a more rigorous fashion than they could muster. "The conduct of a 'conceptual war'," the memorandum explained, "presents a complicated problem in cooperative enterprise. It cannot be carried out by any one, small, isolated group of people no matter how well they are supplied with reference material. Instead," it continued, "hypothetical problems must be taken from time to time to experts in the services, in industry and in science—all of whom would, in time of war, have to deal with problems of the same nature. The synthesis of these individual problems into a meaningful, directed, military research and development effort can be achieved only by placing responsibility and authority in the hands of a competent, full-time, central operational analysis group." It is likely that Shockley's wartime experience in OR influenced the language of the proposal, since R&D had, to that point, never been subjected to "operational analysis." The memorandum went on to recommend that a full 10 percent of RDB personnel be devoted to such long-range analytical efforts, and that they should work in close connection with "already existing service agencies working in similar, but more restricted fields," namely the Navy's OEG, the Air Force's OAD, and Project RAND. The memorandum suggested establishing similar groups under the Army and the Joint Chiefs of Staff as well.[16]

The group's recommendations swiftly made their way into the military bureaucracy. In February 1948, following a discussion of the issue with the Army Chief of Staff, General Dwight Eisenhower, Secretary of Defense James Forrestal made a formal request that a means be established of obtaining "objective and competent" advice on "the technical capabilities and performance of present and probable weapons systems."[17] Amid negotiations between Forrestal and the Joint Chiefs about who would have authority over the new organization, the request stalled.[18] Meanwhile, following the Navy's example of contracting with MIT, the Army began negotiating with Johns Hopkins University in May to administer a "General Research Office." In August, the office began work at Fort McNair in Washington, D.C., and was renamed the Operations Research Office in December.[19] Back at the National Military Establishment, the Weapons Systems Evaluation Group (WSEG, pronounced "wesseg") was established as a part of the civil service staff at the Pentagon in January 1949.[20]

Among civilian military policy research organizations, RAND and WSEG clearly stood apart from OEG, OAD, and the new ORO in that both were dedicated to evaluating potential R&D programs in light of the military's long-term strategic requirements. RAND and WSEG were, however, also very different from each other. As an independent contractor, RAND was able to attract a large staff quickly by building a reputation as an enclave of highly talented and well-paid individuals who were largely free to examine whatever problems they saw fit. Not only did RAND evaluate proposed top-secret R&D programs, its staff performed cutting-edge work in mathematics, and increasingly, computing and microeconomics, and published in the academic literature. Finally, RAND was located in the sunny climate of southern California far from the prying Washington bureaucracy. While it was often difficult to point to precise impacts of RAND's studies—a fact that led to criticisms and even the occasional existential threat—RAND was nevertheless considered an important asset to the Air Force, and a productive and happy place to work.[21]

WSEG, in contrast, was not in RAND's class. It had difficulty recruiting analysts, forcing it to borrow heavily from the services' OR groups, and it was only slated budgets in the hundreds of thousands of dollars, as opposed to the millions made available to RAND.[22] Its small core of a dozen or so permanent analysts were civil servants and were paid accordingly small salaries. The group also suffered from an instability of leadership. Its research directors were all prestigious scientists with wartime experience, who proved eager to return to higher-paying, lower-stress, and more intellectually appealing posts. Against his own inclinations Philip Morse agreed to become the first research director in March 1949, but for only one year.[23] He was followed by Bob Robertson, then at the California Institute of Technology, who left in 1952. Then Bright Wilson came from Harvard's chemistry department and stayed a year. The position then remained vacant for an entire year before William Shockley was recruited. He held the post from mid-1954 to late 1955.[24]

The personnel problems at WSEG adversely affected its work. WSEG was bombarded by research requests from the Joint Chiefs of Staff, most of which were extraordinarily broad in scope, resulting in multivolume reports running into the hundreds of pages.[25] Reports were assembled by collecting technical information, intelligence estimates of enemy capabilities, and strategic thinking being produced by the military and its various contractors, and synthesizing this work into statements about the worth of proposed, massively complex weapons systems. Because of the still rampant competition between services, though, study results were as likely to spark controversy as they were to lend clarity to procurement decisions. WSEG's reports were appreciated as compendia of information, but the reports did not have the intellectual authority to affect policy in the face of determined resistance.

By Wilson's and Shockley's tenures as research director, WSEG's problems were deemed chronic.[26] In April 1953 Wilson and Lieutenant

General Geoffrey Keyes, WSEG's administrative director at the time, visited RAND to see what could be done to improve their organization. Ahead of their visit, the bemused author of RAND's management committee minutes recorded parenthetically, "They assume RAND is successful and can assist them." After the visit, though, the management committee could only marvel at the situation at WSEG: "If RAND has a problem in correlation of its studies, WSEG has several times the problem trying to correlate and sift out the right answers being done by some 30 RAND-type organizations. Some of the basic problems of WSEG are the level in the Defense Department at which they work, inability to obtain competent help, etc."[27] Around that time the RDB was itself shut down and replaced by an office of the Assistant Secretary of Defense for Research and Development. In effect, the department had given up trying to bring a semblance of order to the free-for-all that characterized interservice R&D budget allocation, since it had neither the wherewithal to justify budget allocations rationally, nor the political backing to do it arbitrarily.

When the RDB folded, the Defense Department's leaders decided the department could still lend guidance to the services on those occasions when they agreed to coordinate their efforts through the Joint Chiefs. Therefore, rather than fold WSEG as well, the department and the Joint Chiefs heeded suggestions for turning it into a RAND-like nonprofit entity. In 1955 the department entered into a contract with MIT to recruit analysts. That contract was then superseded by the foundation of the Institute for Defense Analyses (IDA) in 1956.[28] IDA was a nonprofit institution that was supported through a consortium of universities, initially for the specific purpose of aiding WSEG.[29] IDA's head civilian liaison to WSEG doubled as WSEG's director of research, while WSEG itself became a military organization.

This arrangement proved to be a very successful cure for WSEG's immediate ills. Within a few years IDA had reached its goal of recruiting one hundred university scientists to work on WSEG problems. Building on its successes with WSEG, IDA quickly moved into new areas. It sponsored the creation of JASON, a summer study group for elite university scientists that was dedicated to conceptualizing ambitious new military technologies.[30] It also built a working relationship with the Defense Department's new Advanced Research Projects Agency (ARPA), and provided contract research assistance for government blue-ribbon panels such as the 1957 Gaither Committee on civil defense and nuclear deterrence, and the 1958 Draper Committee on foreign military aid.[31]

Even supported by IDA, though, WSEG was never in a position to organize procurement around a unified American defense policy. Not only was President Eisenhower's Pentagon unwilling to attempt the feat, but by 1960 any semblance of an orderly system of military policy analysis had broken down. Over the course of the 1950s contractors multiplied and hybridized into a dense web of military and government offices, and quasi-governmental institutions, gargantuan corporations, and small start-ups. Most contact between military planners and outsiders was, of course, with their industrial suppliers. These suppliers sometimes cleaved off new firms. Much as the Douglas Aircraft Company produced the nonprofit RAND Corporation, in 1953 employees of the Hughes Aircraft Company resigned to form the for-profit R&D consultancy, the Ramo-Wooldridge Corporation, soon renamed Thompson Ramo Wooldridge (TRW), Inc. These organizations joined a number of other independent R&D consulting organizations, such as the Battelle Memorial Institute, established before the war, and the Stanford Research Institute, established just after.[32]

As we have seen, the Army's ORO and the Navy's OEG were administered by Johns Hopkins and MIT respectively, while a consortium of universities administered IDA. Major universities also increasingly hosted research centers to study policy problems and foreign cultures and politics.[33] They also, of course, pursued advanced R&D projects on military contracts. Johns Hopkins hosted the Applied Physics Laboratory, which grew out of the wartime laboratory that developed the proximity fuse, while MIT sponsored a number of military-oriented research projects, foremost of which was the Semi-Automated Ground Environment (SAGE) radar defense system, which gave rise to the MIT Lincoln Laboratory. RAND entered into the field of R&D consulting when it developed a substantial staff of experts in computer systems in support of the SAGE project, which grew to become the largest branch of RAND before it was spun off into the Systems Development Corporation in 1957. The Lincoln Laboratory itself spun off the nonprofit MITRE Corporation in 1958.[34] Universities were also key players in the development of the national laboratory system, including centers for civilian science such as Brookhaven National Laboratory and Ernest Lawrence's Radiation Laboratory at Berkeley, but also the weapons laboratories at Los Alamos and Livermore. All were funded by the Atomic Energy Commission.[35]

The expertise in scientific research, engineering, and military planning flowing through this system of institutions—the extent of which is only hinted at here—made it possible both to create a dizzying array of sophisticated and extraordinarily lethal military technologies, as well as to justify the need for any and all of them. Policymakers, by contrast, had little means of determining what the nation's military requirements might actually be, and which particular technologies and in what quantities could best help fulfill them. Eisenhower created the President's Science Advisory Committee (PSAC) in 1957 in response to Sputnik, and found it useful to have a body of independent experts who could help assess controversial proposals on the White House's behalf, such as a potential treaty to ban nuclear weapons testing. However, PSAC was not, nor was it intended to be, a solution to the general problem of defense budgeting. Eisenhower remained alarmed by the self-perpetuating growth of military suppliers to the end of his presidency, and famously used his 1961 farewell speech to warn of the pernicious influence of what he called the "military-industrial complex," and the possibility that national policy could become "captive" to a "scientific-technological elite."[36]

When John Kennedy replaced Eisenhower at the White House, he installed Robert McNamara as his new Secretary of Defense. McNamara had taught at the Harvard Business School before becoming a part of the Army Air Forces' statistical control organization in World War II, where he helped LeMay plan and administer the incendiary bombing of Japan. After the war he joined the Ford Motor Company, where he and a team of statistical experts helped streamline management and executive decision making. He was eventually named the company's president. At the Pentagon, McNamara moved quickly to institute a new "Planning, Programming, and Budgeting System" (PPBS). PPBS had been developed by RAND economists, and it was administered by staff that McNamara hired from RAND. The Pentagon's budgets no longer simply allocated funds to each service separately, but to particular "programs" that could cut across service boundaries, were funded based on their anticipated contribution to overarching defense policies. These potential contributions were, in turn, assessed by a new Office of Systems Analysis, which initially had only a handful of analysts, but grew to over two hundred by 1966. With Kennedy's, and later President Lyndon Johnson's backing, McNamara used this new PPBS framework to exert unprecedented Defense Department control over budgetary allocation, provoking widespread military resentment toward him and his analytical staff, who were pejoratively labeled the "whiz kids."[37]

17   Wartime Operational Research in Postwar Britain

Of all the ad hoc wartime amendments to the intellectual economy of military planning, only OR was singled out afterward as an especially noteworthy development. Because it was singled out, the terms "operational research" and "operations research" would go on to designate both a new profession and a new body of mathematical theory. Yet, in the first few years following the war, there was no effort to use the term to further either of those goals, nor was any significant outside attention paid to the military's perpetuation of its wartime OR groups. The main reason that OR was considered significant in this period was that it was regarded, and mainly only by a few vocal scientists in Britain, as an exemplar of a newly well-functioning relationship between scientists and military decision makers. As such, OR seemed to augur the beginnings of a new collaboration between scientists and policymakers more generally, which would substantially change the way public policy would henceforth be made.

This idea, of course, had its roots in the rhetoric that scientist activists and bureaucratic reformers had used in 1941 and 1942 to advance their various proposals. As OR groups multiplied rapidly beginning in 1942, the use of that rhetoric waned, only to be revived as the war came to a close. J. D. Bernal was particularly quick to seize the moment. As we have seen, as early as September 1941 he had identified OR with his pre-existing vision of a government that not only consulted scientists, but also was itself "scientific through and through." In November 1945, he picked up on this theme in a lecture to a scientific audience at the Royal Institution in London entitled "Lessons of the War for Science." Inasmuch as scientific reasoning and scientific technology had helped to win the war, their potential in peace was even greater. To him, not only would they become the basis of a planned Marxist society, the year 1945 would mark the beginning of an entirely new epoch in human history.

Significantly, though, Bernal's radical suggestion that humanity stood at the dawn of a new era was buried some two-thirds of the way through his lecture. His intention clearly was not to set aspirational goals, but to portray his ideas as concrete and realistic. He spent the bulk of his time on a sophisticated discussion of the relations between research and development and military planning during the war, and how those relations might translate into the work of peacetime institutions. He called his audience's attention to a number of specific products of wartime research, to the improvement of mundane procedures such as equipment testing, and to equally mundane organizational issues such as alterations to committee structures, all of which had important prewar roots. But he placed a particular emphasis on OR, which he called "virtually an entirely new technique in the application of science." He explained that, through OR, for "the first time immediate, personal liaison was effected between the scientist and the soldier," and that this had led to a "mutual education." This education allowed scientists to see the "real requirements of the [technical] devices as distinct from what they had been imagined to be by the military staffs." Scientists became better able to discern the "operational possibilities" that the devices they were designing allowed, and so began to "influence operations in relation to these possibilities." By integrating the design of equipment and tactics, it also became easier to take into account how crucial "human elements," such as morale, fatigue, training, and enemy reactions, might bear upon choices in the design of both equipment and tactics.[1]

Importantly, scientists' "infiltration" and "integration" into military planning had not entailed any traumatic realignment of authority over either science or military planning. There had been no "dragooning of scientists," which, Bernal noted, was the usual "bugbear of the opponents of scientific organization." On the contrary, enjoying "the greatest freedom and scientific initiative," scientists were able to apply their skills where they would provide the most benefit. "Where there was over-direction of scientists," he pointed out, "it was usually because either the directors were not themselves scientists but were civil servants or military men, or because they were scientists of the old school of government science." Likewise, scientists did not impose their preconceived ideas on military planners. OR, "dealing as it did with actual achievement, provided a quantitative corrective to theorizing." Moreover, OR scientists learned "to appreciate the human qualities of command and performance, and the conditions under which it is necessary to make and adhere to decisions where many of the vital factors cannot possibly be determined in time, or at all." They therefore became adept at selecting only those problems to address where they believed they had something substantial to contribute. Bernal remarked that it was "greatly to the credit of all ranks and branches of the services that this development, which could so easily have been resented as an intrusion, was on the contrary welcomed and encouraged." That spirit, he noted, had not prevailed in Germany and Japan, where "scientists were not trusted with the development and utilization of their own ideas."[2]

For Bernal, the wartime experience clearly invalidated criticisms of the impracticality of scientific planning and socialist governance. "Planned" science would not mean "arbitrary dictation from above," nor would socialism entail "scientists usurping the functions of the state." He assured his audience, "The responsibility for final decisions necessarily taken on incomplete knowledge is the business of those selected by the people to act on their behalf." It was the task of scientists "to see that the information is as complete as it possibly can be at the time and that research is being adequately carried out to ensure the acquisition of more accurate information." In the economic sphere, as the "rigid demarcation" between science and engineering dissolved, previously isolated academic scientists would be able to select topics for research suggested by industrial work, and to pursue those topics wherever they might lead. Because it spanned both worlds, Bernal suggested that operational research represented "a necessary link between laboratory science and industry." He also saw a kinship between OR and the existing field of consumer research. If industry were to serve social needs instead of private profits, policy would need to be guided by social scientific research into what people's needs were, and how well they were being met. Science would thus play a dual role: "The general objective," he declared, "is that of the positive development of scientific ideas in light of scientifically assessed human requirements." He promised his audience that the scientific experience of the war had provided a "demonstration of the possibility of the rational removal" of the evils the war had brought, and augured their "replacement by a better state than we have ever before been able to imagine."[3]

Bernal's ideas about OR were carried into government deliberations by the biologist Cecil Gordon. Also a Marxist, Gordon had served with the OR Section at RAF Coastal Command, where he had been responsible for the development of the command's Planned Flying and Planned Maintenance scheme. In 1946 he was appointed to lead a small Special Research Unit at the Board of Trade, comprising mainly ex-wartime OR scientists, and vaguely tasked with studying problems bearing on industrial productivity, though also, following Bernal's emphasis, consumer research. Later, in 1948, Gordon would join the working party of an ad hoc government Committee on Industrial Productivity, chaired by Henry Tizard, where he would preach a similar gospel.[4]

Ultimately, though, the impacts of early postwar rhetoric surrounding OR were more discursive than practical. Bernal's linking of consumer research to OR was repeated in a chapter dedicated to consumer research in the 1947 book *Science and the Nation*, assembled by the Association of Scientific Workers (AScW), then under the presidency of Patrick Blackett.[5] Blackett himself hailed the wartime development of OR, though his vision was perhaps less ambitious than Bernal's. He used OR mainly as a rhetorical cudgel. For example, in his May 1947 presidential address to the AScW, he chastised the Ministry of Fuel and Power for its handling of scientific matters. He spoke about a meeting he had with ministry officials—"at which, incidentally, no scientist or engineer from the Ministry was present"—wherein he was assured that they would appoint a robust scientific and technical staff. But Blackett felt that nothing was, in the end, done, and so asked, "Is it likely, to take but one example, that with a technologically informed Ministry with an Operational Research Section, this country should have been so short sighted as to take off purchase tax from electrical appliances, and make an unnecessary increase in train services, a few months before a winter during which it was certain that coal stocks would be the lowest ever?" He did not take this to be an isolated failure, deciding, admittedly "flippantly," that it represented a postwar "counter offensive of the administrative civil service" against the wartime incursion of "technologically educated but sometimes rather uncouth outsiders."[6]

Although Blackett remained highly interested in his experience with OR, he did not concentrate on its example exclusively. The peculiar virtues of Henry Tizard also continued to be a crucial point of reference. In his AScW address, Blackett lauded the decision to name Tizard the chairman of both of the government's new top-level science advisory bodies—the Advisory Council on Scientific Policy and the Defence Research Policy Committee—as "decisive evidence" of the failure of the "counter offensive" he detected. Blackett recalled:

One of the main origins I think of the brilliance and soundness of [Tizard's] own judgment on matters of warfare ... was that he made it a policy to seek out the young men, in the establishments and in the squadrons, who were actually doing the real jobs. Few, for instance, who attended it, will forget a Conference in London, the day after a heavy blitz, where he assembled fighter pilots and Air Marshals, radar designers and administrators to thrash the problem of the defence of London. This is the spirit in which the tasks of peace should, but are not always, being tackled.

Tizard had uniquely the quality of knowing what he could not know himself; and of going to the man on the job to find out.[7]

Of course, Blackett did not suppose that Tizard's appointment was a solution to Britain's policy troubles. Much work remained to be done.

Meanwhile, other invocations of the wartime OR experience began to identify it with practically any activity that might in some way inform policymaking. At the August 1947 meeting of the British Association, a session was held on "Operational Research in War and Peace," featuring comments from, among others, Bernal, Robert Watson-Watt, Solly Zuckerman, and the first postwar Scientific Adviser to the Army Council, Owen Wansbrough-Jones. Zuckerman remarked, "Such Cabinet planning bodies as [the Treasury's Economic Planning Board] presided over by Sir Guy [sic, Edwin] Plowden may perhaps be regarded as operational research sections that have had to be established at a strategic level in the national economic field…. So, too, can the [ACSP], presided over by Sir Henry Tizard." Watson-Watt argued that OR was "an important contribution to the still unsolved problem of the relation between the administrator and the victim of that opprobrious epithet 'the expert.'" Such broad definitions created an immediate need to deny OR's novelty. Zuckerman stated that OR "was no more than the application of ordinary scientific methods of thought, or controlled commonsense, to the analysis of complicated and dynamic problems. It should be recognised as no new thing." Wansbrough-Jones similarly argued that "the sole good reason for retaining the term 'operational research' in place of the term 'method research,' which was in general use in the War Office, was to keep the goodwill of the trade name."[8]

Because OR was being conflated with broader concerns over science policy and sound governance in postwar Britain, the fate of the rhetoric surrounding it became closely tied to the outcomes of the political debates of the time, particularly those surrounding "planning."[9] The journal *Nature* began the year 1948 with an editorial on "The Deployment of Scientific Effort in Britain," in which the question of "the limits" of planning "brought to the fore a question which was already under discussion, namely, how far the methods of attack on urgent objectives adopted during the War, and generally known as 'operational research,' can be applied in the present [economic] emergency, or even as a regular feature of our peace-time system." The editorial went on to summarize Bernal's view of OR as a way of selecting research priorities, and to quote Zuckerman's equation of OR with the work of government committees. Hewing to its usual moderate line, *Nature* vaguely supported the leftist-endorsed possibilities that OR represented, but waxed pragmatic about its ultimate potential, warning that "the new method" might be "discredited before it has been fairly tried."[10]

Nature's position mystified geneticist C. H. Waddington, who had led RAF Coastal Command's OR Section for a period during the war. In a letter to the journal, he denied that OR represented any special "technique." He reckoned, "It is, perhaps, because operational research scientific workers had to depend so largely for detailed information on the generous collaboration of their specialist colleagues that some people appear to think that they invented new techniques of teamwork." The world needed such inventions, but that, he argued, was a matter for "the sociologists." More to the point, he was startled to "find *Nature* implying that operational research is a method of planning." He countered: "Research by itself can never produce a plan; its function is to prepare the basis on which a rational plan can be founded, by identifying qualitatively the factors in the situation, and, as far as possible, by evaluating them quantitatively." By this definition, the idea that OR could ever be dismissed as an essential part of rearranging "the scientific effort of Great Britain" was absurd. Naturally, it would be necessary to investigate the demands on various scientific fields, why scientists decided to undertake the research they did, what demands would be made on educational institutions, and so forth. This sort of investigation was a requirement for *any* sound policy planning, but the necessity of it had no bearing on the question of the possible extent of planning and its inherent limits.[11]

By this point, though, it was becoming less important to control the definition of OR in public discourse, as its rhetorical high tide had already begun to recede. It was at this time that Cecil Gordon took up an academic post in genetics at the University of Edinburgh, leaving the Special Research Unit at the Board of Trade to die a quick bureaucratic death. His efforts to push OR through the government's Committee on Industrial Productivity likewise fell by the wayside. Feedback on his proposals had observed that his ideas failed to distinguish themselves from "the technique of good management, cost accounting, work layout, etc." Even Henry Tizard, the committee's chair, felt that "the attention paid to operational research may be overdone."[12] When the sense of peculiar significance surrounding OR passed away, the subject was reduced to the status of an episode in a longer drama playing out between science and the British state. For example, in 1951 the Chief of the Royal Naval Scientific Service and DRPC member, Frederick Brundrett, assembled a history of science in the civil service, which led from the prewar establishment of the directorships of scientific research, to Tizard's chairmanship of the Committee for the Scientific Survey of Air Defence, to operational research, to the establishment of the scientific advisory posts in the services, and ultimately to the DRPC.[13] The same year, in a speech on "Science and the Machinery of Government," Tizard himself applauded how scientists had become "more and more intimately concerned with military thought," adding that it "did not occur without some resistance." He felt this integration was at that time superior in Britain than even in America, and he expressed his hope that it had "come to stay." However, with an eye toward future progress, he acknowledged that in the areas of civil science and civil policy there seemed, as always, to be a long way left to go.[14]

18   The Operational Research Club

As the significance of OR in rhetoric about relations between science and the British state began to wane, it appeared in a more modest form in discussions about research and management in British industry. In this context, wartime OR was invoked by a group of people who had been impressed by its successes and who wanted to replicate them outside the military. Their efforts were primarily responsible for establishing a handful of OR groups in industry, and for instituting an "Operational Research Club," which was soon renamed the Operational Research Society.[1] They did not suppose OR was a novel thing, nor did they attempt to reinvent it as a profession. For them, the term "operational research" served mainly as a rubric for collecting and promoting a scattered array of practices designed to improve industrial efficiency and managerial effectiveness not otherwise captured by the concept of research and development.

Aside from Cecil Gordon, the first figures to promote OR outside the military were Charles Ellis and Charles Goodeve. As the wartime Scientific Adviser to the Army Council, Ellis had overseen the War Office's overseas OR groups, and had been in close contact with the Britain-based groups administered through the Ministry of Supply. After the war ended he elected to remain an administrator, and was named the "scientific member" of the National Coal Board (NCB), the governing body of the newly nationalized coal industry. There he called for the establishment of new research directorates to handle "fundamental laboratory work, operational research on coal-getting and coal preparation, [and] research into matters involving human aspects, e. g. health of the miner."[2] The OR function he mentioned became included in the work of a new "Field Investigation Group," which was eventually renamed the Board's Operational Research Branch in 1962.[3] Meanwhile, as Deputy Controller for Research and Development at the Admiralty, Goodeve had used studies produced by Patrick Blackett's OR group to inform his work on the allocation of the Admiralty's R&D budget.

When the war ended, Goodeve resigned his commission to become the director of the newly founded British Iron and Steel Research Association (BISRA), an organization dedicated to conducting R&D on behalf of the entire industry. There, in addition to championing technical modernization, he created an OR group and encouraged the adoption of OR throughout the iron and steel industry.[4]

Ellis and Goodeve shared with people like Blackett and J. D. Bernal concerns about the need for collaboration and coordination. However, as insiders, they had no need to gesture toward abstract symbols, such as the wartime OR experience, to indicate what essential but missing virtue needed to be cultivated. For example, Ellis's staff at the NCB was sensitive to "the tendency to relegate scientists to the back room (the 'boffin' complex)," but, for them, it sufficed to observe, "impeccable administration and staff duties will help to overcome it."[5] Similarly, shortly after becoming the director of BISRA, Goodeve addressed the Sheffield Society of Engineers and Metallurgists, and defended the industry against charges of "conservatism," supposedly evidenced by its prior lack of a research association. He pointed to the industry's long history of "co-operation between the research worker and the other parts of the mechanism of industry" as well as cooperation "between separate firms and the suppliers of their raw materials and equipment and the users of their products," which was only augmented by the industry's foundation of BISRA.[6]

While Ellis never became a major figure in the OR community, Goodeve quickly became the foremost voice on the subject in Britain. In doing so, he fostered a newly concrete discourse on the topic. In an article on OR he published in *Nature* in 1948, he echoed others' claims that the idea was not novel, describing it simply as a "thinking process," but one which had "already to some extent been used in industry under other names." Examples included work in "market problems, cost accounting, quality control and works efficiency; in other words the operations of industry." This vision of OR cast it as the inheritor of an established legacy, while still allowing the term to be used to describe new breakthroughs. For instance, Goodeve saw a potential for a new "generalized Second Law of Thermodynamics" in industrial operations. His main intent, though, was to use the idea of OR to promote methods that already existed. Pointing to the case of statistics, he observed that while the "application of the statistical tool is still on the increase," there was "no need or likelihood of the tool itself being developed further."[7] (This claim drew a quick rejoinder from the eminent statistician Frank Yates, who assured the readers of *Nature* that innovation in statistical methodology was alive and well.)[8]

A month after his *Nature* article appeared, Goodeve became the convener of the new Operational Research Club. Meeting periodically at the Royal Society's headquarters in London, the club fashioned itself as "a small informal group of people who are working in or are concerned with problems associated with Operational Research." It had a limited membership, initially capped below 100, allowing only one member per organization represented. Its activities primarily involved arranging lectures and informal discussions. Its promotional work was initially very passive. In 1950 the club began publishing *Operational Research Quarterly* (henceforth *OR Quarterly*). Although available by subscription, it was more or less a newsletter. Early issues were between fifteen and twenty pages in length, and featured editorials, club news, an invited article (often ruminating on the meaning of OR), and, most importantly, abstracts of articles appearing elsewhere likely to be of interest to club members.[9]

Club membership partially comprised people who had been involved with OR during the war and had since moved into civil research. Reuben Smeed, for example, had served in the RAF Bomber Command OR Section during the war, and then moved to the government's Road Research Laboratory, where he worked on problems of traffic regulation and road safety. Although the club did not concern itself much with military issues, a few members of the postwar military OR groups, such as Ellis's successor at the War Office, Owen Wansbrough-Jones, did become club members. The club also quickly began to include people working in government and industry who had no experience with wartime OR. In part, these newcomers were people hired by the OR groups created by Ellis and Goodeve, or in groups modeled on those original groups. Donald Hicks and Patrick Rivett, who worked at the NCB, were both important figures in the 1950s British OR community. Another major figure was Albert Swan, who worked for the OR group at the United Steel Companies (USC), which was modeled on the BISRA group. In 1951 Swan left USC to head a new group established by the Courtaulds textile company. USC's Stafford Beer, another well-known member of the British OR community, attempted to tie the OR rubric to his own peculiar "cybernetics"-oriented technocratic vision. Finally, some club members had no formal association with OR at all, but were simply attracted to its goals. The well-known statistician Leonard Tippett was an active member. He had been working since the 1920s on OR-type problems for the Shirley Institute, otherwise known as the British Cotton Industry Research Association.[10]

The intellectual interests of club members were scattered. Early issues of *OR Quarterly* were filled with reports and abstracts of activities in areas as diverse as agriculture, textile manufacture, air traffic control, construction, retail, government surveying, and military operations. The particular activities addressed did not have an intellectually novel dimension, since OR, by definition, was reckoned to be the practical application of ideas already developed in fields such as statistics and economics. Some such applications were explicitly designated as OR when they first appeared; others were identified as examples of OR post hoc. For example, even before the foundation of *OR Quarterly*, G. E. Bell, who led a small OR section at the Ministry of Civil Aviation, published an article in the *Journal of the Royal Aeronautical Society* on the application of statistics to air traffic control problems, which he labeled as being "operational research."[11] Conversely, Reuben Smeed's first full-length article for *OR Quarterly* had originally been a talk given, without any reference to OR, to the psychology section of the British Association.[12]

The diversity and pervasiveness of activities that could conceivably be described as OR almost immediately led the editors of *OR Quarterly* to wonder what kind of policy-related research would *not* be considered OR.[13] In a guest essay in the first issue of the *OR Quarterly*, Patrick Blackett tried to introduce the bureaucratic definition of OR familiar from the war, which distinguished work done directly for policymakers from work that merely had relevance to policy, such as nutrition scientist John Boyd Orr's influential study, "Food, Health, and Income."[14] However, this definition ran aground on the fact that there was more interest in OR in industry in the 1950s than there were bona fide OR groups.[15] Meanwhile, some OR proponents detected a creeping tendency to reduce the heterogeneity of wartime OR to an intellectual essence. In 1951 the Admiralty's Director of OR wrote in a review of the published version of the final report for the United States Navy's wartime OR Group that he hoped it would "destroy the lamentable identification of operational research with statistics which threatens in this country."[16]

Although its conceptualization of OR remained amorphous and its work was modest, the OR Club succeeded in using the rubric of OR to rally interest in the modernization of industrial operations. Hundreds of people became interested in the subject, and in 1953 the club was obliged to reconstitute itself as the Operational Research Society, and to open its membership to as many applicants as were deemed qualified.[17] However, during this expansion, the basic goals of British OR proponents did not change, as reflected in their choice of leaders. Like Charles Goodeve, they were all promoters rather than practitioners of OR.[18] When Owen Wansbrough-Jones was named the first "chairman" of the OR Society, he had advanced to the very high position of Chief Scientist of the Ministry of Supply. William Slater, the second president of the OR Society, was the secretary of the government's Agricultural Research Council. The third president, Maurice Kendall, was a noted statistician at the London School of Economics. When Tony Giffard, the Earl of Halsbury, became the fourth president in 1960, he had just stepped down as the director of the National Research Development Corporation, a government agency created after the war to transfer R&D projects to the private sector. When Patrick Rivett was elected the fifth president of the OR Society in 1962, he was the first true OR practitioner to assume the post.[19]

By the time Rivett became president of the OR Society, British OR had changed substantially, and had begun to assume the mantle of a profession.[20] The clear source of this professionalizing trend was America. There, a number of pedagogical programs in OR had already been established, and a canon of valuable mathematical techniques had become associated with the idea of OR. British OR proponents increasingly took up these techniques as the 1950s progressed, until discussions of them began to dominate the content of the expanding *OR Quarterly*. The first British educational programs in OR appeared in the 1960s, beginning with Rivett's appointment to Britain's first chair in OR at the University of Lancaster in 1964. Precisely why American proponents adopted a distinct view of OR in the 1950s, and why this view proved so influential, is a question that cannot by answered by any simple appeal to gross cultural or economic differences between the two nations.

Between 1945 and 1948 the American reaction to the wartime OR experience remained muted in comparison to the frequent and often enthusiastic invocations of it in Britain. When Americans did begin to develop OR outside the military beginning in 1948, the concurrent establishment of the OR Club in Britain provided an important model for institutionalizing the activity. Yet, America's OR proponents also took a page from the earlier rhetoric casting the advent of OR as a historic development. For this reason, when the Americans decided to civilianize OR, they touted it as a new and important thing. Adopting this strategy forced them to make clear exactly what it was about OR that made it different from existing fields, such as management consulting and economics. In attempting to do so, they followed Patrick Blackett in placing a heavy emphasis on OR practitioners' placement within institutions. But they also leaned heavily on practitioners' background in science and mathematics. Their particular strategies of advocacy opened the door, first, to OR becoming a new profession, and second, to it being identified as a new science with its own intellectual content.

The notion that OR could have its own intellectual content had roots in certain wartime documents, such as Patrick Blackett's 1943 memorandum, "A Note on Certain Aspects of the Methodology of Operational Research." Then, in 1946, Philip Morse and his deputy, the Columbia University physical chemist George Kimball, wrote the final report of their group with an eye turned explicitly to civilian applications. Thus, looking to define OR in some general way, they entitled their report "Methods of Operations Research" and used the bulk of it to detail mathematical techniques used during the war, which were mainly applications of probability theory. However, mindful of the importance of their relations with the Navy to the integrity of their work, they also wrote extensively about the need to build a competent group, to integrate it effectively into an organization, to collect appropriate data, and so forth.[1] Before the subject of OR gained momentum in the United States, the report circulated privately among people interested in its future.[2]

Yet, given the general lack of interest in OR in America, there was little opportunity for any particular view of the subject to take root. This situation began to change when the Princeton statistician Samuel Wilks took up OR's torch. In 1942 Wilks had worked part-time as a member of Morse's OR group before joining the Applied Mathematics Panel. In 1947 he decided to assemble a pair of sessions on OR at the year-end joint meeting of the American Statistical Association and the Institute of Mathematical Statistics. The speakers at these sessions were a heterogeneous selection of scientists, mathematicians, and engineers who had been involved both in wartime OR, or had had close contact with it.[3] Then, in March 1948, Wilks circulated a draft of a proposal for a small conference on OR, to be convened by the National Research Council (NRC). The proposal noted that there had been "some controversy in England and in the United States" as to what exactly constituted OR, and if wartime work could be extended to industry and business, or if it would merely replicate existing activities. In view of the relative intensity of British discussions, he thought it was important for Americans to begin to work these issues out.[4]

This first NRC meeting on OR was held in May 1948, which led to a second exploratory meeting held in June 1949.[5] The second meeting was attended by Wilks; Philip Morse, who was at that time head of the new Weapons Systems Evaluation Group; Ellis Johnson of the Army's new Operations Research Office; Jacinto Steinhardt of the Navy's Operations Evaluation Group; the University of Chicago mathematician Marshall Stone; the research director of the Office of Naval Research Alan Waterman; John Coleman, who was a member of the NRC's Committee on Undersea Warfare; and Clifton Gibbs, the head of the NRC Mathematics and Physical Sciences Division. The meeting also included Arthur A. Brown and Horace Levinson. Levinson, who was retired and living on a farm in Maine, had received a PhD in mathematical astronomy from the University of Chicago in 1922 and taught mathematics at Ohio State University before heading research departments at Bamberger's and Macy's department stores. Brown, who was Steinhardt's deputy, had received his PhD from the prestigious mathematics department at Princeton in 1940, and had worked at Bamberger's with Levinson before becoming a member of Morse's wartime OR group. He was presumably responsible for bringing Levinson into the OR circle.[6]

Discussion in the meeting ranged over the questions of what OR was, whether it had applications in industry, how this activity could be initiated, where OR staff would come from, and whether a pamphlet should be published to define what OR was and to promote it. Among the meeting participants, Wilks appeared the most concerned about precedents. Others, however, sensed that OR was indeed novel because it entailed investigation in support of management decision making that was rigorous and routine. This perception was premised on the view that existing information gathering and analysis practices in business were haphazard at best. Thus, although Ellis Johnson credited the management of some organizations such as Bell Telephone, he was dismissive of a general managerial culture that made decisions "by having a conference and then making a decision without much investigation—an arbitrary handling of a situation." Levinson, an industry insider, agreed, stating: "There is no broad planning. It is generally thought enough to have an accountant's picture of the business. In ordinary business, figures are accumulated over a period of a year and that picture satisfies a man in charge."

The meeting's participants settled on a consensus similar to what prevailed in Britain: OR would serve as a kind of rallying flag to unite scattered efforts to modernize business decision making. However, the result of the meeting was not the formation of another club, but a new NRC committee to be chaired by Levinson, tasked with promoting OR to industry.[7] This decision led the American OR proponents quickly, if unintentionally, away from British precedent. Where the British OR Club provided a venue and a publication, however modest, for discussing existing industrial practices, the American NRC committee had no such mechanisms. The best that it could do was to mount a publicity campaign for the idea of OR itself. In late 1950 Levinson and Alan Waterman visited MIT to encourage the establishment of an "Operations Research Center" to train OR practitioners.[8] Morse published an article in the December 1950 issue of *Physics Today* encouraging physicists to take an interest in OR.[9] He and George Kimball also reworked their final report into a successful book.[10] In 1951, Levinson and Brown published an article on OR in *Scientific American* in order to reach a more popular audience.[11] Committee members also took their case on the road to academic conferences.[12] Most importantly, though, in the spring of 1951 the committee sent a brochure to thousands of companies to try to generate interest among the only people who could make industrial OR a reality.[13]

In promoting OR as an important new development, the Americans knew they would have to answer critics who claimed that OR was indistinguishable from such activities as market research, quality control, and time and motion studies. So, the NRC committee's brochure allowed that, while it was "tempting" to suppose that OR was nothing new, apparent precedents were only piecemeal approaches to improving managerial decision making. Naturally, such precedents would become part of OR, but, the brochure assured its corporate readers, OR was "more than the sum of scattered activities, even though each activity is one of its parts." The keys to OR's transcendence were its practitioners' scientific training and their close relationship with an organization's senior decision makers. The brochure noted OR scientists' ability to deploy advanced quantitative techniques, but emphasized their awareness of a wider array of analytical methodologies, and their ability to match them to problems appropriately. It was this essential catholicity and flexibility in OR that made it misleading to describe it as simply an outgrowth of prior developments.

If, however, OR encompassed some amalgam of whatever techniques might contribute to good decision making, then the question of its limits immediately arose. At a February 1951 meeting of the NRC committee, the unfailingly ambitious Ellis Johnson proposed that "operations research should be broadened to include social and political phenomena and that an operations research group should be placed in the Executive Office of the President to attack broad problems of 'National Political Objectives.'" Bob Robertson, who was attending as the new research director of WSEG, replied that it would be best to limit OR work to "the more concrete problems of control of national defense production"—WSEG's own ambitious mission—"to which it can make contributions in quantitative or at least semi-quantitative form." He suggested, "Political decisions should await the further development of operations research techniques." The committee members then discussed this point in some depth, and ultimately sided with Robertson.[14] However one was to define OR, it did in fact have limits, even if they were set only by the exigencies of practicality.

If the NRC committee members could satisfy their own concerns about OR's novelty and limits, outsiders sometimes proved harder to convince. In September 1951, for instance, an audience member at one of the committee's conference presentations remarked,

I was struck in this presentation by the sort of impression given in some of the examples that operations research is made by people who have on their chests a label: "I am an operations research man." As you were going over the list of studies that have been made, I could think of countless similar studies that have been made for the last ten or fifteen years... There is a whole field, which has been developed in the field of economics, of rational replacement policy... I think economists have done a certain amount in that field and will continue to do so, and it seems to me in general you have left out economists completely.[15]

A similar question arose earlier that year when the work of the OR Committee and a nascent OR group at the Arthur D. Little consulting firm attracted the attention of one of *Fortune* magazine's writers, Herbert Solow, who evaluated OR's prospects in light of the work of what he called "old-line" management consulting firms. He gave credit to OR practitioners' ability to use sophisticated methods, and supposed they could present a challenge to existing activities. But, where OR's proponents felt that OR encompassed and transcended its precedents, Solow supposed that established consulting firms were more likely to simply hire OR practitioners for whatever it was they could do. On the whole, though, the article was flattering, and the OR Committee integrated it into their promotional campaign.[16]

Whatever doubts were being raised, by 1952 and 1953 the American OR movement could be said to have succeeded in its goals. Hundreds of scientists, mathematicians, and engineers had begun to identify with the term, and new institutions dedicated to fostering OR began to be established. In the first half of the 1950s, educational programs in OR were established at the Naval Postgraduate School, MIT, and the Case Institute of Technology. Groups interested in OR at other universities soon forged alliances with related endeavors in management and engineering schools around the country, such as at the Carnegie Institute of Technology and Johns Hopkins University.[17] In line with Solow's predictions, people identifying with OR were hired by consulting firms, as well as by companies in major industries, such as aeronautics and oil refining.[18] In 1952 the American OR movement created a new hub for its activities by establishing the Operations Research Society of America (ORSA), with Philip Morse serving as its first president.[19] The next year ORSA began publishing the *Journal of the Operations Research Society of America.*

At the end of his one-year presidential term, Morse seemed confident that OR's identity issues were reaching their end. He remarked, "We should no longer have trouble explaining the scope and methods of operations research to the layman. We can already say: Operations research is the activity carried on by members of the Operations Research Society; its methods are those reported in our JOURNAL."[20] When he said this, in 1953, he likely expected that ORSA's journal would be publishing reports and case studies not unlike those abstracted in the British *OR Quarterly.* However, within a few short years, mathematical theories of decision making would begin dominating the pages of the American journal, and, within several years, the very identity of the new profession he and his fellow proponents had created.

V Rationality and Theory

20   Theories of Decision, Allocation, and Design

As we have seen, a major interest of mathematicians during World War II was the problem of measuring and improving the effectiveness and efficiency of military operations and logistics. Generally speaking, operations, such as searching for submarines, followed a certain logic. Because this logic was often implicit to the thinking of experienced officers, mathematicians had to work closely with them to determine what that logic was, and whether it could be improved upon. In other cases, though, mainly in engineering and industrial production, practices were already based on formalized theories. In these cases intuition could reveal that the principles employed were in some sense incomplete, and that consideration of meta-calculative issues could improve upon them. Thus, for example, the mathematician Abraham Wald was able to formulate the powerful statistical theory of sequential analysis, which incorporated into the design of testing procedures information to be gained from the tests. Mathematicians' experiences of learning about decisions by analyzing them formally, and of expressing implicit meta-calculations in a formal language suggested that these were areas capable of expansive theoretical development.

The end of the war brought the opportunity to undertake this development. Wald published and expanded on his wartime work in his 1947 monograph *Sequential Analysis*. Three years later he and his wife died in an airplane crash while on a lecture tour in India, but his work became a foundation stone for new research in mathematical statistics and probability theory.[1] This research, in turn, quickly joined with other areas such as economics and the new mathematics of game theory to form a broad tradition of inquiry often referred to as decision theory. Decision theory addressed a variety of dilemmas involved in deciding among an array of possible choices; in allocating resources among a number of possible uses; and in designing technologies, procedures, and policies that were more effective or efficient in achieving their goals than others. These dilemmas were often ones where it was not difficult to make a sound choice, but where it could be very difficult to be certain that a correct or an optimal choice had been made. Even in cases where good choices could intuitively be made, such problems could be a matter of intense practical concern since substantial economies could accrue or be lost if the decision had to be made repeatedly. At the same time, these problems could also have a distinct academic appeal, since formal treatments of them often revealed unique logical structures.

The Traveling Salesman Problem (TSP) is an important case of a problem that was ostensibly practical, but was initially pursued mainly for its academic appeal. If a traveling salesman must visit a certain set of cities, never visiting the same city twice, and returning, ultimately, back to where he began, the problem is to find the shortest possible route satisfying these criteria. Formally articulated, one must find a permutation $P = (1\ i_2\ i_3\ ...\ i_n)$ of the integers from 1 through n that minimizes the quantity

$$a_{1i_2} + a_{i_2i_3} + a_{i_3i_4} + ... + a_{i_n1}$$

where $a_{\alpha\beta}$ are a given set of real numbers. So, in terms of the scenario outlined, there are n cities to be visited (here pre-labeled 1, 2, ..., n; unlike the map in figure 20.1, where cardinal numbers were applied sequentially to the stops in the optimized route). The mileages, a, between every possible pair of cities (i.e., $a_{1\,2}, a_{1\,3}, ..., a_{1\,49}, ... a_{48\,49}$) are known and given on a schedule. The object is to construct a permutation of stops (labeled 1, i_2, i_3, ..., i_n; e.g., 1, 15, 6, ..., 49), which corresponds to a selected set of mileages from the table (e.g., $a_{1\,15}, a_{15\,6}, ..., a_{49\,1}$), which add up to a lower sum than any other set constituting a valid route. "More accurately," the mathematician Merrill Flood wrote in 1956, "since there are *only* $(n\text{-}1)!$ possibilities to consider, the problem is to find an efficient method for choosing a minimizing permutation."[2]

It was not difficult to find a reasonably good solution to the Traveling Salesman Problem using only visual instincts. However, it was almost impossible to find, or to be certain that one had found, an optimal solution to the problem if a sufficiently large number of cities were involved. Further, if one wanted to automate the process of solving the problem using a digital computer, or one wanted to solve an equivalent problem without a strong visual component, one was faced with the problem of finding some systematic algorithm of testing solutions against each other. For his part, Flood had originally heard about the TSP in the mid-1930s while he was a professor at Princeton University, and found himself ensnared by it when he was working on the routing of school buses in

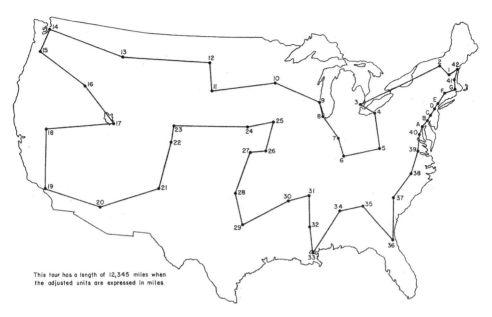

This tour has a length of 12,345 miles when the adjusted units are expressed in miles

Figure 20.1

An optimal solution to a Traveling Salesman Problem. From Dantzig, Fulkerson, and Johnson 1954. © The Institute for Operations Research and the Management Sciences.

1937. The TSP remained unsolved through the war, and in 1949 Flood moved to the RAND Corporation. There, apparently at the instigation of John Williams, the head of RAND's mathematics department, he began soliciting more interest in the TSP. Happily, at that same time, the development of the techniques of linear programming began to make the problem more tractable. [3]

Linear programming originated in the postwar work of the mathematician George Dantzig on the logistical problems surrounding the organization of Army Air Forces' mobilization plans called "programs."[4] While Dantzig had been the head of the Combat Analysis Branch of the AAF's Statistical Control Office, he gathered and processed tremendous amounts of data relating to the sorties flown, targets attacked, bombs dropped, aircraft lost, and so forth, and he became familiar with the vast clerical infrastructure that the AAF had established to coordinate the even vaster logistical infrastructure behind bombing operations. He had also been privy to the 1943 creation of a "program planning" function in the AAF under Edward Learned, a professor at the Harvard Business School. Program planning aimed to coordinate bureaucratically separated functions such as procurement, supply, training, and deployment into an efficient process that was also consistent with an overall war plan.[5]

After the war ended, the AAF consolidated statistical control, program monitoring, and budgeting into the office of the Comptroller. Dantzig was made a mathematical consultant to the Comptroller and tasked with systematizing program planning, and mechanizing the process using newly developed digital computer technology. Working under his fellow Statistical Control alumnus Marshall Wood, Dantzig gathered a group (dubbed Project SCOOP[6] in 1948) to attempt to work out better ways not only to formulate coherent programs—a difficult enough task—but also to compare any of the countless possible configurations of programs to find more efficient ones (see figure 20.2). To approach this problem, Dantzig took his

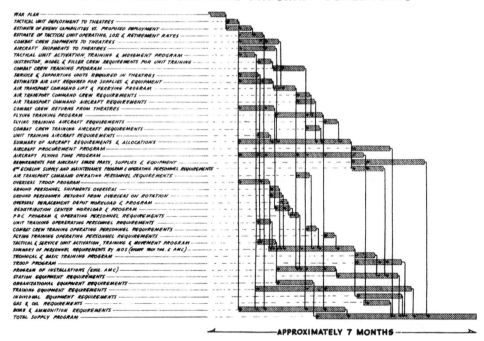

SCHEMATIC DIAGRAM OF MAJOR STEPS IN AIR FORCE WARTIME PROGRAM SCHEDULING

◄──────────────APPROXIMATELY 7 MONTHS──────────────►

Figure 20.2
A wartime program showing the interrelation of different steps in the fulfillment of an overall plan. From "Scientific Planning of Military Programs," Project SCOOP, Report 4-PU, May 20, 1948. Copy courtesy of the Air Force Historical Studies Office.

cues from Harvard economist Wassily Leontief's prewar "input-output" model of the American economy. This model depicted the economy as a matrix describing the connections between inputs of material and labor, and outputs of goods and services, which could themselves be used as inputs. Dantzig and his group at Project SCOOP worked out a similar matrix formulation for combining staged program elements of an overall wartime program, representing the program as an interconnected system of linear equations (representing the dependence of program activities on ones preceding them) and inequalities (representing practical constraints, such as materiel and personnel availability).

Once satisfied that programming activities could be adequately formulated in this manner, Dantzig set out to find a way to efficiently optimize the arrangement of the complicated system. This led him in the summer of 1947 to visit the economist Tjalling Koopmans in Chicago at the Cowles Commission for Research in Economics. During the war, Koopmans had formulated a practical shipping problem wherein the object was to select an efficient set of routes transporting products from possible supply centers with finite resources to destinations where they were required. That work had resulted in the creation of a table presenting "exchange ratios" giving the marginal opportunity cost of using different routes for a shipment, thereby allowing the most efficient route available at any given time to be selected (provided certain conditions, such as safety of the routes, applied). Recognizing in Dantzig's time-staged programming problem the logical structure of his wartime problem of resource allocation at a given point in time, Koopmans became deeply interested in the matter as a means of reformulating problems in neoclassical economic theory. However, he did not have the answer to Dantzig's immediate problem of finding an optimizing algorithm. So Dantzig returned home and employed his own knowledge of the mathematics of convex sets to arrive at a solution that soon became known as the simplex algorithm.[7]

In the wake of the simplex algorithm, theoretical work on what soon became known as linear programming expanded rapidly and into a remarkable variety of research areas. Koopmans's wartime problem became a classical example of a linear programming problem called the "transportation problem" (figure 20.3). Dantzig's team at Project SCOOP tested the efficiency of the simplex algorithm on a similar problem of fulfilling dietary requirements at minimum cost using various foods of varying nutritional properties, which had been proposed in 1945 in the *Journal of Farm Economics* by the economist George Stigler.[8] It became known as the "nutrition" (or "diet," or "housewife") problem. Meanwhile, when

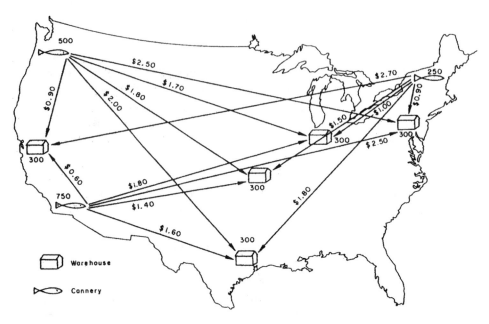

Figure 20.3
An example of the transportation problem. From Dantzig 1963. © The RAND Corporation.

Dantzig paid fellow mathematician John von Neumann a visit to discuss linear programming, von Neumann immediately recognized that it was mathematically equivalent to the two-person, zero-sum variation of the theory of games. Neither Dantzig nor von Neumann published on the subject, but the Princeton mathematician Albert Tucker and his students Harold Kuhn and David Gale proved and published the equivalence in short order.[9]

The first conference on linear programming, entitled "Activity Analysis of Production and Allocation," was held by the Cowles Commission at the University of Chicago in June 1949, and featured papers dealing with a remarkable range of applications and problems in linear programming that had already been developed in the field's short existence. Subjects included the fundamental mathematical relationship between linear programming and game theory and the theory of convex sets; general theoretical problems relating to Leontief's input–output economic models; and more practical optimization problems, such as Koopmans's transportation problem and crop rotation scheduling. All of these areas would continue to provide grist for the mathematical research mill for many years.[10]

Research on the fundamental mathematical aspects of linear programming moved most immediately into what was called the "duality" between it and game theory, as well as into what Tucker and Kuhn called "nonlinear programming" in their 1951 paper introducing the subject.[11] Tjalling Koopmans, meanwhile, would later share the Nobel Memorial Prize in Economics (with the Soviet economist Leonid Kantorovich) for his development of resource allocation models, which, along with game theory, became influential in theoretical economics, especially later in the century. In socialist circles linear programming would be viewed as a path to effective economic planning, but it could also be used to model the ways that markets allocate resources and set prices.[12] At a more mundane level, linear programming became an important technique in industry and logistics. For example, the oil industry quickly adopted it to optimize the blending of gasoline. And, of course, Project SCOOP continued to conduct research related to its original mission of developing applications for military logistics, and was soon joined by the Office of Naval Research and the RAND Corporation, the latter hiring Dantzig away from the U.S. Air Force in 1952.[13]

However, research on linear programming and associated problems was motivated as much by theoretical challenge as practical need. Articles published in scientific and mathematical journals focused on defining and finding solutions for manifestations of linear programming that were constrained in ways that mathematicians found enticing. The transportation problem, for example, was constrained by the finite requirements and availability of goods to be transported. Stigler's nutrition problem had no such constraints, but each possible food contained a fixed quantity of a variety of nutrients. The "assignment problem"—where the problem was to assign a fixed number of people to a fixed number of jobs based upon quantitatively evaluated skills and requirements in various subject areas— is constrained by the fact that there can only be one person for every job. In 1949 mathematician Julia Robinson began formulating a new approach to the Traveling Salesman Problem at RAND. The peculiar challenge in the TSP was that all cities not only had to be visited, but also in a cyclical order where the cycles could not comprise subcycles or disconnected cycles (which were feasible numerical solutions in structurally related problems, such as the assignment problem). In 1954 Dantzig coauthored a seminal paper successfully applying linear programming to the solution of the TSP, which appeared in the second volume of the *Journal of the Operations Research Society of America*, thereby helping to cement the nascent journal's status as a forum for path-breaking mathematical research. When Merrill

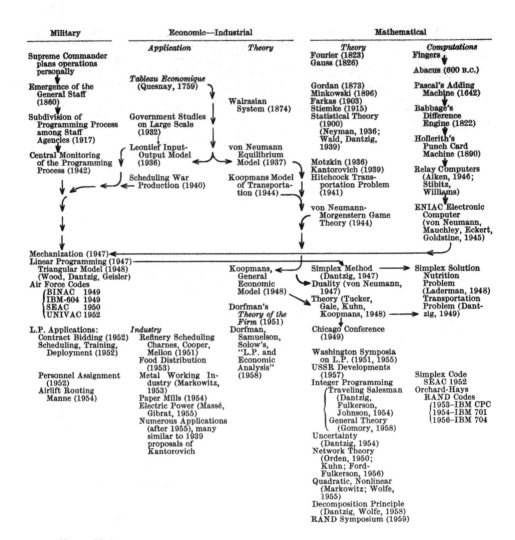

Figure 20.4

Origins and influences of linear programming. From Dantzig 1963. © The RAND Corporation.

Flood (who had moved to Columbia University following his stint at RAND) published a review of work done to date on the TSP in 1956, it was in the same journal, which by this time had been renamed simply *Operations Research*.[14]

The remarkable thing about the postwar growth of decision theory is that it seems to have arisen from a variety of disconnected roots. John von Neumann and Oskar Morganstein's highly influential 1944 opus, *Theory of Games and Economic Behavior*, traced its various origins back to the 1920s.[15] While postwar statistical decision theory was unquestionably spurred by the wartime development of sequential analysis, it also had deeper roots in mathematical statistics, and the development of that field in America by such figures as Abraham Wald, Harold Hotelling, and Jerzy Neyman, which historians have only begun to trace and unravel. And, naturally, there are a host of longstanding practical problems in economic analysis, computing, and industrial management, which provided important motivations and contexts. But, in spite of this variety of prewar roots, World War II and the Cold War unquestionably played a catalytic effect in spurring cross-fertilization between practical problems and different research traditions, as George Dantzig's 1963 sketch of the "origins" and "influences" of linear programming very clearly illustrated (figure 20.4).[16]

21 The Anatomy of Decision: Inventory Theory

Theories of decision became a major research tradition in the postwar period because they led to the development of valuable practical techniques, while at the same time sustaining academic theoreticians' interests. These theories were compelling to academics because, once the logical structure underlying the success of intuitive decisions was articulated, it proved capable of sustaining almost indefinite extension and exploration. Theoreticians could find solutions to explicitly formulated problems of decision and optimization. They could define and constrain these problems in ways that made nontrivial changes to the forms that solutions took. To be considered rigorous, all solutions had to be proved. They might also be proven to be unique or not unique, or generalized to describe the forms that solutions to a variety of related problems had to take. The conditions under which solutions were valid had to probed. Different solutions, if they existed, had to be compared to each other, and different formulations of the same problem had to be proven equivalent or non-equivalent. If solution was deemed impossible, that impossibility would also have to be proved.

Inventory theory, alongside linear programming, became one of the most important branches of the mathematical theories of decision to be developed in the 1950s. The problem of inventory is deceptively simple on its surface. Storing an inventory entails taking on certain costs. The cost of purchasing and managing storage space is only the most obvious one. Storing goods too long also risks costs associated with depreciation, obsolescence, and spoilage. Summer clothes will not sell in August and September without a discount; spare parts might be rendered useless if a new machine is bought; rotten fruit must be thrown away. There are also opportunity costs associated with holding goods in a space that might be devoted to other, more profitable goods. However, the costs of not keeping an inventory can be very steep. If a merchant cannot sell to customers on the spot, or if the anticipated delivery time for ordered goods is too long, customers might decide to make not only that purchase but also future purchases from another merchant. The question to merchants, then, is: how much inventory should one keep, and what policy should be enacted for replenishing an inventory? Depending on assumptions made about the inventory problem, these calculations can be quite difficult to formulate and even more difficult to solve. And it is this difficulty that proved irresistible to some of the greatest mathematical, statistical, and economic theoreticians of the postwar era.

Inventories had, of course, been an issue for business managers long before the middle of the twentieth century, but they had never been subjected to intensive research. Following World War II, the American military services became interested in inventory management as part of the same concentration on logistics that gave rise to Project SCOOP and linear programming. One branch of this work was a bibliographical summation of prior thinking about inventory, sponsored by the Office of Naval Research, and undertaken by Princeton economist Thomson Whitin and Louise Haack of George Washington University. Whitin's work was eventually published in 1953 as *The Theory of Inventory Management*, which included a history of thought related to inventories. According to Whitin's history, the ability of nineteenth-century industrial economies to produce truly massive, and thus expensive, inventories had shifted perceptions of them from being a sign of opulence to a potential liability. At the same time, though, it became easier to transport goods to alternative markets, which meant that profits shrank due to wider competition, and that a premium was placed on even marginal increases in efficiency. Thus, inventories became at once a more prevalent, more burdensome, but also a more controllable and economically exploitable aspect of industrial life.[1]

Of course, inventory size could not always be controlled. The inability to rid oneself of excess stock had become associated with broader economic downturns, especially in the aftermath of the Great Depression. Among some business managers, inventories had taken on the aura of a bad omen, leading to articles in management journals with titles such as "Your Inventory a Graveyard?"[2] Some entrepreneurs began to market inventory "plans" to companies. The "Wilson Inventory Management Plan," for example, was already being employed by Westinghouse Electric and General Foods.[3] Meanwhile, economists began to scrutinize the role of inventory fluctuations as a driver of business cycles. Most notably, the chain leading from unsold inventories to reduced factory orders to higher unemployment to reduced consumer demand, and, thus, to more undiminished inventories had been recognized by, among others, the recently deceased British economist John Maynard Keynes. Whitin observed that Keynes analyzed inventories in terms of a distinction between "liquid capital," denoting surplus stocks, and "working capital," denoting goods in production and transport and being held as a hedge against interruptions of service.[4] He, however, preferred a taxonomy employed by other authors that distinguished "planned" from "unplanned" inventories, because it clearly connected inventory fluctuations to managers' decision making. Yet, he admitted, "To estimate the volume of planned and unplanned inventories … is no simple matter because of the lack of any objective standard for indicating what plans have been."[5]

In order to integrate decisions about inventory management into economic theory, some means was necessary of articulating what would constitute a rational level of inventory to hold. Businesses themselves seemed to be making decisions less often "on the basis of intuition" and were transitioning toward "'scientific' inventory control," thanks to an increasing sophistication in business training, and the increasing role of engineers and cost accountants in business management.[6] The literature on inventories was certainly increasing, and the bulk of Whitin's book was occupied with a review of it. And, just before the book went to print, Whitin inserted last-minute references to two new papers appearing in the journal *Econometrica*: "Optimal Inventory Policy" by Kenneth Arrow, Theodore Harris, and Jacob Marschak; and "The Inventory Problem: Case of Known Distributions of Demand" by Aryeh Dvoretzky, Jack Kiefer, and Jacob Wolfowitz.[7]

Kenneth Arrow had begun his graduate studies in economics at Columbia University before World War II, and served as a weather officer in the Army Air Forces during the war. He completed his PhD over the course of the late 1940s, but spent much of this period as a research associate at the Cowles Commission for Research in Economics, which was based at the University of Chicago. At Cowles, Arrow got to know Jacob Marschak, who was the commission's director from 1943 to 1948.[8] Arrow moved on to an academic post at Stanford University in 1949, but since 1948 he had also been spending summers in Santa Monica at the new RAND Corporation. In the summer of 1950, RAND hosted a conference on inventories. Marschak, who was attending, invited Arrow to join the discussion. Surveying the bibliographical work done by Whitin and Haack, the two of them focused on what they took to be the clearest logical problems bearing upon the formulation of rational inventory policies.[9]

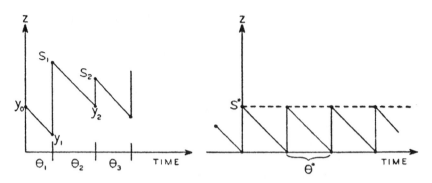

Figure 21.1
Models of ad hoc and optimized inventory replenishment policies in a case of constant demand. From Arrow, Harris, and Marschak 1951. © The Econometric Society.

When they published their work, Arrow and Marschak developed their reasoning through a series of steps so as to define their problem in the clearest way possible. They began by considering a very simple "case of certainty" where demand is known ahead of time, meaning that one could order exactly enough stock to last until the arrival of the next shipment (figure 21.1). Within this model, the problem was simply to balance the cost of "carrying" large, progressively diminishing inventories with the costs incurred by making more frequent orders. They established the existence of a single optimal maximum level of stock, S*, as well as the existence of a single optimal interval between orders, θ*. Establishing these quantities logically precluded the possibility of obtaining profits by varying the time between replenishment or the amount ordered—a point, they parenthetically remarked, that was "usually accepted intuitively."[10]

Next, Arrow and Marschak considered a case limited to only one interval with variable demand. In this case, because there was a finite chance that demand would be unusually high during the interval, one had to make a decision about how much carrying cost one wanted to absorb to avoid the potentially high costs associated with running out of stock. (The paper quoted "A horse, a horse, my kingdom for a horse" from *Richard III* to emphasize the point.) Expectation values were the key: one would behave during the interval as though one were going to pay a fraction of the costs of running out of stock every time one reordered based on what fraction of the time one expected to run out of inventory. Thus one needed to estimate the statistical distribution of demand; one needed at least to estimate how important it was, in dollar terms, that one not run out of stock; and, of course, one needed to know the holding costs and the sale price of the good in order to calculate the optimal policy.[11]

These two models seemed to represent the extent of formal thinking about inventory problems to that point. Arrow later remembered that he and Marschak "quickly saw that the interesting problem was the combination of the two models. That is, the realistically important model was one in which inventory is durable, so what is not used in one period has value in meeting future demands, but in which the demands are random."[12] Suddenly the problem became much more complex, because it was necessary not only to guard against stock depletion, but to take into account the costs associated with holding stocks that were not depleted. To attack this problem Arrow and Marschak decided to use a common, but largely unexamined inventory management strategy, the "two-bin" or "(s,S)" system (figure 21.2). In this system one did not calculate a priori when one would reorder. Rather, opportunities to reorder occurred at regular intervals, at which point one had to decide whether to refill an inventory to its maximum level, S,

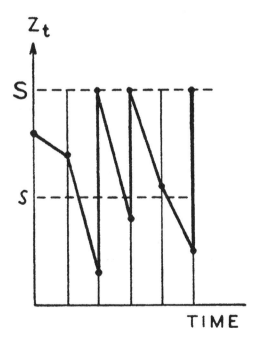

Figure 21.2
The (s,S) model of inventory replenishment policy. From Arrow, Harris, and Marschak 1951. © The Econometric Society.

based upon whether or not stocks had fallen below some predetermined level, s.[13]

Arriving at expressions defining s and S turned out to be far from trivial. To approach the problem, Arrow and Marschak made a number of simplifying assumptions such as neglecting the time lag in reordering, supposing that the cost of restocking to level S did not vary with the size of an order, neglecting any relationship between the cost of depletion and the duration of depletion, and so forth. Needless to say, there would be few instances where the derived values of s and S could genuinely be said to be optimal, but immediate application was not the goal. Arrow and Marschak's aim was to arrive at a more sophisticated, or perhaps less arbitrary, consideration of the inventory problem, compared to all prior considerations of it. Going as far as they did into the issue, they discovered that the problem could be treated as a Markov branching process. This meant that the decision to reorder depended only on the inventory state at that moment, and not on its prior history. It also meant that to calculate optimal levels, it would be necessary to develop expectation values for costs given all possible outcomes of all possible decisions taken at future reordering opportunities. So, one had to determine both the probability that one would have to reorder at the end of the first interval, as well as the probability, in the event one did have to reorder, that the inventory would actually be depleted. It was then necessary to calculate the same probabilities for the second interval, *given* the initial inventory levels defined by the probability distribution of outcomes for the first interval. And that was not the end of the problem. This calculation extended in theory for an infinite number of future intervals, although the costs incurred in distant intervals were discounted through a factor, $\alpha < 1$, which compounded exponentially with each new interval, thus ensuring a convergence at a fixed value, S.[14]

The mathematics needed to determine the optimal levels resulting from this branching process, even with the simplifying assumptions, were not only beyond the abilities (and no doubt the needs) of the average shopkeeper, they were beyond the abilities of Arrow and Marschak as well. They therefore recruited a third author to their effort, RAND mathematician Theodore Harris, who had written a dissertation on such processes at Princeton with William Feller, who had confronted similar problems from the mathematics of predator–prey models in population dynamics and the related economic theory of "industrial replacement."[15] Arrow, Marschak, and Harris's main accomplishment in their paper was that it established a basic model that could be criticized and further developed. At the end of their paper, they themselves pointed out avenues for further investigation.

For example, they took a page from sequential analysis and suggested that the assumed demand distribution, and thus the levels of S and s, might be modified at each reordering opportunity based upon information gathered during preceding intervals.[16]

The two-part 1952 inventory paper by Jack Kiefer, Jacob Wolfowitz, and Aryeh Dvoretzky used intellectual strategies related to those used by Arrow, Marschak, and Harris. This is unsurprising given their related intellectual genealogy. When his wartime work with the Statistical Research Group at Columbia University concluded, Wolfowitz moved to the University of North Carolina before returning to Columbia, before he ultimately landed a position at Cornell University in 1951. Kiefer had been a PhD student of Abraham Wald's at Columbia, and joined Wolfowitz at Cornell. Dvoretzky was a visiting mathematician from Hebrew University. The three of them took up inventory theory as part of the same Office of Naval Research effort that sponsored the 1950 RAND conference. They, too, were primarily interested in defining the logical structure of the problem rather than arriving at practical solutions to it, but took an even more archly theoretical approach to the task. Where Arrow, Marschak, and Harris had focused on finding solutions to the specific (s,S) formulation of inventory problems, they focused on defining what general form optimal solutions to inventory problems might take under varying assumptions, not necessarily adhering to the (s,S) form.[17]

This difference in strategy was highlighted by RAND Corporation mathematician Richard Bellman in his review of a 1958 book on inventory theory written by Arrow with his Stanford University colleagues, Samuel Karlin and Herbert Scarf. In the mid-1950s, Bellman had formulated the mathematics of "dynamic programming," which addressed problems involving ongoing processes of gathering information and make sequences of decisions.[18] He observed that inventory management was among "the most interesting classes of dynamic programming processes." He suggested that, following the "classic paper of Arrow, Harris and Marschak," theoretical development had forked. The trend established by Arrow, Harris, and Marschak was to study "sub-optimal policies," meaning analysis which would allow one to choose optimally between policies adhering to a specific logical structure, such as the (s,S) formulation. However, the trend of theorization begun by Dvoretzky, Kiefer, and Wolfowitz attempted to answer "essential questions pertaining to the existence and uniqueness of the nonlinear functional equations." The settlement of these questions made it possible to "be sure of a one-to-one correspondence between the optimal policies of the original process and the solutions of the defining equations." In other words, it was only by pursuing generalized investigation that it could be made clear whether, in finding an optimized solution to a formulation adhering to a certain structure, one was safe in bracketing out entire classes of alternative structures from consideration.[19]

Each of the trends Bellman identified represented complementary efforts to address the central philosophical problem of whether or not it was possible to say that one inventory policy was truly more rational than another. The only way of being able to do so was to define rigorously the conditions that would have to apply for such a determination to hold. While inventories represented an important problem for managers, and occupied the attention of economists interested in understanding business cycles, the main reason these theories attracted interest from elite theoreticians was that they defined an important new class of logical structures. As Dvoretzky, Kiefer, and Wolfowitz pointed out in the introduction of their first paper, the assignment of the term "inventory" to the problem was merely a matter of promoting "simplicity of exposition and ease of understanding." In reality, their paper dealt with "stochastic processes of a fairly general nature for which, at various times, one may exert some degree of control over the distribution functions in question in order to achieve a desired result." Their analysis could be broadly applied to topically unrelated problems, as in controlling animal populations or in physics "by altering suitably the nomenclature used here."[20] And similar stories were unfolding across the various branches of decision theory. Yet, as enticing as these problems were, they were not necessarily central to the concerns of established fields such as economics and mathematics, leading the people who worked on these problems to seek new intellectual homes for their work.

22   The Rise of Theoretical Operations Research

On the somewhat infrequent occasions when a mathematician looks at the practical problems of operations research, his first impression is apt to be that only known and elementary parts of mathematics are needed. But if he dwells with the subject longer, his reaction tends to be quite the opposite: he begins to realize that, when rightly conceived and formulated, these practical problems often turn on mathematical questions deep enough to go beyond existing knowledge and to require for their answer research into new mathematical fields.

The practical worker in operations research is apt to have the counterpart of this experience; at first he seems to be able to do what he needs to do with only the simple mathematics that is familiar to him. But with increasing experience he sees that some of his most important problems require powerful and sophisticated mathematical instrumentation.

Bernard Koopman, "New Mathematical Methods in Operations Research," 1952[1]

As a member of the U.S. Navy's Operations Research Group, Bernard Koopman was responsible for overseeing its primary theoretical accomplishment, the development of search theory. Yet, when OR began to expand in the postwar period, there was no sense that such theoretical content might define the field or portend its future. In 1952, at the first full meeting of Operations Research Society of America, Koopman was still able to portray the complementarity of mathematical theorization with ordinary OR work as a far-from-obvious realization, and the historical record seems to bear out his observation. In 1951, when Horace Levinson of the NRC's OR Committee was asked directly whether research that was applicable to "two thousand organizations which have roughly the same problems" would qualify as OR, he replied "very definitely" in the negative. It was, he felt, the responsibility of a director of an OR team to be aware of the work "in half a dozen outside fields including engineering, psychology, personnel studies, and so on," and to call upon it as necessary.

However, the idea that the group itself should be "expert engineers, expert psychologists and so on" was "nonsense."[2] Similarly, when Alvin Karchere of the Army's Operations Research Office reviewed Thomson Whitin's book on inventories for ORSA's journal, he deemed the chapters on economics and business management approaches to be particularly relevant to OR. Yet, it was the chapters following those that contained discussions of such topics as game theory and linear programming, which would be considered inseparable from OR only a few years later.[3]

In fact, by 1955 the situation had already changed drastically. That year Philip Morse gave a talk to a national meeting of ORSA, in which he welcomed the development of new theories because they gave operations researchers new techniques to use in their work. But he also expressed uneasiness with the rapidity of the blossoming of theory. Referring back to the wartime experience, he reminded his audience, "Operations research is an experimental science, concerned with the real world. It is not an exercise in pure logic."[4] Since, for him and others of his mindset, the legitimacy of OR hinged on its ability to contribute to practical decision making, it was important not to give theorization pride of place over empirical investigation.

The interesting thing about the placement of a canon of theory at the heart of OR's identity is that it does not seem to have resulted from any deliberate campaign, nor was it seriously resisted, sporadic grumbling notwithstanding. Some historians and commentators have supposed that the ascendancy of theory derived from its ability to make OR seem more scientific, which made it more appealing to academic scientists and enhanced its "authority and prestige" in the broader world.[5] This is a kind of temptation narrative, where OR, per Morse's fears, strayed from an obvious source of legitimacy to a potentially more powerful source of it only to find that it was chimerical. However, I would not be so quick to suppose that theory carried much authority and prestige among businessmen, government officials, and other mainstream decision makers. Moreover, it is unclear that advanced theory held any peculiar allure for most scientists. If these points are granted at least provisionally, it becomes necessary to find a more subtle, and, indeed, plausible explanation for the rise of theoretical OR.

One major problem faced by OR's proponents was the fact that it was unclear what sort of work they might do within commercial and industrial settings. Within the wartime military, OR found room in the processes of doctrine development and military planning, simply because the problems the military faced were new, complex, deeply technological, and changing rapidly, and officers were heavily occupied by pressing duties. Logistical planning, however, was handled by entirely different branches of the military services, and OR scientists' participation in logistics design was minimal, even though logistics were changing no less radically than operations. Industry, meanwhile, had nothing analogous to military combat operations. There, operations were more or less defined as the economic and logistical tasks of manufacturing, distributing, transporting, setting prices, and selling goods. Proponents of a civilianized version of OR had little experience with such problems, but they had close contacts with people who did.

At the RAND Corporation, engineers and mathematicians were tasked with studying the feasibility and comparative desirability of competing proposals for new weapons systems. Their studies necessarily incorporated analyses of the logistical support these systems required. It was for this reason that RAND sponsored conferences on topics such as inventory management, hired people like George Dantzig, and, ultimately, in 1955, devoted an entire department to logistics.[6] As we have seen, RAND was also considered closely allied to the military's own OR groups. Therefore, when RAND mathematicians began publishing on the logistical aspects of their work, it is unsurprising that they often chose to do so in ORSA's new journal.

The addition of cutting-edge techniques in logistical analysis, I would argue, proved a fortunate boon to those developing OR as a new profession. However much enthusiasm OR's proponents managed to generate, questions persisted as to what exactly OR was, and how it could improve upon existing practices in managing industrial operations and in making business decisions. Sometimes these questions even came from within. For example, when the RAND statistician Alexander Mood reviewed Philip Morse's and George Kimball's *Methods of Operations Research* for the ORSA journal, he took the opportunity to argue that OR proponents' routine portrayal of industry as neglectful of analysis was a myth—many businesses, in fact, commonly employed people one might call "operations analysts." If, for example, the task was "to know whether and where to locate a new filling station," then the "operations analysts of Standard Oil" would "have a long list of factors you should consider, complete with weights." He urged, "Next time you inherit a decadent chain of grocery stores, rush out and hire a couple of operations analysts from A and P. Their theorems on inventory control, product display, loss leaders, and whatnot, will amaze you." There was, furthermore, nothing deficient about this work. "It is," he insisted, "quantitative analysis; it is objective analysis; it is useful analysis. Furthermore it is fairly comprehensive analysis."[7]

In a marketplace replete with expertise, OR's proponents could not simply declare that it encompassed all kinds of expertise that might be pertinent to industrial management—they had to offer something more specific. Within this marketplace, developments in such areas as inventory theory and linear programming were genuinely unique, and, in their more practical manifestations, valuable to industry. Thus, it would seem most appropriate to say that theory did not so much consume OR as it was the only thing within the OR rubric that proved capable of substantial growth.[8] In return, OR proponents developed a genuine interest in the new lines of theory. While sequential analysis quickly came to be regarded as a fundamentally important topic within the existing field of mathematical statistics, game theory, linear programming, and inventory theory all occupied an uncomfortable niche between statistics, mathematics, and economics. RAND provided one important home for this work, and a crucial source of support, but to develop more broadly it required journals, and, preferably, a profession dedicated to it. *Naval Research Logistics Quarterly* was an important example of the former, but ORSA and its journal provided both.

The question remains why theoretical OR not only became a part of OR, but essentially dominated its identity. Certainly the increasing dominance of ORSA's journal by theoretical content encouraged identifying OR with theory, as did the need to define the content of the courses and degree programs dedicated to the subject at an increasing number of universities. Moreover, OR was often defined as the application of "scientific method," and mathematicians, peculiarly, tended to identify a statement as "scientific" if it was grounded in rigorously formulated axioms. Thus, they, at least, would have tended to associate OR's claims to being scientific with its theoretical content. Finally, at institutions such as RAND, the broad sorts of studies that OR's proponents associated with OR were usually simply called "studies." For this reason, as we will see, RAND became one of the first places where OR was identified more narrowly with mathematical optimization within the broader ambit of a study. But, whatever the reasons behind OR's turn to theory, it was perhaps not such a great leap as one might suppose to see research intended to enhance the rationality of decision making become identified with theoretical research that clarified what it meant for one policy to be considered more rational than another.

23   Systems Analysis: The Challenge of Rational Engineering

I take it that every person in this room is fundamentally interested and devoted to what you can just broadly call the rational life. He believes fundamentally that there is something to this business of having some knowledge, and some experience, and some insight, and some analysis of problems, rather than living in a state of ignorance, superstition, and drifting-into-whatever-may-come. I take it we all fundamentally believe in the rational life. I rather carefully did not say the logical life, because I am not as exclusively strong for the logical life as I am for what I mean by the rational life. I think there are some things that we need to talk about that are not very logical, but which are still awfully important; and I would like to include an intelligent interest in alogical aspects within what I mean by an enthusiasm for the rational life.

Warren Weaver, Opening Remarks, Project RAND Conference of Social Scientists, September 1947[1]

In the late 1940s, the RAND Corporation began developing a methodology it called "systems analysis," which was essentially an extrapolation of wartime air warfare analysis. The initial application of systems analysis was to identify a preferred configuration for the Air Force's next-generation bomber, the B-52. It was to do so by combining possible design specifications with probabilistic representations of the expected performance of a bomber designed to those specifications in the combat conditions expected to prevail during its intended mission: the large-scale nuclear bombing of the Soviet Union. The resulting model would provide explicit expressions of anticipated cost and mission results associated with different designs, allowing clearer comparisons between designs to be made. Importantly, it was not imagined that the method could provide reliable quantitative predictions of mission outcomes. Rather, the idea was to bring together the best knowledge and assumptions available concerning aircraft and mission design, and to articulate their implications. Since any design decisions would be restricted by limits in knowledge and assumptions, systems analysis was thought to provide the most comprehensive and coherent, and thus most satisfactory, means of comparing options within those limits.

The Origins of Systems Analysis

By 1960 "systems analysis" had long been an integral part of the RAND Corporation's identity. Around this time, RAND staff began regularly describing systems analysis as an extension of wartime operations research work to large-scale policy problems. This description made sense to them because the terms "systems analysis" and "operations research" had significantly shifted during the 1950s. By that time OR essentially connoted mathematical optimization, while systems analysis had escaped an early confinement to mathematical analysis and had become a more heterogeneous form of policy analysis. Thus, it seemed natural to suppose that systems analyses incorporated, contextualized, and were genealogically descended from OR.

Subsequent commentators have accepted this pedigree, and have used it to interpret the history of systems analysis in such a way that it conveys important lessons. According to this history, while wartime OR confined its mathematical treatments to relatively narrow problems, postwar systems analysis constituted an outrageously ambitious attempt by RAND to use mathematics to construct a complete "science of warfare." It was, the story goes, naively hoped that such a science would carry an intellectual authority that could put an end to the politics-ridden disputes that traditionally plagued the procurement process. Meeting a crushing and inevitable failure, RAND reformulated systems analysis into a form that was still scientistic, if somewhat more demure.[2] RAND's obvious naiveté in this scenario has been explained by invoking the influence of a miasmatic postwar-era faith in science and technology, as well as the psychology of the key figure in the development of systems analysis at RAND, Edwin Paxson. The journalist Fred Kaplan has provided us with some enduring images of Paxson: he was "ingenious, rude, abrasive, a driven man, hard-working, hard-drinking, chain-smoking." At internal RAND briefings he was a brutal critic of others' work, driving one briefer into a faint. He was "the numbers-cruncher *par excellence*. He loved to devise and try to solve equations of gargantuan dimension, the more numbers and variables and mathematical complexities the better." Quantifying "every single factor of the strategic bombing campaign" was "his dream."[3] From such observations we are supposed to know who Paxson was well enough to explain his work. He did not get on well with humans. His comfort was in vice, and his trust was in numbers.

In reality, however, our knowledge of Paxson is scant, and our comprehension of his work remains limited. Unlike some other RAND employees and affiliates he did not publish extensively in the open literature, and his internal RAND memoranda and working papers provide little commentary on his work. Nevertheless, with an improved understanding of the theoretical tools he wielded and some additional information on his wartime experience, his immediate postwar work, and his activities at RAND, there is no need to resort to the idea that his systems analysis work derived from a sort of hermetic mental dysfunction. There is ample evidence to demonstrate that systems analysis was a well-accepted extension of wartime air warfare analysis. Though certainly overly ambitious in its own way, it was more an attempt to grapple honestly with real conundrums in engineering complex technologies than a radical rationalistic effort to reduce defense procurement to a methodical science.[4]

During the war, it had been clear that air warfare analysis was analytically constrained in important ways. When L. B. C. Cunningham first outlined his theoretical ideas in 1937, he clearly stated that his theory could provide expectation values for combatants' likelihood of combat victory, provided they "adhere to their tactical programmes." However, it was often the case that a particular design could not be evaluated simply on the basis of any single tactic, not least since enemies could adapt their own tactics to it. In October 1944, in the middle of the AC-92 contract, the mathematician Merrill Flood suggested that John von Neumann and Oskar Morgenstern's theory of games could be used to derive a "pattern of action" that mixed different tactics at random. He also argued that in bombing operations the existence of multiple desiderata for evaluating operational success—accuracy and volume of bombs dropped, aircraft and crew survival, and number of enemy fighter aircraft destroyed, for example—prevented the outcome of particular aerial combats from being the sole criteria for determining the value of a particular configuration of equipment. Instead success gained meaning only by taking into account the higher strategic goals of the war.[5]

As we have seen, in his final report Warren Weaver embodied these considerations in his fantastical Tactical-Strategic Computer. However, it was Paxson who was most intent on making practical use of them. After the war, Paxson went to Europe to consult for the United States Strategic Bombing Survey before being named the co-head of the mathematics department at the Naval Ordnance Test Station (NOTS), located in the Mojave Desert near the small town of Inyokern, California.[6] He was only there for a short time, but, while there, he produced a working paper entitled "The Role of Mathematical Research in the Airborne Fire Control Development Program," spelling out a vision clearly deriving from the AC-92 project.

According to Paxson's paper, the advanced planning of fire-control equipment development had to take into account two points: first, what strategic and tactical situations were likely to be faced at future points in time, and, second, what technology was expected to be available to address those situations. "Airborne fire control development of the past," he claimed,

has consistently violated both principles. In the first instance, design proceeded on the basis of naive assumptions of rectilinear target courses, and low angular rates; and, in the second instance, design presumed that any improvements in projectiles, in radar inputs, in barbettes, in speed and maneuverability of the parent aircraft, and stabilization or second order sight corrections could be simply adjoined to existing equipment.

During World War II, Paxson had regarded cobbling together "stop-gap" solutions to inadequacies in basic design to be a proper role for mathematicians. Now that R&D schedules were somewhat more leisurely, he viewed it as wasteful to relegate mathematicians to the status of "patch up" specialists who "had little contact with functional specifications and design." If mathematicians were allowed to conduct thorough theoretical analyses preceding "rigid commitment to a particular design," they could help ensure that the assumptions built into devices would be less arbitrary. This would minimize the "real dangers of obsolescence and inadequacy" associated with changing technical and tactical realities.[7]

Much as decision theorists were often more interested in using formalism to explore the logical anatomy of decisions than they were in calculating actual quantities, so Paxson felt equipment design would benefit more from mathematicians' "feeling for generalization, structure, and alternative" than their "specific knowledge of differential equations, function theory, statistics, geometry, mathematical physics, and computing techniques." He outlined their role at each of four stages of the design "problem": long term, auxiliary, short term, and operational. In the long-term stage, design had to incorporate estimates of what "ranges, range rates, angles off, angular rates, target course curvature, tolerated sight settling time, [and] motion of gun platforms" were likely to characterize future combat.

Mathematicians could then "distill" available estimates into models of various combat situations, and recommend plausible armament configurations. For instance, if bomber speeds were assumed to be sufficiently high, it might be wise to place barbettes (remotely operated gun turrets) on the rear of engine nacelles to guard against a preponderance of rear attacks. But nacelle barbettes precluded the use of jet engines, so it was important to weigh these factors against each other. Then, on the basis of proposed configurations, the next step was to place "reasonable bounds" on "mass reaction" effects, that is, to estimate the range of likely outcomes of sizeable air battles. Paxson noted that during the war these effects had been measured empirically by operations analysis groups. As empirical investigation was not possible in peacetime, the theoretical evaluation of these long-term issues had to make use of "auxiliary" studies, such as evaluations of the probabilities that different gun configurations had of scoring a hit on distant moving targets.

The thorough analysis and measurement of long-term and auxiliary problems addressed the fire-control problem "in the large." Only once these problems had been considered could the problem "in the small" be intelligently addressed. According to Paxson, the "short term" stage was the time for the fire-control engineer's "rugged job of invention, design, construction, and extensive bench testing of pilot models," guided by mathematicians' earlier studies. Then, in operational contexts, the mathematician would reappear to help evaluate airborne tests, to aid in gunnery training, and to "serve in an operations analysis section" in times of war. This last job meant offering advice on field modifications, devising "statistical theorems of performance," and extracting information from operations that could be fed back into long-term and auxiliary studies for the development of future equipment.[8]

Paxson's memorandum addressed the importance of formulating fire-control equipment designs in view of stated aircraft designs and tactical scenarios. However, in a letter to John von Neumann, he also pointed out that, in principle, those contexts need not be so arbitrarily constrained. Instead, technologies and tactics could be combined together from the ground up in all the different ways in which it was feasible to do so, and the anticipated mission results for each combination calculated. Von Neumann's game theory was essential to this process of combination. For example, if one were calculating the optimal armament configuration for a fighter aircraft attacking a bomber, the calculation would depend on if the fighter was expected to open fire early or late in an approach. That choice, in turn, depended on the bomber's armament configuration and its choice of tactics. Since the bomber was likely to change tactics to cope with whatever armament configuration and tactics the fighter employed, the real solution was to find a mix of tactics and a corresponding armament configuration that would be most effective against the bomber's correspondingly optimal mix of tactics. In game theory this was what was known as the "minimax" solution, which maximizes gain or minimizes loss. In a game of rock-paper-scissors, to take a trivial example, the minimax solution is to select randomly between the three choices, weighting the frequency of each evenly. Paxson supposed that such logic could be taken further, and applied, for example, to the assignment of preferred targets in a large melee, or even to choices of missions and campaigns.[9] He did not suppose that these formulations could cut through all the inherent uncertainties and eventualities that characterized war. Rather, he supposed that military officers always made their decisions based on some equivalent but tacit logic, which, no doubt, harbored inconsistencies and fallacies that formal analysis could ferret out. NOTS, though, was not the place to pursue inquiries into all the interrelations linking equipment design with tactics development and strategy.

In early 1946 John Williams briefly joined Paxson's staff at NOTS. Finding that there was no adequate housing available on the dusty base near Inyokern, he arranged to work instead out of an auxiliary NOTS office in the Los Angeles suburb of Pasadena. As Williams settled there, Project RAND was just being set up in nearby Santa Monica, and its director, Frank Collbohm, was looking for staff. Impressed by AMP's war work, Collbohm contacted Warren Weaver, who recommended he hire Williams. According to Williams's later recollection, Weaver had told Collbohm that Williams was "the laziest man that he had ever met" and "therefore could be relied upon to find an easy way to solve hard problems." And RAND's problems were extremely hard. If the tactical use of the B-29 had proven too open-ended for AMP to digest, then RAND's open-ended remit to study ways of conducting intercontinental warfare was still more daunting. Yet, as with the AC-92 study, it was an irresistible chance to go beyond the piecemeal studies that were continually hampered by a lack of time and information, and constrained by commitments to equipment that could not be changed, and to plans that were already unfolding. So Williams sheepishly informed Paxson and NOTS that he had accepted a job working on a project for their rivals in the Army Air Forces.[10]

Project RAND was divided into various sections, later called "departments" in the independent RAND Corporation. Most of these were straightforwardly technical: nuclear energy, rockets, communications, electronics, and airplanes, with others added later. The "military worth" section, created by Williams, dealt with the problem of integrating studies into a comprehensive analysis, and evaluating them in view of overarching military goals. Military Worth soon became the "Evaluation" section, and, later, when economics and social science departments were added, it became "Mathematics," but continued to focus on the same broad intellectual problems.[11] Paxson himself left NOTS to join RAND in March 1947, where, independent of sectional affiliation, he took control of its first major research project, which would occupy him for the next three years.[12]

The Work of Systems Analysis

It was probably no accident that, where the AC-92 contract was devoted to a study of the B-29, the Strategic Bombing Systems Analysis quickly became oriented around selecting a design for the new B-52. Boeing had already won the contract to design and construct the bomber, but the company's engineers were being sent repeatedly back to the drawing board as their designs failed to live up to the U.S. Air Force's constantly changing specifications for its range, speed, flying altitude, and size.[13] These were, of course, precisely the kinds of issues that Paxson felt could be averted by developing operational requirements and equipment designs in tandem by employing mathematical analysis in the early conceptual stages.

The basis of Paxson's work would be information gathered from aircraft companies and military offices, including factors such as what budget was likely to be available, how many people could be trained and in what skills, what targets would need to be hit, and with how much force. Pertinent data was usually found in existing documentation, or obtained through direct inquiries. In a 1947 memorandum, Paxson urged RAND to develop contacts with the new Air University in Alabama, writing, "one may suppose that the major RAND function is to supply technical and scientific numerical implementation of the qualitative [Air War College] type thinking."[14] This was not to say that the military work was devoid of quantification, only that that quantification did not typically extend to problems such as tactical design.

Of course, obtaining the necessary information was not always a straightforward matter of asking for it. After the conclusion of the Strategic Bombing Systems Analysis, Paxson participated in an analysis of a potential Soviet offensive through West Germany called Exercise STYX, and had to estimate how many casualties could be expected, given varying assumptions about attacking and defending forces at the Rhine River. Paxson wanted to model the situation using casualty estimates given in military officers' planning manuals and Frederick Lanchester's World War I-era differential equations, which yielded expectation values of attrition of opposing forces of determined strength. First, however, he made a visit to Fort Leavenworth in Kansas to discuss the "physiology of ground operations" with military planners. There he was informed that the manuals' figures were meant to be used for planning logistics, not for estimating the outcome of any particular battle. When planning actual battles, they told him, it was necessary to consider "terrain, weather, the degree of organization or disorganization of the forces on a given side at a given stage of the battle, ... the number of days of combat which the various forces have been submitted to and, most particularly ... the quality of the replacements coming up." These factors could have a surprisingly radical impact on battle outcomes.[15]

Paxson was skeptical that officers actually made calculations anywhere near so intricate, "and still felt that there were a few magic little numbers which an officer estimating the situation was really using." To test his hypothesis, Paxson asked the officers to carry out a simple simulation of a battle on a map. The demonstration convinced Paxson of the officers' point—he remarked in a memorandum that a graphical model of such intricate force movements could be animated beautifully by Walt Disney. As a result of his meeting he became "rather discouraged about the entire approach to the problem of ground operations through differential equations." Instead, he asked, "Why shouldn't we use professionals as professionals in connection with the estimation of the results of ground operations?" He proposed to prepare a matrix of special situations versus relative ground forces, and then to bring in a small panel of officers to fill in numbers of "approximate losses in personnel and materiel" for the various scenarios and then use a distillation of their combined opinions to fill out the analysis.[16]

The sensibility underlying the construction of Paxson's analysis was to defer to expertise where that expertise could be found. However, where it was absent, it was not considered responsible to simply exclude a factor from consideration, thereby not only neglecting its impact, but also leaving that neglect hidden. In such cases it was considered acceptable for analysts to employ theoretical placeholders, such as using Lanchester's equations to model "mass reaction" effects.[17] In some cases almost totally arbitrary assumptions had to be made: the audacious invention of the so-called "Paxson Packing Factor" made a lasting impression on Paxson's colleagues.[18] In other cases, it might be possible to bring in nonmilitary expertise. Cognizance that "human factors" could have a real impact on battle prompted RAND to develop an interest in the social sciences.[19]

In assembling a highly technical analysis, the insights of the social sciences had to be included if essential but intangible factors such as "morale" were to be taken into account. John Williams later recalled that RAND's leaders were skeptical of including less technical areas in analysis:

As I remember it, their attitudes were: well, we've had a little contact with economists, and we're not exactly transported by the experience. As for the rest of the people in the social sciences, we know nothing about them, but they seem to be at least an order of magnitude worse than the economists, so far as really having their fields in hand is concerned; we're skeptical, and it appears that if we go blundering into this business, as is suggested, it doesn't seem to us to be a very conservative thing to do.

Still, according to Williams's recollection, he insisted that to be complete, and thus valid as a schematic for action, RAND's analyses had to be "catholic." He agreed that the social sciences were "in the fourteenth century" compared to the physical sciences and engineering, but, he explained, "I thought that was our situation: fourteenth century, tenth century, who cared what century; we need them to the extent that they can contribute. We can't do the job we want to do, leaving these factors out—these social, economic, and political factors."[20]

Addressing a conference RAND assembled in 1947 to recruit social scientists, Warren Weaver framed the issue in terms more flattering to them, appealing to their belief in the "rational life," which he contrasted with the "logical life." In making this contrast, he allowed that social science did not have a strict logical relationship to engineering problems, but that, nevertheless, there was certainly some relationship, which had to be considered in an "intelligent" way.[21] It was necessary to have "some knowledge, and some experience, and some insight" into the impact of human factors, which the social scientists were best equipped to provide.

Social science certainly did not fit into systems analyses easily. In a 1950 memorandum RAND sociologist Herbert Goldhamer reaffirmed the importance of social science, noting that human factors were "not trivial additions intended to lend a supererogatory breadth and elegance to the functions of the systems." When quantitative variables denoting human intervention were incorporated into analyses, "Quite reasonable variations in these values and in the manner in which the parameters are incorporated into the payoff function produce radical differences in the payoff variables." It was, therefore, necessary to take these parameters seriously, whatever "degree of intrepidity above and beyond the ordinary call of scientific duty" was entailed that might "embarrass [the systems analyst's] intellectual efforts." Goldhamer noted, in particular, that to that point human factors had usually been reduced to a "degradation factor," indicating an expected deviation in the efficiency of a mission function from some theoretically obtained ideal.[22] He urged that a more sophisticated formulation of the impact of human factors was badly required to ensure usable results.

With so many difficult-to-define elements inhabiting a systems analysis, Paxson employed various methods of testing the reasonableness of assumptions. Experience gained in World War II was sometimes useful: in one case an analytical model of an aerial combat scenario predicted a success rate thirty times greater than what had actually been experienced.[23] As new technologies made wartime experience obsolete, analysts had to rely on tests and exercises to check their assumptions. Reasonable-sounding coarse estimates could also be compared with the results of more detailed analyses. What Paxson called "first order" theories of bomber attrition involved making a fairly arbitrary estimate of what percentage of bombers was likely to survive the trip to a bombing run, and what percentage of those survivors was likely to make the trip back. A "second order" theory incorporated information from the potential "history of a mission" which had to "be given in idealized form (i.e., a mathematical model)." One could estimate how likely a bomber would be to encounter flak or fighter interceptors, and then estimate using methods familiar from warfare analysis how vulnerable a bomber would be to such attacks.[24] Disagreement between perspectives forced the analyst to make efforts to establish where unreasonably arbitrary assumptions underlying the discrepancy resided.

Between 1947 and 1950 Paxson and a large fraction of RAND's staff assembled the expected anatomy of a bombing mission against the Soviet Union to take place some time in the mid-1950s. In the end they concluded that a most effective system design would deploy a large number of relatively inexpensive and slow turboprop bombers rather than a fewer number of expensive jet-propelled bombers. This recommendation, however, was not heeded, and the B-52 was developed as a jet aircraft. RAND's management was livid at the Air Force's response to their recommendations. When, for instance, the Air Materiel Command suddenly produced new figures indicating that a turboprop bomber would be twice as expensive as initially proposed, they felt certain that the figures had been manipulated to support a prejudiced preference for jet-propelled aircraft.[25]

However, the Air Force's rejection of RAND's recommendation should not be taken as an overarching opinion of RAND's systems analysis work. As RAND systems analyst Edward Quade recalled, "Everyone liked the method, really."[26] The trouble was that the three years of work that went into the systems analysis produced a single final result that did not integrate well into the ongoing design process. Nor did it mesh with the Air Force's evident interest in designing a forward-looking aircraft, whatever time-horizon it had explicitly placed on RAND's analysis. In a 1953 memorandum Quade argued that Paxson's analysis, and others performed since, had suffered from "an excessively narrow formulation of the problem or the omission from the criterion of factors the Air Force considers of overriding importance."[27] These factors were no doubt at least partially Air Force prejudices, but those prejudices were not necessarily irrational. There was no particular reason why the study had to be limited by the budget constraints that prevailed prior to 1950, nor by the assumption that bombers would be used almost immediately upon delivery. Since war fortunately never came on any timescale, it is difficult to judge whether RAND or the Air Force was in some sense correct. The jet-propelled B-52 became a successful, versatile, long-serving aircraft, which remains decades from its scheduled retirement. However, the Soviets' equivalent, but more economical, turboprop-propelled Tu-95 "Bear" bomber was also successful and long serving.[28]

Systems analysis, logistical studies, operations research, and closely associated branches of statistics, mathematics, and economics all combined in the 1950s to form an intellectual matrix in which new theories of decision making developed. While intricately intertwined, these theories varied substantially in their purpose, and in the basic strategies that theoreticians used to develop them. Some theories were attempts to express what form an ideally rational decision would take, given strictly defined criteria governing the worth of a decision. Many, but not all, of these idealized theories were intended to serve as tools for producing actual numerical results. Other theories were conceptually heuristic attempts to express the tacit considerations that governed intuitive forms of decision making. Still other theories were intended only as approximate models of real decision making, which, when considered in aggregate, could test or generate explanations for observed economic phenomena. And still other theories were more abstract ruminations on the logical structure of decisions, or even on the fundamental mathematics underlying those structures.

In order to understand the enthusiasm that developed around different lines of theorization, it is essential to appreciate the distinctions between these lines, and the proclivities of different individuals to different varieties of theory. Some theorists had very practical interests. Others were interested in exploring abstract spaces. A few proved highly adept at moving between the practical and the abstract. But, in any event, postwar decision theory was a broad church that accommodated the wide variety of theories under development, and encouraged traffic among them. How theorists perceived their place within this heterogeneous culture of theorization informed their understanding of what their theories accomplished, the strategies they used to develop their theories, and their beliefs concerning how their work should be used and by whom. In OR, for instance, theorists had to decide, at any given moment, whether a theoretical model successfully articulated an optimal solution to a problem, whether it heuristically articulated some suboptimal decision as measured against some as-yet-unarticulated optimal decision, or indeed whether it satisfactorily described the rationality already implicit within seasoned decision makers' intuitive choices. Similarly, in economic models, theories of optimal decision making were increasingly used as placeholders for economic actors' choices. In some cases the theorists understood these theories to idealize those choices, but in others they understood them to be only rudimentary models that had yet to account for all the rational choices that economic actors in fact make.[1] Of course, these sorts of distinctions were by no means obvious, and debates surrounding them were often contentious. Nevertheless, in the 1950s, it was by drawing such distinctions that very different lines of work could each be regarded as intellectually legitimate and very much worth the effort put into it.

The Theorization of Practical Decision Making

The single constant that ran throughout all the various facets of decision theory was the conviction that there existed certain kinds of decisions that inevitably had to be made, whether that decision was an active or passive choice, and whether it was made by reason or by an arbitrary, even unwitting selection among alternatives. Further, even if decisions were made unconsciously, they always at least embodied some rationale, which could be expressed, at least imperfectly, in some explicit language, preferably mathematics and symbolic logic. Moreover, since all decisions are made to further one or more goals, their underlying rationales could, in principle, be optimized to further those goals most effectively. But, in any event, it was considered generally desirable that some attempt be made to express the rationales underlying decisions on the grounds that it made it easier to understand, critique, and improve upon them.

As we have seen, this conviction motivated the wartime development of search theory and sequential analysis, Warren Weaver's vision of the Tactical-Strategic Computer, the postwar development of inventory theory, and Edwin Paxson's Strategic Bombing Systems Analysis. One of this conviction's most remarkable expressions occurred in Aryeh Dvoretzky, Jack Kiefer, and Jacob Wolfowitz's first paper on inventories in 1952. In the middle of an otherwise deeply abstract and technical inquiry into the general form of solutions to inventory problems, they paused to offer a lengthy defense of a particular function, labeled W, representing the expected costs associated with inventory depletion. These costs included such elusive quantities as the sales expected to be lost due to customer dissatisfaction. They wrote:

It may be objected that our method requires one to specify the function W and that this function may be unknown or difficult to give. We wish to emphasize that the need for a function W is inevitable in the sense that any method which does not explicitly use a function W simply uses one implicitly. Thus one who selects a method of solving the inventory problem which ostensibly has the advantage of not requiring the specification of W is simply relinquishing control of W, and may be implicitly using a W of which he would disapprove (if he knew it). It is difficult to see what advantages can accrue to the ordering agency from deliberately burying its intellectual head in the sand. Even if the function W is very difficult to obtain it seems preferable to make some attempt at an intelligent decision about it. A rough approximation of a greatly simplified version of the underlying W may be preferable to completely ignoring this fundamental datum of the problem.[2]

In this passage, the deep parallels in sensibility between their articulation of the abstract logical structure governing inventory policies and the inclusion of semi-arbitrary assumptions in Edwin Paxson's ostensibly practical systems analysis become evident.

A major philosophical difficulty in constructing decision theories was to determine whether or not they adequately captured all the considerations governing the rationality of decisions made in practical situations. In some circumstances the problem was trivial. For example, in applying linear programming to gasoline blending, it would be immediately clear whether or not the application of advanced methodology improved the efficiency of the process. However, in other cases it could be profoundly difficult to determine whether or not some decision-making process was actually improved through the application of a new methodology. For instance, an ostensible improvement over existing practice might neglect some concealed, and perhaps unsuspected, virtue in the existing practice that guarded against some rare but catastrophic contingency. Such reasoning could, in principle, be used to argue for an absolute form of conservatism, wherein, since no reform could be conclusively demonstrated to be superior to the tacit wisdom of an existing policy, no reform could be rationally justified. However, since no one took such a stultifying philosophy seriously, it was generally accepted among theoreticians that the best that could be hoped for was to articulate the rationales underlying existing policy and the case for reform in the most thoroughly developed way possible, and to make comparisons and plans of action on that basis.

In any case, theoreticians who understood their work to be conceptually exploratory generally experienced few anxieties about whether their work matched the wisdom implicitly inhabiting established practices. Because their work involved the construction of an explicit conceptual vocabulary, their only immediate concern was whether their work augmented an existing lexicon archived in professional journals. For them, improvements to the existing lexicon could come, as sequential analysis had, through consideration of real decision makers' commentary and behaviors. But they might just as easily arise by following one's own intuition, or the logical implications of a particular line of abstract reasoning. In the decade following World War II, all these channels of theory development proved extraordinarily fruitful, and some theoreticians did not restrict themselves to any particular source of inspiration.

No theorist was more attuned to the ideas driving this theoretical literature, or was more influential in giving concrete shape to those ideas, than Kenneth Arrow. Arrow's foundational work on inventory theory was perhaps the least of his accomplishments in the 1940s and 1950s, but it was, in many ways, of a piece with his more widely celebrated achievements. These achievements came in a period that saw him shuttling back and forth between influential institutions and coming into contact with a variety of individuals considering new and apparently profound intellectual problems. His PhD work at Columbia University was conducted under statistician-turned-economist Harold Hotelling, and, during the course of his studies, he also took courses with Abraham Wald. During the war, he had been privy to Wald's development of sequential analysis. Then, when he became a research associate at the Cowles Commission, he met Abraham Girshick.

Like Arrow, Girshick had been a student of Hotelling's at Columbia. Upon finishing his PhD in 1937, he moved to the United States Department of Agriculture (USDA) in Washington, D.C., where, in 1939, he became the principal statistician at the Bureau of Agricultural Economics. During the war he returned to Columbia to join the Statistical Research Group, and worked with Wald in developing sequential analysis. After the war, Girshick returned to Washington where he met David Blackwell. Blackwell had received a PhD in mathematics from the University of Illinois in 1941 at the age of twenty-two, and then received a one-year fellowship at the Institute for Advanced Study in Princeton. However, Blackwell was an African American, so, despite his prodigious talent, the only long-term employment opportunities that proved open to him were at segregated black colleges. In 1944 this trail led him to a position at Howard University in Washington, D.C. In 1945 he was inspired to take

Figure 24.1
Left to right, Kenneth Arrow, David Blackwell, and Abraham Girshick in Santa
Monica in 1948. From DeGroot 1986. Courtesy of the Institute of Mathematical
Statistics.

up mathematical statistics by a talk that Girshick gave at the busy local
chapter of the American Statistical Association.

Meanwhile, one of the junior researchers under Girshick's direction at
the USDA was Selma Schweitzer, who soon left to take up graduate study
at the University of Chicago, where she met and married Arrow. On a visit
to the Cowles Commission, Girshick met Arrow through Schweitzer, and,
after leaving the USDA for RAND (before moving to Stanford University
shortly thereafter), Girshick extended Arrow and Blackwell an invitation
to come to Santa Monica in the summer of 1948, where the three of them
collaborated to elaborate upon Wald's formulation of sequential analysis
(figure 24.1).[3] This work entailed incorporating conditional and subjective
factors into sequential testing procedures.

Recall that the question at the heart of a sequential testing sequence is
to know when enough information has been gathered during the testing
process to satisfy the test's requirements so that the sequence can be ter-
minated. That formulation assumes that the cost of testing is non-zero so
that there is some value in terminating the test as early as possible. It also
assumes that the requirements for passing the test are fixed, perhaps rationally, but perhaps arbitrarily. In their paper, Arrow, Blackwell, and Girshick

probed these assumptions by considering how the decision to terminate a sequential test of a hypothesis would be affected by taking into account certain anticipated costs and benefits of testing. The costs included those associated with deciding incorrectly between hypotheses, as well as the costs of undertaking additional trials. These costs were then to be balanced against the anticipated value of the information to be gained from the trials. In this way the requirements for passing the test would be more clearly defined as a function of the ultimate goals of the decision makers undertaking the test. Arrow, Blackwell, and Girshick published their paper in *Econometrica* in 1949, but they were beaten to the punch by Abraham Wald and Jacob Wolfowitz back at Columbia, who derived many of the same conclusions in a paper published in the *Annals of Mathematical Statistics* in 1948.[4]

The coincidence of Arrow, Blackwell, and Girshick's and Wald and Wolfowitz's work was indicative of a broader interest permeating decision theory concerning the meta-calculative considerations that established under what circumstances one decision could and could not be considered more rational than another. In the case of sequential testing, the essential insight was that no decision as to when to terminate the test could be said to be more rational than another, unless the costs and value of testing, as dictated by the higher purpose that the test was intended to serve, were taken into account. A similar principle applied with inventory theory, in which the rationality of an inventory policy was inextricably tied to the ability of the inventory to balance the benefit of having stock on hand with the cost of holding the inventory, as measured in terms of each factor's contribution to the higher goal of profitability. Since much of the point of maintaining an inventory was to maintain good customer relations, Wolfowitz, Jack Kiefer, and Aryeh Dvoretzky naturally insisted that W, the relationship between having stock on hand and the quality of customer relations, be articulated, since, otherwise, no inventory policy could actually be said to be superior to another.

Another issue permeating postwar decision theory was that the value of the outcome of a decision-making process was not merely subjective, but also dependent upon decisions being made by others at the same time. The value of deciding to undertake an action might depend on whether or not someone else took a complementary action, or the value of making a decision quickly might depend on what one stood to lose if someone else made that decision first. Notably, once introduced to the RAND Corporation, David Blackwell quickly became an expert in applying game theory to aircraft duels, wherein the value of the decision to open fire at a certain range not only depended on the calculated chance of scoring a hit, but also on the prospect that the other airplane would soon start shooting as well.[5] However, game theory could also be thought of in more generalized terms, and applied to scenarios that were not strictly competitive. A final section in Arrow, Blackwell, and Girshick's paper united sequential analysis with game theory by supposing that the process of hypothesis testing could be formulated as a game played against "Nature." In the game Nature selects a hypothesis according to an unknown but fixed "mixed strategy," while the statistician selects a test procedure in an attempt to determine both efficiently and effectively what hypothesis Nature has selected.[6] Further work along these lines considered scenarios where tests were undertaken in situations where conclusions had to be drawn in view of competition with other decision makers.

All of these sorts of theoretical considerations were incorporated into Abraham Wald's 1950 monograph, *Statistical Decision Functions*, published just prior to his untimely death.[7] These considerations also augured the rehabilitation and development of Bayesian probability. Bayesian probability frames hypothesis testing as a problem of making a decision between possible hypotheses based on the strength of a subjective belief that they are true. Initially, these probabilities are assigned some a priori value, which might be an even distribution between possibilities or a representation of the decision maker's vague intuition. The probabilities are then modified on the basis of subsequent experiments. Perhaps the key figure in this development was Jimmie Savage. Savage had received his PhD from the University of Michigan, and had served in the Statistical Research Group (SRG) at Columbia during the war. Afterward, he and fellow SRG statistician Allen Wallis founded a department of statistics at the University of Chicago. In 1954 he published *The Foundations of Statistics*, which later became informally known as the Bayesians' "Bible."[8] The same year Girshick and Blackwell published *Theory of Games and Statistical Decisions*. Girshick would die prematurely the following year, but Blackwell, having joined the statistics department at the University of California, Berkeley in 1954, was on his way to becoming an authority on Bayesian statistics. In 1965 he would be the first African American to be elected a member of the National Academy of Sciences.[9]

Because the rationality of a decision was deemed by decision theorists to be contingent on the value ascribed to the various outcomes of not only the decision but the decision-making process, some theorists began to examine how these values could be defined. At the RAND Corporation this discussion centered on the concept of "military worth," which ostensibly was defined as victory in warfare. However, since not all victories were created equal, choices had to be made concerning what kinds of wars one intended on fighting and when, and what kinds of victories one wanted to pursue. This issue presented Kenneth Arrow with the problem he finally alighted upon for his PhD work. In his first summer at RAND, when he wrote his paper with Girshick and Blackwell, he also became embroiled in discussions of military worth, which, it was agreed, had to be somehow defined in terms of "national interest." This problem, it seemed, was effectively identical to the more longstanding problem of "social welfare" in economics, which centered on how policy goals could be defined by a collective entity such as a society. Arrow supposed that the only way a social welfare function could be defined, outside of the arbitrary declaration of a dictator, was through the summation of the "preference functions" of all the individuals comprising a society. Following the logic of the so-called "voting paradox" outlined by the Marquis de Condorcet in the eighteenth century, Arrow established that the aggregation of preferences was virtually guaranteed to result in an incoherent ordering. For example, it would be found that, in aggregate, a goal, A, would be preferred to another, B, which would be preferred to another, C, which, in turn, would be preferred to A. This result compelled theorists to reflect more deeply than they previously had on the basis of the legitimacy of state action. Arrow did not offer any easy way out of the paradox. Rather, after being published as the book *Social Choice and Individual Values* in 1951, his so-called "impossibility theorem" proved rich enough to establish its own class of logical and philosophical problems, which political scientists would pursue with great zeal over the course of the following decades.[10]

Modeling across the Limits of Computability

Decision theory was, in its essence, a theory of practical decision making. It was intended to establish whether one decision could be considered more rational than alternative decisions. This rationality was defined in terms of the decision's ability to achieve various goals set by the decision maker. Moreover, it was defined not in terms of the transcendental correctness of the decision, but in terms of what decision would be most rational in view of the knowledge possessed by the decision maker at the time of the decision. This proviso derived from the fact that the process of gathering knowledge itself carried costs that detracted from a decision maker's overarching goals, meaning that all decisions would be made under conditions of greater or lesser uncertainty. The ultimate—and, it was admitted, unachievable—objective of this line of theorization was to define the rationality of a decision in such a way that it would be impossible for the decision maker to object that a formulation did not truly define a rational decision on account of its failure to account for some hidden meta-calculative criterion that bore upon how a decision had to be made in view of all relevant practical exigencies.

Needless to say, the history of the development of theories *of* practical decision making should not be confounded with the history of the development of theories *for* practical decision making. Incorporating all of the meta-calculative considerations that bore upon the rationality of a decision invariably made the formulation of the decision problem so complex that, ironically, it became impossible to use it for many practical computational purposes. The obvious response to this objection was clearly that some alternative logic must exist that describes how practical computations may, in fact, be rationally undertaken. Some theoreticians would follow that line of inquiry down its own philosophical rabbit hole. However, most spent their time developing theories in the service of alternative goals.

As we have seen, scientists interested in developing the field of operations research took an interest in new theories primarily as a means of deriving actual numerical results that could be directly implemented in military and industrial operations. They relied on close contact with established decision makers and on the continuous monitoring of operational results to ensure that theory-derived amendments to plans, policies, and procedures resulted in real improvements. While more sophisticated versions of theories could have applications, particularly in more engineering-oriented situations, overelaboration tended to make theories less directly useful. It did not, however, mean that elaborate theories had no use.

Obtaining conceptual clarity from a model was considered by many theoreticians to be at least as valuable as obtaining numerical results. Kenneth Arrow discussed this issue in an address at the 1956 annual meeting of the Operations Research Society of America entitled "Decision Theory and Operations Research." While Arrow allowed that theories could certainly be used to produce numerical results, he believed that that goal should not preclude their use as heuristic devices for exploring their own inadequacies. For instance, he pointed out, an inventory model might well yield an optimized solution to a well-formulated inventory management problem, but it was the *process* of modeling inventory problems that yielded the most edification. This assertion led him to argue that the "open and tentative character of operations research, and of all scientific analysis, can itself be discussed in terms of decision theory." He remarked:

I believe that the notions of decision analysis in sequential situations ... can, if properly understood, be very revealing of the tentative and approximate nature of any particular solution to an operations-research problem. This shows that, on the one hand, solutions to particular problems are less definitive than their formal statement suggests, and, on the other, formal over-simplified models which are not yet capable of being quantified may nevertheless be of great practical value.

Modeling could, for example, elucidate what elements of a problem were most sensitive to choice, and thus "the value of certain types of information," which bore upon decisions concerning what kinds of records would be valuable to keep, and what kinds of models would be valuable to develop in the future. Just how much a model had to be developed before it stopped being useful was an important question. "Naturally," he noted, "better results will be obtained by broadening and deepening the problem... But there is always a point where we must stop." The decision to terminate the modeling sequence depended on the theoretician's goals in undertaking the act of modeling. It was perhaps the one meta-calculation that could never be incorporated into the model one was building.[11]

How one built models according to one's specific goals in model building was a major driver of various twists and turns in the early history of game theory.[12] John von Neumann and Oskar Morgenstern regarded their book *Theory of Games and Economic Behavior* as offering an axiomatically grounded discussion of a new field of mathematics, from which could be derived a conceptual vocabulary for discussing competitive economic behaviors.[13] However, as we have seen, the theory was initially valued for its ability to provide explicit solutions to problems concerning the selection of preferred combinations of armaments and tactics.[14] At the same time, the abstract mathematical structure of game theory, its relationship to linear programming, and the existence and form of game-theoretical solutions were also considered ripe subjects for academic inquiry. In 1950, following this last line of inquiry, the Princeton PhD student John Nash established that any game, zero-sum or not, and involving any finite number of players, will contain a solution concept, that is, an optimal mix of strategy for all players.[15]

Many game theoreticians quickly came to regard the "Nash equilibrium" as a concept of great importance. By definition, it encompassed all possible meta-calculations that a game's players might conceivably make, meaning that there are no strategies or incentives available to them except for those defined by the rules of the game. Moreover, unlike in applications of game theory to combat analysis, there were no fixed concepts of "friend" or "enemy" within the game: cooperative alliances could be made or broken by the players in any way that maximized their payoff. Naturally, in more complex games the likelihood that real players might find and abide by the solution concepts defined by the Nash equilibrium decreased dramatically, and many kinds of game defied numerical solution altogether. But the reason that the equilibrium was important was that it was a seemingly fundamental formal result implied by the mathematical structure of the theory. However, as theorists continued to develop game theory, the meaning and significance of Nash's work would prove ambiguous, revered by some, and dismissed as irrelevant by others.[16]

One major question that faced game theorists as time passed was whether they should be interested in solutions to game-theoretical formulations at all. Increasingly, game theory came to be understood as a way of evoking, defining, and, above all, organizing ideas about the possible forms that competitive human behavior could exhibit, whether it was in some sense optimized or not. The "Prisoner's Dilemma," which was formulated at RAND by Merrill Flood and Melvin Drescher in 1950, proved particularly appealing and enduring in this regard. In the problem, two captured conspirators are separated from each other, and have contradicting incentives to protect and betray each other, the worst results coming to players if they protect while their partner betrays. The Nash equilibrium of the problem dictates that mutual betrayal is the optimal solution concept, even though it is clearly a less desirable outcome than mutual protection. However, the point of the Prisoner's Dilemma formulation was not to find an "optimal" solution to it, but to explore its inadequacies, and what strategies might exist external to the stated structure of the problem. Games with indefinite rules, strategies, and number of repetitions were useful for conceptualizing behaviors such as reconnaissance and trust building. These concepts, defined in game-theoretical language, then provided a means of identifying different strategic dilemmas in the real world, and describing different options for facing those dilemmas.

One of the first public presentations of this more conceptual use of game theory was John Williams's popular exposition, *The Compleat Strategyst*, published in 1954. Shortly thereafter, conceptual models in game theory proliferated into the social sciences through new journals such as *Behavioral Science*, which was founded in 1956, and *The Journal of Conflict Resolution*, founded in 1957. The subject found its first scholarly textbook in R. Duncan Luce and Howard Raiffa's *Games and Decisions*, published in 1957. Thomas Schelling's book, *The Strategy of Conflict*, appeared in 1960 and quickly became a landmark work in formalized strategic theory.[17]

Rationality Assumptions and Economic and Cognitive Theory

In von Neumann and Morgenstern's original articulation of game theory, prescriptive and descriptive applications were closely intertwined. The work of Luce, Raiffa, Schelling, and others began to draw a clear delineation between numerical, prescriptive applications and conceptual, descriptive applications. Of course, descriptive applications could have prescriptive implications, but only through the derivation of conceptual clarity from the act of theorizing, not from the results of a calculation deriving directly from a theoretical formulation. However, this delineation did not dispel the issue of the extent to which theoretical depictions of apparently optimally rational decisions could be used descriptively.[18] Nowhere was this issue more contentious than in economic theory.

Keynesian economic theory employed the presumption that economic actors make decisions that seem rational to them, but it also showed how, in aggregate, those decisions could have pathological consequences for national economies. While Keynesian models were intended to be descriptive, it was anticipated that the conceptual understanding of economic mechanisms that they provided would permit appropriate remedies for economic ills to be chosen from a range of possible policies. Other economic models had a distinct goal: to develop models of aggregated decision making that would provide insight into how market mechanisms worked in the first place. While not, in any strong sense of the term, descriptive of real economies, these models were also not intended to be, in any precise sense, prescriptive, except insofar as some authors took them to demonstrate why, by virtue of their complexities, centrally planned economies were practical impossibilities.[19]

The so-called Austrian School of economists was deeply interested in the functioning of market mechanisms, and Oskar Morgenstern, who had been associated with that school, was interested in game theory as a means of providing a new language for describing that functioning. Meanwhile, in the postwar period, linear programming and the related mathematics of game theory were viewed, particularly by members and associates of the Cowles Commission, as a new means of reformulating the basic model of an economy as an auction originally articulated in the nineteenth-century "pure economics" of Léon Walras. A crucial step in the development of this line of inquiry came in 1954 with the publication of Kenneth Arrow and French mathematician and economist Gérard Debreu's paper, "Existence of an Equilibrium for a Competitive Economy." Their paper employed the Nash equilibrium as a means of proving that a Walrasian economy formulated in game-theoretical terms could, in fact, be guaranteed to establish a set of price points that cleared the market of goods and services.[20]

In the immediate postwar era, the interests of the majority of economic theorists leaned toward the more active policy orientation of the Keynesians, but many were also interested in what insights more sophisticated models of economic decision making might yield. The highly mathematized and neoclassical but nonetheless Keynesian economics developed by Paul Samuelson, Robert Solow, and others in the rapidly expanding postwar MIT economics department was exemplary of this view. Samuelson, notably, became interested in linear programming following a visit to the RAND Corporation in January 1949. After considering and rejecting the possibility that the empirical study of market mechanisms might actually yield a more powerful optimization procedure than the simplex algorithm, he began to focus on using linear programming as a model of market mechanisms. However, rather than focusing on the Walrasian model, he turned linear programming back on the economic formulation that inspired it, Wassily Leontief's input–output models of a national economy.[21]

Where Leontief's model had represented a static economic structure, linear programming held out the promise of rendering the model dynamic that could show how actors in different sectors of the economy were likely to adjust their production, prices, and consumption in light of shifting conditions in neighboring market sectors. But, rather than focusing on a fully fleshed-out input–output model, Samuelson preferred to work with a highly constrained model in order to clarify very particular points. At Tjalling Koopmans's "Activity Analysis" conference, he offered a relatively simple logical proof that showed that when substitution is allowed between inputs—a possibility allowed with linear programming, but implicitly disallowed by Leontief's model—changes in the labor market or in the final composition of output goods would not produce market incentives to substitute between inputs. Koopmans, Arrow, and Vanderbilt University economist Nicholas Georgescu-Roegen offered similar proofs.[22]

For Samuelson and Solow, linear programming was useful as a means of approximating rational decision making, but only in cases where prior models did not already take that decision making into account. However, they did not suppose that economic actors' decision making was necessarily optimally rational. In a book on the relations between linear programming and economics that they cowrote with Harvard economist and OR theorist Robert Dorfman, they explicitly denied that managers made decisions about production inputs by "formulating the production function mathematically and finding its optimum by the methods of the differential calculus." Rather, linear programming sufficed as a means of approximating the effects of decisions that were made by some indeterminate means, perhaps through "trial and error or survival of the fittest." Their relative lack of interest in general equilibrium models was probably linked to a belief that optimal decision models were sufficiently inaccurate that they could not be fruitfully aggregated to encompass whole economies. But judging the ability of a model to produce useful insights was a matter of perception. For Arrow and Debreu, the Walrasian model constituted "a reasonably accurate description of reality, at least for certain purposes."[23]

It is important to appreciate, though, that theorists' perceptions of the utility of incorporating models of rational decision making into economic models had less to do with their various beliefs concerning economic actors' actual rationality, than it did with their views about the proper heuristics of model building.[24] All economic models are abstractions, and in the postwar period it was certainly possible to understand that abstraction as over-idealization. However, the development of decision theory and the application of linear programming to economic analysis contributed to an overarching sense that the problem with most economic models was that they were too rudimentary, and did not yet take rational behaviors sufficiently into account. This attitude toward model making could potentially be elevated to the status of a fundamental precept. In principle, the inadequacies of *any* model could be explained in terms of its failure to take into account meta-calculative considerations or economic actors' undeclared goals. Thus, the object of future models would be to incorporate these aspects of the problem. According to this view, if one deemed, for instance, a Nash equilibrium to a game-theoretical problem to be an inadequate representation of reality, it was not because the Nash solution concept did not apply, but because the rules and stakes of the game had not been adequately described.

The obvious problem with such a philosophy was that it led to the creation of models that could not possibly reflect economic actors' actual thought. This issue was the subject of substantial philosophical reflection. Before World War II, Paul Samuelson argued that economic actors' goals could be best discerned not by asking them directly, but by ascertaining what goals their behaviors suggested they were rationally pursuing. Following the war, he rearticulated this idea as the doctrine of "revealed preference." This approach to the problem of rationality reflected a fundamental ambivalence toward it, since, according to it, any behavior could be deemed rational simply by redefining the ends that it appeared to accomplish, rather than by assessing actors' ability to accomplish specific ends.[25] Another approach, which better preserved the utility of the concept of rationality, was to suppose that economic actors arrived at rational decisions, but without undertaking any conscious act of calculation. UCLA economist and RAND affiliate Armen Alchian suggested that "profit maximization" models should be replaced by ones that describe the existential pressures of the "economic system" as an "adoptive mechanism," akin to Darwinian natural selection, "which chooses among exploratory actions generated by the adaptive pursuit of 'success' or 'profits.'"[26] As we have seen, this was just the view that Samuelson, Solow, and Dorfman ended up taking toward linear programming as a model of economic decision making.

Still another approach was to dilute the rationality in theories until those theories accurately described observed behavior. This approach was championed by Herbert Simon, who began his career in the late 1940s as a student of administrative thought and practice. His goal at that time, as articulated in his nonformalistic 1947 book *Administrative Behavior*, was to reform the field of public administration by overcoming the inadequacy of its "linguistic and conceptual tools for realistically and significantly describing even a simple administrative organization." He wanted to clarify how it was that organizations went about "getting things done" by describing the way administrators set goals, balanced conflicting goals, motivated employees to achieve those goals, and coordinated their efforts.[27] Around the time Simon moved to the new Graduate School of Industrial Administration at the Carnegie Institute of Technology in 1949, where he would spend his career, he adopted the formalistic vocabulary of economists, statisticians, and the new decision theorists.[28] His facility in moving between different styles of formalized theorization rivaled even Kenneth Arrow's. He contributed a paper on the application of linear programming to economic analysis at the 1949 "Activity Analysis" conference, and published a paper in *Econometrica* applying servo engineering theory to problems of supply-chain management just as the new inventory-control literature began to develop.[29] He would remain interested in management and economics, but his main interest soon gravitated to the description of individuals' observable decision-making behaviors, such as those exhibited in games of chess.

As with other decision theorists, Simon was interested in how constraints in resources, ideas, time, institutional politics, and so forth, forced decision makers to adopt expedient decision-making strategies. However, where other theorists factored these constraints into optimization calculations, Simon concentrated on articulating semirational rules that more realistically described, step by step, how decision makers coped with complexity. The object of such rules was not to arrive at a choice that was optimal, but rather one that "satisficed," that is, was deemed satisfactory by the decision maker. To the dissatisfaction of many theoreticians, this notion of "bounded rationality" had no fixed intellectual foundations, and was not very useful in building up well-organized bodies of theoretical knowledge. Simon, however, was unbothered by such concerns, and increasingly, with his collaborator Allen Newell, turned his attention from theory to the development of digital computer simulations of cognitive behavior.[30]

Simon's rejection of fundamental ontologies of behavior and any metaphysics of theorization left a collection of freestanding computer models to constitute the epistemological core of his and Newell's work. These models had a defined structure, which could be transplanted from OR to economics to cognitive psychology, insofar as the structure successfully described observed decision-making behaviors. Although intended to be more realistic than economic or decision-theoretical models, he was well aware that even his digital simulations could make no claim to perfect realism. Much later, in his 1969 book *The Sciences of the Artificial*, he argued that even if digital logic did not correspond to the underlying ontology of a brain, a business, or an economy, the construction of artificial models was not only a valid means but also an inevitable means of inquiry. There was, he urged, no approach to any science, natural or social, which did not employ artificial metaphors of one kind or another in order to proceed. Digital computation was simply the most flexible metaphor that he knew.[31] But there was also a further advantage to modeling in a digital medium: one could reverse the process of inquiry so that understanding became engineering. In addition to his many other interests, Simon was also a pioneer in the field of artificial intelligence.

Coda: The Management of Theory

From 1950 onward, debates and pitched battles over what sorts of theories and models are the most valid or useful continued unabated. Individual assessments of a model's worth often hinged on an implicit sense of the uses to which it was to be put, and a subjective experience of the insight it helped to develop. Such criteria for assessing the value of a theory were considered legitimate because the persona of the theorist dictated that theorists' only responsibilities were to acknowledge and build on the work of other theorists. It was generally agreed that the burden of ensuring the legitimacy of the deployment of a theory fell not on the theorist, but on the theory's users. This view was illustrated by a story that George Dantzig liked to tell about one of his first presentations of linear programming at a conference. Decades after the fact, he could still recall how the eminent and physically imposing Harold Hotelling had stood up and curtly informed him that "the world is nonlinear," and then sat back down. The virtually unknown Dantzig was left speechless, but he was saved by the even-more-eminent John von Neumann, who stood up and replied on his behalf: "The speaker titled his talk 'Linear Programming.' Then he carefully stated his axioms. If you have an application that satisfies the axioms, use it. If it does not, then *don't*."[32]

An important point that Dantzig's story missed was that the practical validity of theories was itself a subject of intensive theorization at the time, and that many of Hotelling's own colleagues and students were at the forefront of that movement. The sheer complexity and meta-calculative abstractness to which this line of theorization led amply demonstrated that the relationship between theories and their legitimate application was a far from trivial matter. As new theories multiplied across numerous disciplines in the second half of the twentieth century, the management of the relationship between theory and practice would prove an enduringly difficult problem. Enticing theoretical problems were likely to be inspired by real problems faced in industry, the military, and politics, but it was generally understood that such theorization might well proceed along tangents from those problems. But, for those economists, operations researchers, and engineers who were interested in solving practical problems, the work of theorists could appear tantalizingly concrete and useful in one instance, and maddeningly removed from reality in the next. They fretted, with reason, that the theoretical work that increasingly dominated their professional journals would move unfiltered into practical work and sour relations with finicky clients and employers. In the next section of this book, we will examine how theory was managed in practice.

VI The Intellectual Economy of Theory and Practice

25 The Civilianization of Operations Research on the Charles

During the 1950s operations research and systems analysis merged together with the new multifaceted corpus of decision theory, creating a potent intellectual and professional force. On the one hand, OR, in particular, provided an important professional space beyond the existing fields of economics and mathematical statistics, where new theories could develop and cross-fertilize. It also provided crucial institutional access to the military and industry, beyond what any single organization, even the RAND Corporation, could provide. This access naturally presented theoreticians with opportunities for sponsorship, but it also yielded practical problems, consideration of which could lead into new theoretical terrain. On the other hand, new theories gave OR proponents a valuable advantage in a crowded professional marketplace populated by statisticians, economists, cost accountants, management consultants, and others hoping to reform industrial practice.[1] The critical problem was to balance the clear value of theory with the danger that its more abstract and impractical manifestations might be injudiciously applied to practical problems, thereby threatening OR proponents' credibility. This problem could only be solved by creating a culture, consistently reinforced through networks of institutions, that could strike and maintain such balances in a variety of milieus.

The Arrival of OR at MIT

Cambridge, Massachusetts, was among the places where this essential problem was first addressed, as OR proponents attempted to craft a civilianized version of wartime work.[2] As early as 1944, suggestions that the Massachusetts Institute of Technology establish an educational program in OR flowed back from the field. These suggestions began to grow more insistent once MIT contracted to administer the Navy's Operations Evaluation Group (OEG).[3] Physicist Charles Kittel, a Guggenheim Fellow at MIT who had been a member of the Navy's wartime group, also published an article in *Science*, calling attention to British efforts to expand OR beyond the military.[4] In late 1947 MIT administrators even invited Patrick Blackett to travel from Britain to consult with them on the subject. Blackett did not encourage the venture, arguing that OR was not such a novel thing in industry.[5] However, MIT's leaders were not discouraged and set up an experimental full-year graduate-level course on the subject in 1948–1949, taught by OEG staff visiting from Washington and mathematician George Wadsworth. It was repeated in a one-semester version in the spring of 1950.[6]

A full description of the second offering of the course survives in the form of an OEG report, which made the aims and approach of the course abundantly clear. The report observed that "a strong effort was made to convey certain basic principles and viewpoints of operations research." These primarily involved the need to strike a balance between the utility of special mathematical techniques and making valuable contributions as members of an organization. "Real problems," the report stressed, "do not fall neatly into one or another of the usual categories of knowledge, but cut widely across boundaries." Success demanded that the "wholeness of the problem" not be "artificially suppressed." Developing a qualitative understanding of a problem was a "prerequisite to useful quantitative work." The report noted that a "crude solution of the real problem" could have great benefits, "whereas a precise solution of an unjustified idealization of the problem is at best worthless and, if its misleading conclusions are acted on, may do great harm." It pointed to the factor of "urgency" in determining what kind of a study should be designed, and to the "utmost importance" of the "relation of an O/R group to the organization which it serves." It also observed the potential for conflict between "the requirements of scientific freedom" and "the discipline and authority necessary to the effective functioning of a large organization, military or nonmilitary."[7]

The course covered the technical methods of statistics, search theory, and game theory extensively, but also emphasized problems such as the "formulation and solution of real problems," to which it devoted six full sessions. These priorities were reflected in the course's written assignments. Some were simply mathematical exercises: one advanced problem, for instance, asked what an optimal mixed strategy would be in a game-theoretical problem where "each player chooses a real number; if X chooses x, and Y chooses y, then the payment to X is sin(x+y)." But the course's final written assignment was simply: "Make an operational study of the taxicab business. Submit results in the form of a report to the executive of the company." The course organizers pointed out that they kept such problems purposefully vague in order to reduce the artificiality of the classroom experience and to simulate the unformed nature of a problem when it arrived in the hands of an OR group.[8]

The instructors were also on the lookout for signs students were not treating the material in the spirit they wished to instill. The report noted that "many of the students, especially those majoring in mathematics ... appeared to be disturbed emotionally by the uncertainty and large scope of some of these problems, and too quick to seize on some neat set of assumptions and then quickly reduce the discussion to a problem in pure mathematics." These students tended to spend only a few lines formulating the problem and then working out pages' worth of analytical solution. The report went on: "It was necessary to combat this tendency vigorously by emphasizing the importance of qualitative considerations and crude approximate analysis, and by de-emphasizing elaborate analysis by showing that in many cases the essential features of such analysis are understandable in very simple terms and, where not, are not reliable anyway." In the end, apparently, "all but two or three fell into the spirit of the assignments and produced very creditable reports." The report supposed that those who did not fall into the proper spirit might instead prove to be "a valuable member of a mathematical service section."[9]

OR and Consulting at Arthur D. Little

While MIT was implementing its first OR course, its neighbor on the Charles River Basin, the Arthur D. Little, Inc. consulting firm, began setting up an OR program of its own. Over decades the firm had built a formidable reputation in research and development problems, but had not taken part in the prewar boom in management consulting.[10] Following the war, Bruce Old and Gilbert King suggested that operations research could provide a springboard for the company's scientifically skilled consultants to make an entry into management issues. Old, a metallurgist, had served in the Navy's Office of the Coordinator for Research and Development, and had had some experience with the Navy's wartime OR work. King, a research associate in chemistry at nearby MIT and an associate of the firm, had become involved with OR in the later years of the war.[11] Together they convinced Raymond Stevens, a vice president of the firm, to establish an industry-directed OR group on an experimental basis.[12] Stevens, in turn, assigned Harry Wissman, one of the company's few holders of a business degree, to take on the task. In the fall of 1949 Wissman hired as an assistant John Magee, a recent graduate of the Harvard Business School. However, he recruited the lion's share of the OR staff from the Navy's OEG. George Kimball, the deputy director of the Navy's wartime OR group, consulted part-time, and later became a vice president of the company.[13]

One of Arthur D. Little's existing clients, Sears, Roebuck & Company, agreed to serve as a test case for the new activity. Sears had collected punch-card records of the names, addresses, and ordering histories of some ten million of its mail-order customers. Because of the prohibitive costs of mailing the bulky spring and fall catalogs, Sears managers knew it was only worthwhile to send them out to those customers most likely to make a purchase. Over the course of decades, the company had established an elaborate set of rules to manage catalog distribution, and had even conducted experiments wherein they sent catalogs out to everyone on the list in certain markets to see how reliable their rules were. The OR group was given the task of improving these rules.[14]

The OR group members began their test by trolling through the data that Sears had collected, seeking out hidden regularities. Their quantitative analysis revealed certain facts, such as that the frequency of ordering was a far better guide to future ordering habits than order size. Kimball suggested that the predictive value of customer information would decay exponentially with the data's age, and turned out to be correct. In the end, although the OR group's suggested rule changes were not radical, they did result in millions of dollars in extra revenue every year. In 1953 Magee published the results of the study in the second issue of the new *Journal of the Operations Research Society of America,* disguising Sears as a coffee distributor that was trying to decide how to allocate its promotional aid to stores. For its part, Sears continued to employ Arthur D. Little's OR services for many years afterward.[15]

The Sears study managed to balance a perspective of OR as a field defined by the use of advanced quantitative methods with the wartime connotation of OR as the scrutiny of existing policies. This balancing act was also evident in a 1953 article that Magee cowrote for the *Harvard Business Review* (*HBR*) with Cyril Herrmann, a professor of management at MIT, to promote OR to businesses.[16] They described OR generically as something that would "single out the critical issues which require executive appraisal" and provide "factual bases to support and guide executive judgment." At the same time, they embraced OR's connection to science and mathematics. A two-page sidebar worked through examples of mathematical OR problems likely to prove forbidding to most managers. However, Magee and Herrmann did not suppose that science and mathematics conveyed any authority among managers. To the contrary, they were sensitive that OR could be regarded as a gimmick. To overcome such suspicions, they conveyed their full awareness of the failures of prior "scientific" approaches to business, such as "efficiency engineering." They also sought to allay fears that "regular employees might resent 'outsider' investigators dipping into the internal operations of the company," by suggesting that managers work with "reputable consulting organizations who are trained to approach their work with integrity and tact."[17]

Magee and Herrmann's notion of tactful work drew heavily on the idea, familiar from military OR, that operations researchers' work was most successful when it was well integrated with existing managers' thinking. While OR staff would undertake the heavy lifting of an analysis, executives would shape OR work around their organizations' unique values, concerns, and practices. In fact, Magee and Herrmann attempted to maintain a fairly stark line between managers' and operations researchers' intellectual responsibilities. "Some practitioners," they wrote, "take the rather broad point of view that operations research should include rather indefinite and qualitative methods of the social fields." However, they agreed with what they claimed was the bulk of "professional opinion," which restricted OR's meaning "to the quantitative methods and experimentally verifiable results of the physical sciences." This delineation of OR appealed to the notion that operations researchers were akin to technical specialists who worked successfully within professional cultures, rather than the more ambitious idea that OR should automatically have a privileged place within an organization. Accordingly, they also urged that an OR group should begin "modestly, at a lower point within the organization; and, as it proves itself, to develop and grow to a more prominent position."[18]

The way that Arthur D. Little's operations researchers integrated their work with management was reflected in how they approached the problem of inventory management, in which they quickly developed a specialty. The firm began working on inventory problems in 1951 with a study for Johnson & Johnson led by George Kimball.[19] In 1956 Magee published an unusual three-part series in *HBR* on inventory policymaking distilling Arthur D. Little's expertise on the subject.[20] Unlike his previous article, this one made only passing mention of OR, concentrating instead on how managers could view inventory management as an opportunity for profit rather than as a necessary evil. He framed the problem as essentially a managerial one of mediating between different departments' demands on a firm's inventory policy. He wrote that each departmental manager "fails to recognize costs outside his usual framework. He tends to think of inventories in isolation from other operations. The sales manager commonly says that the company must never make a customer wait; the production manager says there must be long manufacturing runs for lower costs and steady employment; the treasurer says that large inventories are draining off cash which could be used to make a profit."[21] In order to assess the problem from an overarching managerial perspective, one first had to ask the question of what roles an inventory served in one's specific company. Were they the result of economies achieved through large production runs or buying in bulk? Did they guard against volatile shifts in demand? Were they being built up for an anticipated busy season? Once one asked such questions, one could begin to formulate a rational response to them.

Magee's treatment of rational inventory policymaking had many parallels with the growing body of inventory theory in the OR literature. For example, on the always-vexing problem of how to calculate the costs of inventory depletion, he stressed that such calculations were inherent to inventory policy, whether consciously considered or not. However, rather than force companies to correlate depleted inventories with lost sales directly, he suggested that a company might instead fix what it supposed was a "reasonable" standard of customer service, and then set inventory policy around that standard. He noted that inventory policies were always formulated in view of such cultural contexts. "Fluctuation stocks," he wrote, "are part of the price we pay for our general business philosophy of serving the consumers' wants (and whims!) rather than having them take what they can get. The queues before Russian retail stores illustrate a different point of view."[22]

How sophisticated a particular inventory system needed to be was also a function of the needs of individual companies. In some cases establishing relatively arbitrary standards would suffice. One could chart expected inventory levels versus actual inventory levels on a graph and then determine one's inventory policies on the basis of the efficacy of that graph in satisfying the company's cost control needs. Alternatively, in more cost-sensitive systems, one could actually calculate optimal policies using more rigorous mathematical standards, which had by that time become familiar elements of the OR canon. In a 1958 book on inventory and production control, Magee likewise pointed out that in many cases trial and error could be trusted to converge on acceptable solutions, but that in certain cases, such as where inventories needed to be coordinated among multiple product lines, advanced techniques like linear programming could yield significant savings. However, the book's contents offered only basic inventory formulas, while using the footnotes to point to more advanced theoretical treatments.[23]

The lack of emphasis on theory in Magee's paper, and, indeed, its lack of emphasis on OR in general, were indicative of the fact that Arthur D. Little's OR consultants were beginning to stake out a reputation for themselves independent of the OR profession. And Magee, at least, began to feel estranged from the increasingly theoretical OR literature. At one point he submitted an article, which, to his chagrin, was rejected because it was deemed "too pragmatic."[24] Magee's own fortunes were certainly most closely tied to the OR group's success within the firm and within the consulting market in general. He eventually rose to become the head of the OR group, and went on to serve as Arthur D. Little's president and chief executive officer for much of the 1970s and 1980s. Yet, Magee and the OR group did continue to draw value from their association with the OR profession. Martin Ernst, who joined Arthur D. Little's OR group from the Navy OEG in 1959, served as president of the Operations Research Society of America for 1960. Magee himself served as president for 1966, and George Kimball was president the following year.[25]

The MIT Operations Research Center

The first experimental course on OR at MIT was only moderately successful. Students in economics and business administration who might be most interested in the subject did not have the requisite mathematical background to approach some of the problems. Moreover, it was difficult to elicit interest in the subject in the absence of a broader pedagogical program. By 1950, though, pressure was beginning to mount to set up such a program. Some members of the Operations Evaluation Group pressed MIT to continue its efforts to civilianize OR.[26] Arthur D. Little had already set up its OR group, and, in the summer of 1950, Philip Morse returned to MIT following his term as research director of the Weapons Systems Evaluation Group. That fall General Motors magnate Alfred Sloan agreed to donate to MIT funds from his foundation for the creation of a new School of Industrial Management (SIM). Shortly after the donation was agreed upon, but not yet announced, the National Research Council's OR committee sent Horace Levinson and Alan Waterman to MIT to advocate the establishment of an "Operations Research Center." After their meeting with MIT's provost, Julius Stratton, OR was placed in the front-running as a field that could define an approach to the work of SIM that drew upon MIT's leadership in scientific and technological research.[27]

In 1951 Pennell Brooks, a vice president at Sears, Roebuck & Company, was chosen as the first dean of SIM. He assigned Tom Hill, a young professor of accounting, to investigate the appropriateness of OR to the new school's interests.[28] In January 1952, after surveying the promotional literature on OR, speaking with members of the Arthur D. Little group, and attending a seminar series on OR being run by Morse, Hill registered his skepticism. Since OR seemed to be defined only in terms of a general "*modus operandi* of the scientists," he concluded, "the validity of any claim to innovation made on behalf of Operations Research hinges entirely on the novelty of applying the scientific method (and the scientific mind) to areas outside the usual purview of the scientist." While, he imagined, the establishment of OR on this basis had been productive within a military context, in business it seemed much less likely to prove important. He argued:

Case discussions in the OR seminar have revealed that OR groups have frequently done no more than to arrive at operating methods which we recognize as corresponding to existing practice in certain well-managed, progressive companies. This fact has been disappointing to those of us who were anticipating dramatic revelations of startling results achieved by new techniques completely foreign to our own experiences.

He blamed OR's supporters for building up "expectations and perhaps subconscious antagonism" by insisting on their work's novelty. He felt that only industrial secrecy prevented more examples of work performed by "groups less naive concerning American industrial practice" from becoming well known.[29]

Hill did not dismiss OR outright, however. He recognized that its proponents' comfort with mathematical techniques was an advantage, and that it was significant that "persons entirely ignorant of certain best practices laboriously developed in business so often arrived at those same practices by independent and perhaps simpler routes." He also felt that the "employment of the truly scientific approach in commerce and industry is spotty; and, viewed in proper perspective, the tentative exploratory efforts of OR groups indicate possibilities for far more generalized application." Ultimately the question came down not to the validity of OR, but to what sort of relationship SIM should establish with respect to it. As Hill put it, "How do we capitalize on the current interest in Operations Research in furthering our educational objectives? To fail to take advantage of the fact that we have a ready-made common meeting ground with our colleagues in science would seem to me unsound. On the other hand, I am equally convinced that neither a slogan nor a technique is any foundation on which to build an educational program." He recommended that SIM steer an independent but friendly course, and that the school simply incorporate quantitative methods into its regular pedagogy where appropriate.[30] For his own part, he became a founding member of ORSA several months later.[31]

Meanwhile, unaware he had a spy in his ranks, Philip Morse was mystified that Pennell Brooks seemed uninterested in establishing a pedagogical program in light of his and his fellow proponents' success in fomenting interest in OR. In July 1952 he wrote to Brooks, "I have had an average of about three inquiries a week, all last spring, from spaghetti factories and from textile mills and from railroads, asking questions about the subject and wanting to know where they could learn more." Brooks waited until September to make his noncommittal but sympathetic reply, whereupon Morse turned to Stratton. He wrote that while he was hesitant to be seen as "usurping" SIM's prerogative, Brooks's latest memo seemed to "throw the ball back" to him. Morse figured that since OR had begun during the war as an "application of methodology of *physical* science into problems of management," the physical scientists "could be actively in the picture from the beginning here at Tech." That fall, with Stratton's authorization, Morse established an interdisciplinary committee on OR to coordinate research and supervise graduate work on OR-related topics.[32]

Morse's committee comprised members of several departments, including physics, mathematics, the new management school, mechanical engineering, electrical engineering, and economics, and its members encouraged students to undertake equally diverse coursework. The committee also maintained a relationship with OEG, and students were assigned local problems at MIT to work on, such as library circulation and parking. They also sometimes worked on projects with the consultants at Arthur D. Little.[33] PhD work done under the committee was supposed to be undertaken on behalf of outside organizations to give students experience in the practical reality of OR. John Little, a student of Morse's in the physics department, received the committee's first PhD in 1955. Initially interested in machine computation, his dissertation studied water management in the system of dams along the Columbia River.[34] The object was to maximize the production of energy prior to spring snowmelt in view of uncertainties in precipitation patterns. He approached the problem by twisting the usual inventory-control scenario so that demand for electricity was a constant while supply was modeled using a stochastic distribution derived from records of past rainfall. However, following in the wartime tradition of comparing results to existing practice, he also compared his model's efficacy with actual results obtained from the "rule curves" that dam managers used to control water flow. Although the curves were not statistically rigorous, Little took the time to remark on why his own models achieved only a 1-percent improvement over them.[35]

The MIT OR committee's emphasis on practicality also informed its efforts to reach out to industry through a series of summer courses. The first course, offered with help from OEG and Arthur D. Little staff, lasted for three weeks in the summer of 1953, with mornings devoted to mathematical techniques, and afternoons to "laboratory sessions." For the afternoon sessions, course participants were divided into small teams, meeting with instructors role playing as executives. Morse played an MIT administrator and a hotel manager. John Magee played an inventory manager, and the head of both a sales force and an advertising department. Others ran restaurants, warehouses, maintenance facilities, libraries, taxicab fleets, railroads, an airport, a power company, and so forth. Actual data were obtained where possible from contacts in industry. Course organizers made sure the data were not "too pre-digested," since the object was to train participants in the realities of OR work.[36] The summer course continued in one guise or another for fifteen years.[37]

By the middle of the 1950s, OR had become better established at MIT and as a profession, and was beginning to be identified more closely with its theoretical content. In 1954 Tom Hill reversed his position on OR, urging with other professors that SIM should push forward with "quantitative analysis in what are now regarded as the 'non-technical' areas of industrial management." Hill admitted that the suggestion replicated work done by the OR committee. But, observing that "a substantial gulf exists between the typical management and the theorist in autocorrelation, linear programming, game theory, and other such areas pertinent to industrial activity," he felt SIM might aid in the "reduction of theory to practice."[38]

By that time, though, OR had already made a strong claim to both theory and practice. Responding to Hill's memorandum, Morse allowed that his OR committee was an interim measure, and that the time might have come for SIM to take the work over, yet he had also become convinced that it might be wise to keep OR interdisciplinary.[39] And, indeed, when MIT established its Operations Research Center soon thereafter, it remained an independent organ, attracting participation from multiple departments, including SIM. Morse also retained control over OR, holding the directorship of the center until his retirement in 1968. At that time he was replaced by John Little, who had, in the interim, left MIT after receiving his PhD, and then returned as a professor at SIM, now renamed the Sloan School of Management.[40]

It's hard to recall how and why I moved my intellectual dwelling some half century ago from epistemology to management. The two questions, "What's wrong with logical positivism's theory of knowledge?" and "How many 15½-33 men's shirts should be kept in a retail store's shelves?" do seem a bit different, don't they?

West Churchman, 1994[1]

Almost all of the impetus behind the professionalization of operations research in the early postwar years originated in the desire to replicate the successes of wartime OR in peacetime contexts. The one major exception to this rule was the OR program developed at the Case Institute of Technology in Cleveland by West Churchman and Russell Ackoff. Churchman and Ackoff were both trained as philosophers of science at the University of Pennsylvania, and were committed to the development of the "experimentalist" ideas of Churchman's mentor, Edgar Singer, Jr., who had himself been a follower of pragmatist philosopher William James.[2] In Churchman and Ackoff's philosophy, the essence of science was to be found not so much in the knowledge it produced, as in its efficacy as a methodology for answering carefully stated questions and solving carefully stated problems. For this reason they viewed practical problems of management to be equally amenable to scientific analysis as the problems of natural science. This vision of science aligned with OR proponents' advocacy for OR as an application of scientific methods to the problems faced by organizations. Consequently, after learning of the proponents' efforts, Churchman and Ackoff came to view OR as a vehicle for putting their philosophical ideas into practice.

In the immediate postwar years, Churchman and Ackoff's work focused on the philosophical problem of defining what constituted science. This approach was, in part, a reaction to the logical positivism of the interwar era, which drew sharp boundaries between well-founded "scientific" knowledge and all other kinds of knowledge. For their part, they preferred to view the distinction as one of degree rather than kind. To them, "science" was distinguished in that it always sought a "best" solution to a problem rather than suffice with some less-than-optimal solution. In their terminology, sufficing with a nonoptimal solution constituted a "lag" behind science. However, they were also aware that a zeal for reform could lead to worse rather than better solutions. They labeled this phenomenon "anti-lag."[3]

For Churchman and Ackoff the most important prerequisite to proper science was the maintenance of a record of stated problems and attempted solutions. How to ascertain the best solution among all solutions constituted a deep problem. To approach it they drew heavily on then-recent advances in statistical methodology. During the war Churchman had worked for the Frankford Arsenal in Philadelphia, where he was assigned problems of assessing production quality very similar to the ones that inspired the development of sequential analysis. After the war, in his 1948 book *Theory of Statistical Inference*, he drew on the problem of testing as a way of illustrating the philosophical problems underlying scientific inquiry. He took, as an example, the industrial problem of deciding whether a new and supposedly improved material is stronger than an existing material. A simple side-by-side comparison might show the new material to be consistently stronger than the older one, but a large variance in the strength measurements of the new material might also hint that the test was actually inconclusive. Fortunately, Churchman observed, advanced statistical methodology could help distinguish rigorously between the sets of data by determining just how likely it was that the new material was actually stronger.[4]

Yet, as decision theory was showing at that time, results of statistical testing could themselves hinge on value judgments relating to how certain experimenters wanted to be that their results were true. Tolerance for error determined just how much testing would be conducted and how conclusive results needed to be before they were accepted. Such tolerance, furthermore, often had an ethical component. For example, ethics could influence the design of a test to determine the lethal dosage of a chemical, including whether it was better to overestimate or underestimate the value.[5] In positing an interconnectedness between knowledge and values, Churchman and Ackoff acknowledged a certain debt to the claim of philosophical relativism that knowledge was contingent upon one's willingness to believe in it. Yet, rather than follow that line of reasoning to the logical endpoint of simple skepticism, they hoped to build on pragmatist foundations. By acknowledging the arbitrary assumptions and values that sustained knowledge, they believed it was possible to identify, head on, how social factors impacted knowledge claims.[6] This prospect, they believed, even offered philosophers an opportunity to amalgamate the natural and social sciences, ethics, and policymaking into a single entity. Churchman and Ackoff believed, for instance, that personality traits could be described and measured in terms of the "efficiency" of the choices that individuals made in various contexts, given the range of choices available to them. They also aspired to create a "science of ethics," which could empirically study how it was that individuals agreed that a particular course of action was justified, ethical, and fair.[7]

Inasmuch as Churchman and Ackoff relied on statistical principles to cut through ambiguous data and to illuminate the assumptions underlying it, they also recognized that the data were themselves the product of experimental design, which also had to be subjected to intensive scrutiny. They acknowledged, for instance, that some hypotheses simply could not be tested because it was impossible to design an experiment capable of clearly correlating the results with the hypothesis. To decide which hypotheses could be reliably tested and how, it was necessary to formulate both hypotheses and the design of experimental tests by making deft use of the entire system of experimental knowledge that had been assembled to that point.[8] This was clearly a daunting problem, which Churchman and Ackoff understood to be badly exacerbated by the rapid expansion, diversification, and, above all, specialization of the sciences. If experiments were only designed making use of the narrow knowledge available to one discipline, then the danger of "stagnation" in science was pronounced. To combat this danger, they urged a "unification" of the sciences. However, unlike the unification urged by logical positivists, which sought to build a solid body of knowledge out of foundational truths, Churchman and Ackoff did not regard any particular truths as more fundamental than others. Rather, at the locus of experiment design, all sciences leaned on each other in the general search for more effective knowledge. They pointed, as an exemplar, to the interrelations between biology, psychology, and sociology in Friedrich Bessel's well-known nineteenth-century development of the "personal equation" as a means of improving astronomical observations.[9]

To handle the enormous effort that would be needed to coordinate different branches of knowledge in experiment design, and to ascertain which results provided the best answers, Churchman and Ackoff looked to an institutional solution: a set of "Institutes of Experimental Method."

These institutes, divided into four sections, would train methodologists—specialists in the issues of scientific methodology—who would serve as consultants to scientists. Members of a "general methodology" section would keep track of scientific knowledge so as to criticize and improve scientists' methods of experimentation and the criteria used to evaluate the results of those experiments. A mathematical statistics section would develop the insights of statistical theory and ensure that theories were properly employed. Experts in a separate section devoted to sampling techniques would scrutinize and help formulate the presuppositions underlying experimental samples. Finally, in recognition of the fact that scientific work does not always retain knowledge of why it has taken the paths that it has, a history of science section would investigate the ways scientific investigations of the past influenced the ways current scientific inquiries were framed, and would help determine what aspects of past scientific work could be revived most fruitfully in the present.[10]

As a first step toward setting up their institutes, Churchman and Ackoff helped organize a series of conferences on issues they saw as harboring rich methodological problems. They held their first conference in May 1945, covering the links among statistics, psychology, and physics. Their second conference, on the "measurement of consumer interest," was held one year later. As they, with their colleague Murray Wax, wrote in the introduction to the conference proceedings, the notion of what it meant to measure consumer interest remained unstable. By bringing together "methodologists," statisticians, survey sampling experts, psychologists, and marketing researchers, they hoped to provide a model for how to "make all fields of research more self-conscious," and to help "determine the most fruitful steps to be taken toward making research scientific." This goal, they remarked as an aside, was "similar to that of 'operational analysis' discussed by Professor Wilks" who had contributed to a panel on the "specifications for consumers' goods."[11]

The Princeton statistician Samuel Wilks, as noted earlier, was a key figure in the early postwar American push to establish OR beyond the military. In his talk, entitled "Research on Consumer Products as a Counterpart of Wartime Research," he professed his ignorance of the field of consumer research, but suggested that the goals of the Navy's OR group and similar efforts had important commonalities with the interests of the conference's attendees. Pointing out that during the war the industrial power of Britain and America had been "geared to the training of men and the production of war materiel," he related how OR groups "studied the effectiveness of these products for the purpose for which they were designed, by going to the theatre of operations and working with the users of the products—the men in combat." Now that peace had arrived, there was "a huge segment of [the] economy geared to manufacturing consumer products." Accordingly, a new emphasis was being placed on the "problem of studying the effectiveness with which the peace-time consumer products, such as refrigerators, automobiles, and electric appliances, are doing the job they are supposed to do." To get a handle on this kind of problem, Wilks thought it important to develop appropriately "scientific" means of polling, which meant understanding, among other things, how the interview process impacted the information received from consumers. He then spent the remainder of his short talk discussing the need for cooperation among businesses, universities, and various "agencies and organizations" to promote the social sciences and improve these methods.[12]

Wilks's talk was not especially illuminating on either consumer interest or military OR, but it did strike a chord with Churchman and Ackoff. The kind of heterogeneity that characterized military OR work mapped onto their ideas about needing to break down disciplinary barriers to develop properly "scientific" conclusions. However, the idea's impact on their careers was not immediate. Following the conference on consumer research, the two philosophers assembled two other conferences under the Institutes of Experimental Method rubric. In 1947 Ackoff took a position in the philosophy department at Wayne University in Detroit, and Churchman followed him soon thereafter. Ultimately, their interests in methodology clashed with the interests of the philosophical community there, and the two of them moved to the Case Institute of Technology in 1951.[13] Case was, at that time, arranging with the Chesapeake and Ohio Railway Company to undertake a statistical accounting and logistical study and to secure sponsorship for a new professorship in OR in the Department of Engineering Administration. Churchman and Ackoff were able to use the opportunity to find places at Case, and to begin setting up a new program in OR. In November 1951 they held the first-ever conference on the subject.[14]

Although the move to Case and into OR distanced Churchman and Ackoff from their original goal of improving scientific methodology, they not only remained committed to their original philosophical vision, they also attempted to mold the nascent OR profession around it. To make a claim to being "scientific," they felt that, beyond applying piecemeal methods to managerial problems, OR would have to help managers make decisions that accorded with the experimentalist definition of the term. For them, managerial problems were even more appealing, from a

philosophical point of view, than scientific problems, because they habitually involved integrating diverse specialist perspectives and mediating
between business goals and ethics.

As a means of parsing the value of management decisions from overarching perspectives, Churchman and Ackoff quickly became enthusiastic
proponents of decision theory and worked to incorporate it into the methodology of OR. This enthusiasm embraced not only the practical deployment of advanced mathematical tools, but also the more abstruse reflection
that typified more academic work in decision theory. For example, they
understood inventory theory to be important not only because it promised
to improve firms' finances, but also because it was a paradigm of the
general philosophical problem of how divergent interests could be recon-
ciled through an appeal to a higher measure of worth.[15] Moreover, they
felt, the very concept of higher worth required substantial philosophical
reflection. Even though they were now primarily interested in assisting
business firms, they maintained, as in Ackoff's 1953 book, *The Design of
Social Research*, that the overarching goals of society still had to be taken
into account.[16] For them, this concern manifested itself most clearly in
problems of business ethics, in which they took a passionate interest.
However, they also pursued the problem in its more deeply theoretical
forms as well. In one of the earliest issues of the *Journal of the Operations
Research Society of America*, Churchman and Ackoff debated with Nicholas
Smith, a member of the Army's Operations Research Office, the relative
merits of experimentalist attempts to measure values, versus the assignment of values from, in Smith's words, a "relative and arbitrary" standpoint
as "constructs of human rationalization."[17]

Churchman and Ackoff's interest in operations research as both a methodologically rigorous activity, and as something geared toward the practical problems of management, was reflected in the diverse content of a
seminal textbook on OR that they published in 1957 with their colleague
Leonard Arnoff.[18] It also informed the program that they built at Case. As
at MIT, students at Case were expected to take a diverse array of courses,
spanning statistics, scientific methodology, computation, and engineering.
Undergraduates in the program were even expected to take a course on the
history of science and technology. As at MIT, the Case program ran a
summer course for people from industry. It also undertook consulting
projects for industry, in which students were expected to participate.[19]

Through their extraordinary initiative, Churchman and Ackoff quickly
became leaders in the new operations research profession. They were
instrumental in encouraging the development of decision theory under

the OR rubric, and Ackoff would be elected the president of ORSA for 1956.[20] And, initially at least, their ambitious vision for OR seemed to closely coincide with the aspirations of the profession's other proponents. Yet, there was also an inherent tension between their philosophical ideas and efforts to establish OR as a new profession. Because they insisted that, to be scientific, managerial decisions had to encompass all relevant interests and expertise, and be optimal from the broadest possible vantage point, their thought necessarily trespassed into managerial problems in which OR had nothing unique to contribute. Where people such as John Magee believed it important to confine the definition of OR, and to leave nonspecialist issues such as personnel management to managerial prerogative, this was precisely the level at which Churchman and Ackoff thought it most important to intervene. Ackoff would find some support for this view in Britain, where some of the broad societal significance attributed to OR in the immediate postwar period continued to motivate some OR practitioners.[21] But most practitioners declined to follow Churchman and Ackoff down this path.

In the military, operations researchers sought to maintain the legitimacy of their work by maintaining an engagement with the thinking of military specialists. As OR's proponents worked to build a new profession, they likewise stressed this sort of engagement at places such as MIT, the Case Institute of Technology, and the Arthur D. Little consulting firm. But, to ensure that the legitimacy of OR would be protected *wherever* it might be established, the leaders of the Operations Research Society of America had to try to maintain control over the broader culture of their new profession. As Philip Morse recalled in his memoirs, "As soon as it began to look as though O/R would become popular, a few quacks began using its name to sell their magic."[1] ORSA did not, of course, have the legal authority to authorize its membership as the medical and legal professions did. Instead, it divided its membership into three echelons: fellows, members, and associate members. Fellows were essentially the elite of the profession. Full members were those whose work was deemed certifiable as OR, while associate membership denoted a status of novice or supporter.

Hierarchical membership made sense from the perspective of its earliest proponents, but it clashed with the ideals of others who were otherwise attracted to ORSA's goal of improving industrial operations and business management. In his memoirs, the Hungarian émigré Andrew Vazsonyi recalled ORSA's stratification as less beneficent: "Full members were only those theorists and mathematicians certified by the core group. Associate members were riff-raff from the real world—guinea pigs from business who could try out the full members' theories." It reminded him of "the good old days of semi-feudal Hungary." Vazsonyi had escaped Europe in 1940—his two brothers and extended family were killed by the Nazis on account of their Jewish heritage—and in 1941 he accepted a fellowship at Harvard's Graduate School of Engineering. In 1945 he moved to industry and later the Navy, working on problems on aerodynamics and control systems. In 1953 he landed at Hughes Aircraft, where he worked with Simon Ramo and Dean Wooldridge, applying his skills to the automation of production management. When Ramo and Wooldridge resigned to form the Ramo-Wooldridge Corporation, Vazsonyi followed shortly thereafter.

When Vazsonyi saw announcements for early meetings of ORSA, he believed that the organization's goals aligned with his own, and joined as an associate member. Aside from his dismay at the hierarchical membership, he also quickly found himself frustrated with the "sense of academic hubris" pervading ORSA conferences. He recalled, "Morse and his disciples seemed too far removed from the day-to-day world of American business." He was also disappointed in ORSA members' apparent lack of interest in applying digital computers to managerial problems. Disillusioned with ORSA, he soon joined forces with a group working toward the creation of another new organization, The Institute of Management Sciences (TIMS).[2] TIMS was, essentially, an outlet for anyone who, for whatever reason, did not find a true home in ORSA. Some, like Vazsonyi, felt a new organization would address practical business and industrial concerns better than ORSA's academic scientists and military consultants. However, others saw ORSA from the opposite perspective: it was altogether too practical for their tastes.

Melvin Salveson had trained in engineering at the University of California, Berkeley, but became involved in the management of submarine construction during World War II. After the war he went to MIT to obtain a master's degree in industrial management. After completing his degree, he began working for the McKinsey and Company management consulting firm, and then taught and did research at the business school at UCLA. From his experiences, he later recalled, he found that managers "had excellent experience and intuition as to what worked." Still, he was dissatisfied with the conceptual amorphousness of the management techniques he had learned and was teaching. So, he went to the University of Chicago to obtain a PhD in business, and arranged to work under Tjalling Koopmans at the Cowles Commission. He finished his doctorate in 1952, having written a dissertation in the burgeoning field of linear programming.[3]

While at Chicago, Salveson began agitating with others whom he had met in the course of his work, including West Churchman, Russell Ackoff, and others at the Case Institute of Technology, Herbert Simon and others at the Carnegie Institute of Technology, as well as employees of the RAND Corporation including George Dantzig and Merrill Flood. Collectively, they hoped to create a new organization dedicated to developing, in Salveson's words, "the bodies of knowledge and all the disciplines that underlie and serve as the base for the processes and actions of those who are engaged in the practice of management." He later remembered, "We considered this overall body of knowledge to include such disciplines as economics, psychology and the behavioral sciences, mathematics and statistics, and the information sciences."[4] William Cooper, a colleague of Simon's interested in practical applications of methods such as linear programming, questioned whether it was wise to compete with ORSA.[5] Others, however, were convinced that the new organization had substantially different aims, and their plan moved forward.

TIMS was established in December 1953. Its founding members elected Cooper to be its first president, Vazsonyi the first "past president," Merrill Flood as "president elect," and Salveson, Simon, and IBM mathematician Cuthbert Hurd as vice presidents. They also convinced West Churchman, who already edited *Philosophy of Science*, to edit their new journal, *Management Science*, as well.[6] Having by this point entrenched himself as an operations researcher, Churchman now attempted to encompass management science within his philosophical vision as well. His hope was that TIMS would create an environment in which research that "lived up to the standards of good science" could be done. OR, in this scenario, would then constitute "the practical application of that science."[7]

In accord with his expansive view of the various perspectives that needed to be included to obtain properly scientific conclusions, Churchman aimed to make the content of *Management Science* eclectic by including articles on issues such as ethics, and philosophical reflection on the nature of "good administration."[8] And, to an extent, the journal did begin eclectically. For example, in its first issue, appearing in October 1954, Vazsonyi published the first part of a three-part series on the production control methods he had implemented at Hughes Aircraft and Ramo-Wooldridge. He argued that mathematical theory was having a transformational effect on production and inventory control, but that the rate of this transformation would "depend on the degree of integration effected between the production men, the management scientist, and the electronic engineer." He lamented that there had been "very little effort expended in 'popularizing' … mathematical concepts," and felt that many people simply despaired of the task. The point of Vazsonyi's series, then, was not to develop new theoretical methods, but to convey how they had been successfully integrated into management through such means as "Gozinto" (i.e., "goes into") diagrams. Vazsonyi liked to joke that they had been developed by the "Italian mathematician" Zepartzatt Gozinto.[9]

Very quickly, however, Churchman's aspirations for *Management Science* were subordinated to Salveson's and other theorists' goal of fostering a formalized body of managerial theory. Churchman later recalled with some bitterness that he had never had strong control over the journal's content. He reflected, "What happened, I believe, was that there were several economists who had written papers on linear programming but couldn't get them published in *Econometrica* or other economics journals that published papers in mathematical economics. So they decided to set up their own journal and society. It wasn't a battle between OR and MS, but a battle inside economics."[10] Although Churchman's recollection correctly identified the important links between formalized economic and management theory, it neglected to account for those who were mainly interested in developing theory for applications in management and engineering. And, of course, some elite theoreticians such as Kenneth Arrrow, president of TIMS for 1963, and his colleagues at Stanford, and Herbert Simon and his colleagues at Carnegie Tech, would navigate the divide between management science and economics relatively freely. Whatever frictions may have existed between theoretical economics and management science, it was never so much an issue for TIMS in this period as was the definition of the relationship between management science and OR.

In 1955, upon finishing his term as the second president of TIMS, Merrill Flood attempted to clarify the purpose of the organization by drawing a clear distinction between management science and OR. Taking the mathematician's view of "science," he defined management science not as a service profession, but as a particular body of theory defined by the peculiarity of its axioms and logical structure. Importantly, this body of theory had no necessary relationship to actual management. Flood pointed to John von Neumann and Oskar Morgenstern's *Theory of Games and Economic Behavior* as a "shattering advance in the science of decision making" akin to what Gregor Mendel's genetics represented in biology and what Max Planck's quantum theory represented in physics. He allowed that "scientists" would be aided in their "creative effort by close association with lawyers, accountants, labor economists, managers, and members of other professions who are close to the problems of interest to the management scientist." But, he also observed that von Neumann had originally begun to work on games in 1928 while he was formulating quantum mechanics in a rigorous axiomatic language. Thus, Flood supposed, "scientific contributions of first importance in understanding management"—meaning contributions of fundamental theoretical interest—"are not very apt to result from any direct interest in management problems."[11]

Recall that in their end-of-war report Philip Morse and George Kimball believed that engagement with actual decision-making problems was what made OR "science" rather than "philosophy." For Flood, OR's practical focus consigned it to the realm of "engineering." This terminological difference should not, however, be read as a battle over what could and could not claim the title of "science." Rather, what Flood was arguing for was the worth of a division of responsibilities. It was, in his view, engineers' responsibility to make sound use of a body of knowledge, preferably "scientific" knowledge, when they undertook practical tasks. Theorists, in contrast, had to dedicate themselves to making sure the body of available knowledge from which engineers could draw was as broad, as deep, as subtle, and as "scientific" as possible. It was this division of responsibilities that, in turn, justified the division of ORSA and TIMS.

In the long run, the differences in sensibility between people like Morse, Kimball, Churchman, Ackoff, Vazsonyi, Salveson, and Flood would prove subtler than any of them or anyone else cared to consider at length. Many people joined both ORSA and TIMS and failed to see any substantial differences in their activities. Flood's speech forced this more ambivalent view into the open. John Lathrop, a former member of the Navy's Operations Evaluation Group who had moved to the Lockheed Aircraft Corporation via Arthur D. Little, responded to the editors of both *Operations Research* and *Management Science* suggesting that the societies be merged. In his view, the concerns of the membership of each society overlapped extensively, and many of the initial complaints about ORSA had ceased to carry weight. Military analysts constituted a smaller percentage of the society's membership, its journal regularly published work on theoretical problems, and it was considering abolishing its stratified membership.[12]

Lathrop's letter, in turn, was met with two others suggesting that the organizations should strive more forcefully to situate themselves with respect to each other. One of those letters was from David Hertz, a former Columbia University mathematician who was at that time working for the Arthur Andersen & Company accounting firm. He argued that maintaining the split between the organizations would allow individuals to choose between supporting professionalization and supporting theory development, and that, naturally, there would be many who would support both.[13] He himself would serve as the president of TIMS for 1964 and as the president of ORSA for 1974. In any event, after the dust settled nothing was done either to distinguish or unite the societies. People continued to join either or both ORSA and TIMS, and operations research and management science grew constantly closer together in meaning. The societies would remain separate until 1995 when they finally merged to form the Institute for Operations Research and the Management Sciences.

The pioneers of quantitative systems analysis at the RAND Corporation understood their work to be valuable because it integrated and harmonized engineers' and military planners' knowledge, and explored the logical implications of that knowledge for equipment design and tactical planning. However, the rationality of designs and tactics was always contingent upon their ability to serve higher-level strategic goals. Yet, understanding of how designs and tactics related to those goals was sufficiently vague that incorporating more or less arbitrary assumptions about those relations into an analysis threatened to render the analysis no more rational than less formal means of approaching the problem. At RAND this conundrum was recognized at least as early as January 1947 in a memorandum, probably written by John Williams, recommending the establishment of Project RAND's Evaluation Section. It observed that "there is no particularly logical place at which to curtail the analysis, short of the limit imposed by inadequate information and understanding," at which point the analysis would have to be considered specious. Therefore, it warned, studies had to be designed based on their "promise to increase information and understanding of the consequences of warfare operations."[1]

How to define the limits of an analysis so that it maximized "information and understanding" became a crucial problem. According to the memorandum, maintaining practicality could entail circumscribing an analysis to establish what *would* constitute a rational course *given* a scenario constrained by certain assumptions. This sort of analysis would be of value "for the reason that it is easier to prognosticate the effect of an operation in a specific setting than it is to derive a universally applicable forecast."[2] As we have seen, Edwin Paxson followed exactly this course in designing his Strategic Bombing Systems Analysis. The failure of Paxson's analysis was particularly discomfiting to RAND's analysts, not simply because it was ignored, but also because it was impossible to say whether or not the Air

Force's analytically simpler decision was actually the more valid. Because Air Force leaders could be said to have premised their decision to pursue jet-propelled bombers on contexts not considered in RAND's analysis, RAND could say very little about the extent to which the design Boeing and the Air Force ultimately produced was either rational or irrational.

Introspection at RAND

What only became evident in the wake of Paxson's analysis was that, by delivering a single model representing a single set of decisions to be made at a single point in time based on a single set of constraints, the ability of such a model to increase information and understanding was severely limited. For such a tightly delimited approach to have any clear value, the Air Force and its contractors would have had to commit to adhering to the constraints they had initially placed on the analysis, and perforce to abide by the analysis's conclusions. This was, of course, a deeply unrealistic requirement. Not only would it have given RAND unacceptable authority over the final design, it failed to respect the way aircraft designs actually unfolded. During World War II, producing a single, integrated analysis of competing design choices for a fire-control device, for example, made some sense. Proposals typically only had to be judged in view of their immediate appropriateness to a well-defined, near-term goal, which was largely dictated by fixed technological and strategic commitments. However, in the Cold War, systems evolved from basic concepts over the course of a lengthy development and design process to accommodate the availability of new component technologies, new mission profiles, new intelligence on enemy capabilities, and new politics.

In the early 1950s, RAND's analysts became increasingly appreciative of the fact that systems analysis had to be valuable to the Air Force and its contractors whatever course of action they might ultimately pursue. Because any course of action might be rational if it was taken within a context or it furthered a goal that made it rational, RAND had to find a way to design its analyses to elucidate which choices would be more or less rational with respect to which contexts and which goals. It would then be up to the study's readers to determine which contexts and goals applied. For example, analyses might have to consider the validity of their conclusions within various time frames concerning when a weapons system would be used, rather than a single time frame assigned arbitrarily. While still at NOTS, Paxson had himself observed that fire-control system designs had to consider ranges of scenarios, where war might occur in six months, six years, or sixteen years.[3] But, when considering major technological systems, simply expanding an analysis to incorporate entire ranges of scenarios was out of the question. The limits of analytical practicality were precisely why overly limiting constraints had been placed on Paxson's analysis in the first place. Instead, RAND's analysts took notice of the fact that the main impact of the Strategic Bombing Systems Analysis had come through the cannibalization of its component studies (see figure 28.1). Henceforth, RAND analysts would take into account the fact that their work would be used by others as resources in their own decision making rather than simply present final conclusions that had to be accepted or rejected wholesale.

RAND's follow-up to the Strategic Bombing Systems Analysis, called the "Air Defense Study," was a detailed evaluation of America's potential defense capabilities against an anticipated Soviet strike. It was directed by Edward Barlow, another RAND analyst, and completed in 1951.[4] The generalized term "study" connoted something larger than a "systems analysis," which continued to mean Paxson's style of appraising competing system design characteristics through an expansive quantitative analysis. In this case, what was called the "Defense Systems Analysis" constituted a substantial "numerical phase" within the larger qualitative framework of the study. Situated within this framework, the Defense Systems Analysis was intended to serve as a rigorous model of the overall problem of defense against large-scale air attack. It was hoped that it would bring to light what aspects of defense were likely to be critical, and what factors were actually less significant than initially imagined, as only mathematical analysis could reveal.

However, departing from the tradition of L. B. C. Cunningham and Paxson, the new study also took pains to explain to its readers how the quantitative aspects of the study had been formulated, explicitly pointing out where its presumptions were especially arbitrary, so as to allow new information or different assumptions to be more easily integrated into a revised version of the analysis.[5] Barlow also related the study directly to discussions being held outside of RAND, pointing out that many of its conclusions simply reinforced those already arrived at by "other agencies." These conclusions were nevertheless included in the RAND study, both to maintain the comprehensiveness of the analysis, but also because the study represented "documentation and corroboration of these other investigations." This confirmation of accepted conclusions through rigorous

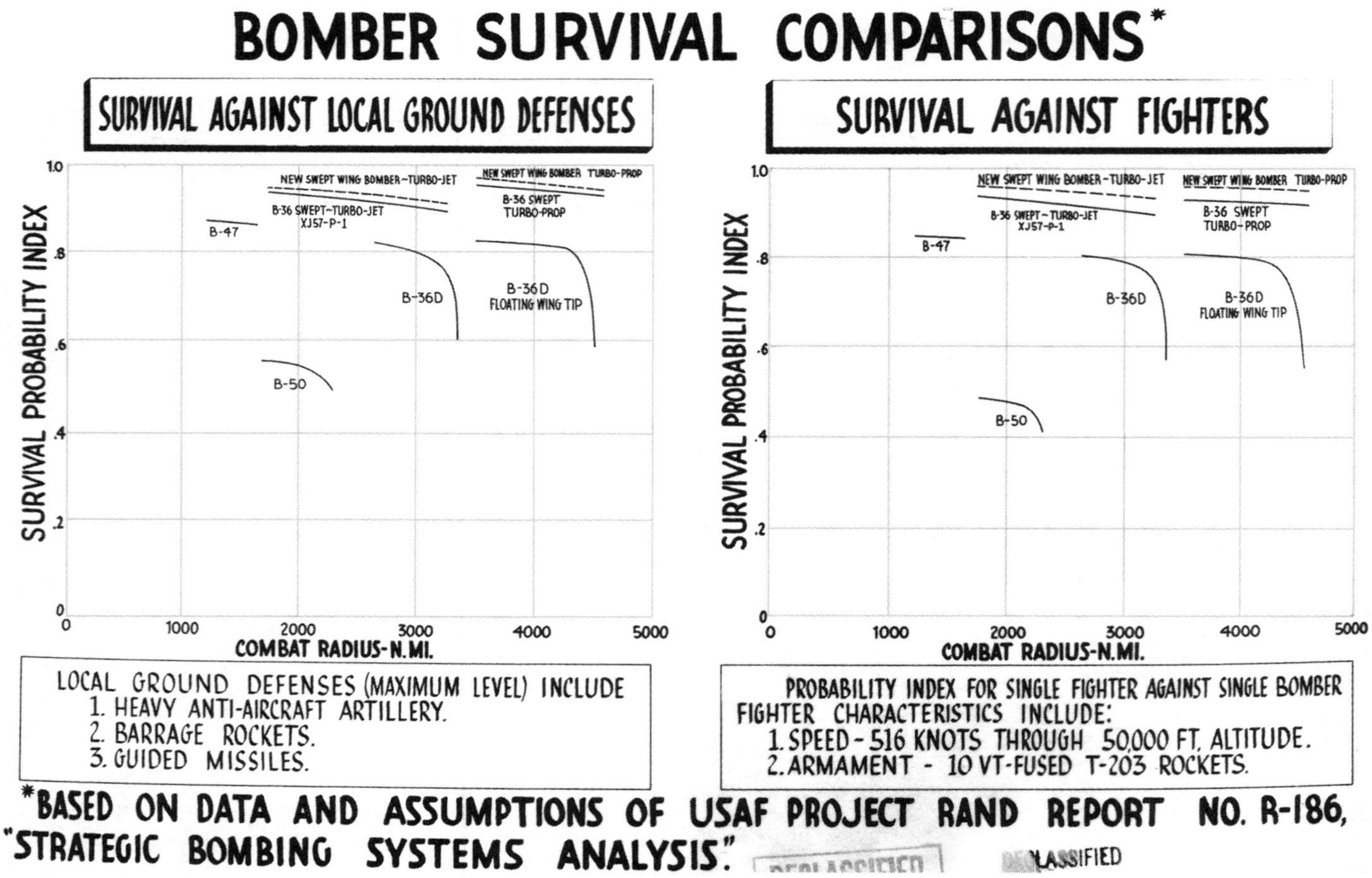
BOMBER SURVIVAL COMPARISONS*
SURVIVAL AGAINST LOCAL GROUND DEFENSES
SURVIVAL PROBABILITY INDEX
1.0
.8
.6
.4
.2
0
0
1000
2000
3000
4000
5000
COMBAT RADIUS-N.MI.
NEW SWEPT WING BOMBER–TURBO-JET
NEW SWEPT WING BOMBER TURBO-PROP
B-36 SWEPT-TURBO-JET
XJ57-P-1
B-36 SWEPT
TURBO-PROP
B-47
B-36D
B-36D
FLOATING WING TIP
B-50
LOCAL GROUND DEFENSES (MAXIMUM LEVEL) INCLUDE
1. HEAVY ANTI-AIRCRAFT ARTILLERY.
2. BARRAGE ROCKETS.
3. GUIDED MISSILES.
SURVIVAL AGAINST FIGHTERS
SURVIVAL PROBABILITY INDEX
1.0
.8
.6
.4
.2
0
0
1000
2000
3000
4000
5000
COMBAT RADIUS-N.MI.
NEW SWEPT WING BOMBER–TURBO-JET
NEW SWEPT WING BOMBER TURBO-PROP
B-36 SWEPT–TURBO-JET
XJ57-P-1
B-36 SWEPT
TURBO-PROP
B-47
B-36D
B-36D
FLOATING WING TIP
B-50
PROBABILITY INDEX FOR SINGLE FIGHTER AGAINST SINGLE BOMBER
FIGHTER CHARACTERISTICS INCLUDE:
1. SPEED - 516 KNOTS THROUGH 50,000 FT. ALTITUDE.
2. ARMAMENT - 10 VT-FUSED T-203 ROCKETS.
*BASED ON DATA AND ASSUMPTIONS OF USAF PROJECT RAND REPORT NO. R-186,
"STRATEGIC BOMBING SYSTEMS ANALYSIS."
LASSIFIED

Figure 28.1

Projected comparative survival rate curves adapted by Convair engineers from components of RAND's Strategic Bombing Systems Analysis. From Convair Swept Wing B-36 Program Summary, November 11, 1950, in LOC CEL, Box B96.

analysis was in turn supposed to lend the study additional credibility in places where its analysis led to conclusions that departed from what was already accepted.[6]

This more flexible vision of systems analysis was complemented in RAND's internal discourse by a critique of systems analysis leveled at that same time by the head of RAND's economics department, Charles Hitch. Hitch had attended the University of Arizona before going to Harvard for a year of graduate study, and then to Oxford University as a Rhodes Scholar. He then remained at Oxford, becoming a fellow of the Queen's College in 1935. In 1941 he joined the war effort, undertaking studies of British wartime materiel controls at the Lend-Lease mission in London, before moving to the War Production Board in Washington, D.C. Shortly thereafter, he was drafted and assigned to the Office of Strategic Services and was sent back to Britain to become the deputy head of the joint American-British RE-8 division of the Ministry of Home Security at Princes Risborough to assess the impact of air raids over Germany.[7] He then joined the Joint Target Group at the Pentagon to assess the effects of air raids on Japan, before finally returning to Oxford at the end of the war.

When John Williams began considering hiring social scientists for Project RAND, Dana Bailey, a RAND physicist who was a friend of Williams from the University of Arizona, told him about their fellow Arizona alumnus. Bailey had met Hitch when he himself had held a Rhodes Scholarship, and thought that he would be a good fit for the organization. Much to Williams's surprise, they managed to recruit him: Hitch became the head of RAND's new economics department in July 1948. Over a decade later, after Hitch had left to become the comptroller in Robert McNamara's Pentagon, Williams reflected, "I'm still so pleased about getting Hitch that I can hardly contain myself. I think that of all the men that have been at RAND, a list which by now has contained hundreds of my peers and betters, to my mind Charlie is still the number one boy."[8]

When he arrived at RAND, Hitch was less concerned with the detailed engineering problems that occupied people like Edwin Paxson and Edward Barlow, and more with the economic problem of how to design a military budget that balanced economy with efficacy. As we have seen, in the 1950s formalistic theories, especially linear programming, that developed around systems analysis were swiftly becoming integrated into economic theory. While supportive of the development of these theories, Hitch believed their utility in solving practical economic problems was constrained. In a 1949 talk at the annual meeting of the American Economic Association entitled "Planning Defense Production," he actively grappled with the implications for defense budgeting of new analytical techniques, including Wassily Leontief's input–output analysis, linear programming, and the "philosophy of game theory." He made clear that, in contrast to the more committed economic theoreticians, he did not wish to propose "a grandiose system of Walrasian equations—game theoretic or otherwise—to solve the whole [budget] allocation problem." In principle, such systems could be used to represent and compare different allocations of a budget, even taking into account possible competing Soviet budgets. But, Hitch reckoned, the construction of such synoptic models "may or may not be a desirable ultimate objective." Instead, he argued, "within the foreseeable future all that we can hope to do is expand the area of 'rational' decisions and reduce the element of 'judgment' or hunch in the planning process."[9]

To address the military budgeting problem constructively, Hitch suggested that the best thing to do would be to solve "a multiplicity of sub-optimization problems at various lower levels" of policymaking. As we have seen, during World War II, mathematicians such as Warren Weaver found strict concentration on lower-level problems dissatisfying because it was never possible to say which lower-level problems needed to be addressed to make a real difference, and whether a lower-level solution might not actually be counterproductive from a higher-level perspective. Hitch was certainly cognizant of this problem, but he was not seriously perturbed by it so long as lower-tier work was situated within a "satisfactory general framework of analysis." To make his point, he drew on the example of the national economy. America, he suggested, out-produced "the world not because we have been exceptionally clever at dividing resources between consumption, investment, and government; or income among individuals and classes." Rather, he argued, "We have managers of firms who solve hundreds of thousands of suboptimization problems. ... And by doing this well, bothered by no philosophic doubts regarding the relation of relative to absolute optima, they have put us streets ahead of other economies."[10]

In the military, some of the best opportunities for suboptimization came at the level of equipment procurement, which led him into a direct engagement with the problems of systems analysis. Hitch supposed that a systems analysis, like a budget, should be treated as a collection of suboptimization problems rather than as one grand optimization problem. For example, in addressing the by-then venerable problem of selecting a "'best' gun sight," one could see how an analysis could quickly balloon outward by "fitting the gun sights into planes, then the planes into fighter groups, then the groups into relevant military operations." At some level systems analysis and economics came together, as the analyst might even "want to ask what tasks or budgets make sense in view of the whole military operation and of political realities." However, one could not "spend one's life seeking [an] *optimum optimorum*." By the time one approached that level, the analysis was bound to break down under limits in investigative resources, the time available to complete the study, computational power, the ability of the analyst to influence decision making, and even perhaps the lack of importance at the highest level as to which gun sight was actually chosen. One had to accept that there would be a point where a line would be semi-arbitrarily drawn defining the "limited context" of the study's validity. The analyst would have to "get on with his inevitable job of sub-optimizing," while simply remaining aware of the "shortcomings and biases" of the analysis "in a general and qualitative way."[11]

In defining systems analysis in this way, Hitch swiftly became interested in what sorts of rules of craft governed the process of amalgamating sub-optimizations together into a legitimate analysis. In a paper given to the first full meeting of the Operations Research Society of America in November 1952, he urged that careful attention be paid to the choice of criteria to optimize, so that the optimization made sense in view of, and did not obviously conflict with, higher-level goals. Pointing to a by-then famous case example from World War II, he observed that increasing convoy size to reduce losses to U-boats and maximize destruction to U-boats had to be considered in light of ports' ability to handle simultaneous loads and other possible limiting factors on the desirability of a larger convoy size. In this case, the correct criterion to be optimized had to be the throughput of goods vital to the war effort, rather than a ratio of vessels lost to U-boats destroyed, as reported in Philip Morse's and George Kimball's *Methods of Operations Research* textbook.

In taking this example, Hitch became one of the first people to define OR as the optimization of components within larger systems. This redefinition of OR vis-à-vis systems analysis violated, and in some ways inverted, the wartime conception of warfare analysis as the mathematical study of combat and OR as the empirical investigation of combat operations, which helped keep mathematical analyses tethered to operational realities. However, as OR came to be more mathematical and systems analysis came to be less mathematical, Hitch's conceptualization of their intellectual relationship with each other began to make increasing sense. By the time Hitch became the president of ORSA for 1959, and he and RAND economist Roland McKean published their touchstone work on defense budgeting, *The Economics of Defense in the Nuclear Age*, the following year, his view had become well accepted.[12]

Systems Analysis Reconceived

RAND's analysts began to reform systems analysis as early as 1951, and Charles Hitch began to offer his prescriptions for its ills around the same time. But the fully worked out ideas detailed in *The Economics of Defense in the Nuclear Age* developed only slowly over the course of the 1950s as RAND's analysts continued to struggle with the tendentious practical problem of aligning increasingly powerful but complicated equipment designs with increasingly uncertain expectations surrounding the conditions in which that equipment would be used. The design of complicated technological systems was an expensive process involving the production of specially built prototypes. In a 1953 memorandum, systems analyst Edward Quade reported that the Air Force's contractors were becoming increasingly agitated that they were not being adequately compensated for development and design work that did not lead to a procurement contract. Whatever its flaws, the explicit evaluation of systems analysis still seemed to be a fruitful approach to this problem, because it promised to help eliminate unfruitful designs before they absorbed too much money. Some industry representatives were beginning to suggest that the Department of Defense perform systems analyses prior to issuing new contracts for prototype development. These sentiments were also shared by many in the Air Force. "Indeed," Quade wrote, "there seems to be a feeling in some parts of the Air Force that the systems approach may provide the complete answer to all questions of development, procurement, and operation as well as those of design."[13]

Quade, to an extent, welcomed the enthusiasm. Systems analysis was initially designed to collect, consolidate, and reconcile expert opinions pertinent to design problems. However frustrating it was that RAND's general conclusions were ignored, the cannibalization of RAND's studies by military contractors for use in their own proposals had placed "RAND in the role of an information collecting and transmitting agency," which was "generally considered to be desirable." Furthermore, "A properly executed systems analysis might not only furnish an 'optimum' choice (subject to the restriction that the requirements and assumptions on which it is based are likely to be extremely arbitrary) but also an effective means of argument to the Air Force (provided the Air Force understands systems analysis!) that the particular choice is superior to those made by competitors. Both the Air Force and the contractor should save money."[14] If everyone could at least agree on what range of assumptions might dictate what new designs would be most valuable, the design process would ostensibly become more coherent and successful through the application of analysis.

There was, however, a danger that components of RAND studies would simply be reintegrated into new analyses without understanding the importance the studies placed on the context surrounding those components. Factors such as expectations of enemy capabilities were based on highly variable information that required significant reinterpretation when incorporated into revised analyses. While industry could obtain people competent in the application of "techniques of operations research"—here connoting mathematical optimization—RAND opinion of others' ability to employ these techniques responsibly was low. Quade lamented, "There is little indication that industry has profited from RAND's mistakes." In addition, he felt that the Air Force had been "oversold" on systems analysis, and worried that "a surfeit of low quality studies may cause not only a complete loss of faith in the product but also in the salesman."[15]

To address the problem, a proposal had been floated that RAND offer a course in systems analysis that would "teach the airframe industry the 'correct' way to do such a study and the Air Force how to understand one." Quade felt it was a good idea. Insofar as RAND could be said to know how to do a systematic study properly, the corporation had never published any sort of report on the methodology itself. "Unfortunately," he observed, all the wisdom that had been gained on such matters was "completely unorganized" and often "buried in highly controversial opinions concerning the range of application, the validity of the results, and the adequacy of the method of systems analysis as a means of solving Air Force problems." The course would not only inform industrial suppliers how to construct their own analyses, it would "disabuse the Air Force as to the technique and range of the method," pointing out areas to which it did not apply, such as in making decisions such as what programs of basic research to pursue. At the same time, the course was expected to give RAND a chance to collect and collate the various ideas being circulated internally about what systems analysis was and what it could and could not accomplish. "Like a university faculty," Quade argued, "the faculty assigned to any systems analysis course could profitably spend the major part of its effort on research about the subject in general. A research effort in this field is called for regardless of any educational effort we might make."[16]

The need to decide on the nature of systems analysis was acute. Work in Charles Hitch's economics department on risk management in budget allocation suggested that RAND's systems analyses might be leading the Air Force down a dangerous path. By offering the illusion that requirements and design configurations could be coordinated into complete designs ahead of time, systems analysis might cause the Air Force and its contractors to believe that the inevitable pursuit of fruitless design paths could be avoided. The economists Armen Alchian and Reuben Kessel took up this line of critique.[17] Responding to Quade's 1953 memorandum, they observed, "Inadequate compensation for development work is the reason developers feel inadequately compensated." There was little that systems analysis could do to remove the facts that research led to waste, and that the journey from design to production was always long and arduous. They argued that only the development of several projects could serve as adequate insurance against the dangers of failed technical development. The Air Force might simply have to offer more lucrative development contracts to offset the inevitable costs.[18]

Alchian and Kessel knew they could be accused of sounding a false alarm. Systems analysis referred only to processes of drafting sensible designs, not to the economists' goal of informing overall R&D funding policies. There was nothing in systems analysis dictating that only a single project should be pursued, and high-level Air Force policymakers had to that date been much more likely to ignore RAND's conclusions than to use them to justify singular lines of development. The economists recognized this fact, but they were still concerned that decision-making bodies in the Air Force might yet be overwhelmed by enthusiasm. They wrote, "Little boys and matches neither logically nor inevitably lead to fires, but the probability is distressingly high, if it's your boy and house." They argued that, as it stood, systems analysis was only really useful if forecasts of future circumstances were not more or less arbitrary. If forecasts were poor, as they so often were, it would be better to develop studies of how to hedge budgetary bets and improve forecasting methods than to work out the analytical conclusions of debased assumptions.[19]

Quade accepted the economists' wisdom concerning poor forecasts, but not to the point of agreeing that systems analysis was a specious exercise. In a more detailed proposal for the course, which he and the economist Malcolm Hoag[20] wrote in 1954, they allowed that systems analysis would be more reliable "if the examples were restricted to a selection from within a family of essentially similar instruments." Yet, it was evident that there was a need to compare rather different options when addressing weapons and defense system requirements. Conceding the difficulty of the task, they pointed out that the economists did not seem to offer any guidance aside from stressing that diversity of investment was a virtue. But it was surely not an unlimited virtue: "To hedge without limit by developing everything is, in view of the vast array of technically feasible choices, economically out of the question." Thus, to focus on the economics was to dodge the crucial point: potential development programs had to be compared, and these comparisons had to be made on some basis. This led Quade and Hoag to ask, "Should [these comparisons] not be made as carefully and as explicitly as possible, with no exclusion of alternatives arbitrarily or on the basis of petty considerations? Is that not Systems Analysis?" They reckoned it was: "To select by analysis is by our definition Systems Analysis, which is emphatically not to say that the techniques and results need be the same for procurement and development decisions nor that unique rather than multiple choices are inevitable." Quade and Hoag concluded their thirty-four-page proposal by reemphasizing that the theme of their proposed course was to discuss "important questions" explicitly, "albeit necessarily imperfectly," and to "treat them as best one can, and end with the analyst being honest with himself and his customers about the arbitrary and uncertain elements of importance in his analysis."[21]

Meanwhile, the problem of how to undertake such an analysis was still unresolved. One way of reforming systems analysis was by simply placing lower-level considerations within comparative strategic contexts. This approach was pioneered in a study led by Albert Wohlstetter, completed in 1953, concerning the choice of arrangements for basing the Air Force's strategic bombers. Wohlstetter, a mathematical logician and a consultant to RAND, had been urged to consider the problem by Hitch.[22] In a departure from prior systems analyses, Wohlstetter began his work alone as he contemplated the structure of the problem, which led him to an important insight. As it turned out, basing bombers overseas was preferable from a cost standpoint, *if* one assumed that the objective was simply to launch a strike on the Soviet Union. However, *if* the objective were to deter Soviet aggression, then the bombers would have to be defended against a surprise attack. From this perspective, it would be more cost effective to defend the bombers from bases within the United States.[23] While the study naturally required detailed comparative cost analyses to reach its conclusions, the answer to its central question turned not on the outcome of advanced mathematical formulations, but on a reconsideration of the policy's larger strategic context.

In the wake of Wohlstetter's basing study, systems analysis began to be recognized as a more generic form of policy analysis, characterized by a methodological heterogeneity. It was used to explore the feasibility, value, and possible ramifications of different policy choices within a range of possible scenarios. Since individual studies were no longer as exhaustive or conclusive as originally supposed, they no longer took years to complete, nor was their validity taken to hinge on the applicability of a particular set of assumptions. The objective now became to assess what sorts of policies and technological systems would be desirable, if not necessarily "optimal," and under what circumstances substantial cost burdens would accrue, or fatal flaws would manifest. In a 1958 memorandum Wohlstetter suggested that the work would more accurately be called "systems design" than "systems analysis."[24]

Reformed methodology accompanied the reformed systems analysis. Logistical optimizations employing OR tools, such as linear programming, continued to prove important for constructing feasible and efficient designs. However, there was now room for other methods more appropriate for exploring ideas rather than calculating optima. The uses of game theory at RAND morphed in exactly this way. Where Paxson's systems analysis made direct use of the solutions to game-theoretical models of gun duels and equipment selection, game theory was now used more as a means of exploring strategic scenarios. For example, game theory was thought to help guide analysts' intuitions on questions such as whether it was better to develop a single technology that could perform adequately against a variety of threats, or to develop many technologies that were specialized to perform against different kinds of threats. As Charles Hitch put it in a 1955 paper, "Suppose you have your defenses deployed as well as you can. Now you get more defenses. How do you deploy them?" He admitted, "Well, my intuition told me (and so did most people's) that you deploy them to protect additional targets ... that you did not previously have enough to defend." Game theory suggested otherwise. "You use additional defenses mainly to increase the defense of targets already defended." Faced with this result, "you think about it and begin to see the rationale."[25]

Even as game theory became more of a background activity at RAND, "gaming" and simulation became more central. War gaming fit easily into the long tradition of military war exercises and equipment tests, which established so many of the parameters that shaped system designs in times of peace. However, more abstract sorts of games could be played on boards like chess, or they could be played on digital computers wherein stated policies could be played out under varying circumstances, or varying policies could be played out under the same circumstances many times over. The results of a game, or even very many games, could never produce demonstrative conclusions in the way an analytical answer to a mathematical problem might. But, when analyzing complicated situations that were not amenable to analytical solution, games could help evaluate the robustness of a particular policy by exploring hidden ways that it might collapse, such as amid unsuspected logistical bottlenecks. More positively, games might bring to light unsuspected strategies, much as the continuous gambit had emerged when the Navy's OR group had moved from analysis to gaming of U-boat search problems during the war.[26] Both games and the conceptual uses of game theory were also linked to the RAND Corporation's burgeoning interest in strategic studies. While RAND hired the political and strategic thinker Bernard Brodie as early as 1951, it was only later in the decade, with the rise of a more systematically developed strategic theory, that RAND became widely famous—and, to many, infamous—as a center for exploratory thinking about arms policy and the conduct of nuclear wars.[27]

As systems analysis became more methodologically eclectic, and as its emphasis shifted, at least partially, from engineering to policy, the burden of managing the intellectual status of RAND's work grew heavier. Engineers had a prerogative to take or leave suggestions for various elements of a system design, based upon their assessment of the suitability of those suggestions. Policymakers possessed a similar prerogative, but tended to have less of an expert sense of whether or not suggestions would actually improve policy. Generally, because it was accepted that policymaking is rife with uncertainty and ambiguity, policymakers understood that they were permitted to make decisions and to draw upon advice as they saw fit, until such time as their superiors or a voting public deemed someone else might do a better job. While many policy analysts were no doubt content to exercise influence according to the favor they managed to curry with policymakers in power, in principle the legitimacy of their influence was tied to their ability to improve policies. Since the ambiguities surrounding policy often made it difficult to be certain that their work did not actually make policies worse, they could occupy an uncomfortable position, both politically and intellectually.

This essential discomfort in policy analysis was felt at RAND, where it was intimately linked to the persistence of a culture of methodological critique. In a 1957 report entitled "Ten Common Pitfalls," systems analyst Herman Kahn addressed this anxiety, writing, "Probably no applied professional group is so intensely and continuously concerned with methodological and philosophical questions as Operations Analysts and Systems Analysts." He partially attributed this preoccupation to "the normal introspection to be expected in any new field." Yet, he wrote, "it is hard to avoid the feeling that much of this self-questioning is caused by a sort of mass inferiority complex or at least a general sense of insecurity." He suspected this feeling arose from the "nebulous and unspecialized nature of most of the work," and from the angst that a lot of it was "not quite passable" due to the prevalence of "common mistakes."

Since there were no methodologies that could guarantee the quality of an analysis, Kahn took a therapeutic approach, warning analysts to avoid certain pitfalls. These included concentrating too much on a model and not enough on the problem ("Modelism"), doing studies too far removed from the first-hand knowledge of experts and policymakers ("Hermitism"), mistaking "real" uncertainty for probabilistic uncertainty, making unrealistic assumptions about enemy capabilities, concentrating too much on one topic, choosing a topic that is too ambitious, or making other mistakes likely to undermine the legitimacy of an analysis.[28] Around the same time, similar suggestions were aired in the OR community by practitioners such as Bernard Koopman, who, echoed by Charles Hitch, warned against "mechanitis" and "authorititis," which stemmed from overreliance on the power of computers and policymakers' formulations of problems, respectively.[29] These sorts of lessons were also pressed through the aforementioned course on systems analysis, which RAND ended up offering in 1955 and 1959. A book based upon the course was published in 1964. The course was given again in 1965, and a revised edition of the book appeared in 1968.[30] Of course, the repetition of sage advice was never any guarantee that it would be followed, nor could it protect analysts from accusations that they were ignorant of the basic wisdom it conveyed.

VII   Epilogue

29 Dr. Strangelove: Rationality, Authority, and Sin

From World War II to the 1960s, proponents of the sciences of policy were acutely aware that their intervention could render deliberations less rather than more rational. To confront this possibility, they continually discussed how various institutional structures and intellectual strategies, embodying principles of sound practice, might help to ensure that what influence they had over policy was constructive and legitimate. The idea that the violation of these principles was a common cause of analytical failure also proved extremely powerful as a polemical tool. In the late 1950s, irreconcilable differences that developed between people associated with the sciences of policy resulted in a rash of public accusations that their opponents' claims derived from violations of fundamental precepts concerning the proper power, scope, nature, and role of their work. Polemicists warned that, unless these violations were exposed, the analysts committing them could exercise an illegitimate influence over policymakers, resulting in systematic policy failures. In the nuclear era, such failures might affect the fate of nations, and even humanity itself.

These polemics were also closely related to a longstanding discourse holding that a blind trust in the power of science, technology, and reason would lead humanity down an uncertain, and likely dangerous, path (see figure 29.1). This anxiety found its iconic expression in the rising director Stanley Kubrick's 1963 dark comedy, *Dr. Strangelove, or: How I Learned to Stop Worrying and Love the Bomb*. The film was an adaptation of a 1958 novel called *Red Alert*, in which an insane Strategic Air Force general exploits a plan that allows him to order his nuclear bombers to attack the Soviet Union if the chain of command from the President has already been destroyed by a sneak attack. However, the film's title character, a wheelchair-bound German émigré scientist and government adviser, was an invention of the film. The figure of Strangelove established a stronger sense of contrast between the logical intricacy

LEADERSHIP

Figure 29.1
A scientific figure exercising an authority deriving from mysterious mathematics. Illustration by physicist and operations researcher Robert Herman, accompanying Charles Goodeve, "The Future of Co-operative Research in the Steel Industry," October 10, 1968, in CAC GOEV, 8/1. Reproduced with the permission of Roberta Herman.

of the government's plans and the persistent reality of the human fallibility that undermined them.

In the film's second half, the problematic relationship between planning and fallibility achieves apocalyptic proportions when it is revealed that the Soviets have secretly armed a "Doomsday Machine" that will irradiate the entire surface of the Earth for a century if a nuclear bomb ever detonates on Soviet territory, or any attempt is made to disarm it. The Soviet ambassador reports that the machine had been built in response to a *New York Times* report that the Americans were thinking of developing their own such machine. Strangelove, appearing for the first time, explains that he had commissioned a "BLAND Corporation" study of such a device, which had found that it was "not a practical deterrent" for reasons that had become "all too obvious." He chastises the ambassador for the Soviets' lack of logic in keeping their device a secret, undermining its entire point as the ultimate deterrent. The ambassador replies that the Soviet premier (a drunk who "loves surprises") was to announce it at the Party Congress the following Monday.

Despite heroic efforts to recall the bombers, the film concludes as one completes its mission and triggers the Doomsday Machine. Strangelove consoles the distraught President by pointing out that there is still time for the nation's leaders and strongest citizens to go into hiding in mineshafts deep beneath the Earth's surface. At a ratio of ten physically alluring women to each man, life—and strategic logic—will continue. Soon the war room is abuzz as the assembled generals and politicians begin to worry about whether the Soviets might stash away a bomb for the day a century off when their progeny return to the surface. The bombastic General "Buck" Turgidson, an avid consumer of nuclear war studies, warns that the Soviets might even attempt an immediate attack to take over American mineshaft space, allowing them "to breed more prodigiously than we do." He insists, "We must not allow a mineshaft gap!" This was an unmistakable reference to the "missile gap" issue from America's 1960 Presidential election, but with an added overtone of race war and demands for *Lebensraum*. In the film's final line, Strangelove, personifying a reborn Nazism, rises from his wheelchair, shouting, *"Mein Führer, I can walk!"*

Although *Dr. Strangelove* is an unmistakable critique of the nuclear arms race and its grim and intricate logic, one of the more remarkable aspects of the film is its ability to ridicule without condescension or petty mockery. Although the characters are certainly deeply flawed, their flaws, save for the insane general's, are not the proximate cause of the film's horrific conclusion. In fact, most of the characters prove capable of transcending their weaknesses, if only briefly. Further, their ideas and actions are never portrayed as blatantly illogical or naive regarding human behavior. The BLAND Corporation, notably, is proven right about the danger of the Doomsday Machine. Turgidson's complaint that the Air Force's human reliability screenings should not be condemned "because of a simple slip-up" is plainly absurd, but his tacit argument that it is unrealistic to expect perfection from the screenings suggests his understanding of their limitations.

It is, I argue, the film's underlying sympathy for its characters that makes it so devastatingly effective. If it were easier to condemn their morality or to poke holes in their reasoning, the film's conclusion would seem preventable simply by enacting a more rational policy, or choosing more rational leaders. But the entire point of the film is that rationality's futility in the face of folly and sin transforms it into the vehicle of the realization of those forces' full destructive potential. It is the inescapability of this problem that allows the film's comedy to feel like an appropriate response to its overarching fatalism.[1] By contrast, scientists' portraits of each other in issues of national policy often showed little such sympathy.

Charles Hitch versus Russell Ackoff

Many disputes over principle revolved, in one way or another, around the limits of useful analysis. West Churchman and Russell Ackoff's philosophy of decision held that for a decision to be considered properly "scientific," all factors of the decision had to be taken into account in as sophisticated a way as possible. Since they envisioned the new field of operations research to be the aegis under which decisions were to be validated as scientific, they urged people working in OR not to constrain their vision. Accordingly, in 1957, when Ackoff ended his term as president of the Operations Research Society of America, he gave a speech at the society's annual meeting entitled "Operations Research and National Planning." In it he implored the society's members to take an interest in problems of national policy, and in particular the economic planning of India, which was at that time a frequent beneficiary of postcolonial developmental assistance and advice. Ackoff did not imagine that the mathematical techniques of OR he had helped foster would prove particularly valuable in the task. But he did believe that operations researchers' "knowledge of system design, control processes, and the structure of decision-making" made them particularly well suited to aid in the establishment of effective government organizations.[2]

The RAND Corporation's Charles Hitch responded to Ackoff's speech in the journal *Operations Research*. While he was sympathetic to Ackoff's goals, he argued that he knew "of no evidence ... that operations research has (as yet) much to offer at the level of national planning." He recited his credo, now well sharpened from his critiques of systems analysis at RAND, that OR was the "art of sub-optimizing, i.e., of solving some lower-level problems," and that "difficulties increase and our special competence diminishes by an order of magnitude with every level of decision making we attempt to ascend." None of this was to say that "some operations *researchers*" could not "make excellent advisers on some high-level problems." He acknowledged, "The profession contains some men with a combination of first-rate minds, broad interests, wide and relevant experience, and excellent judgment."[3] His concern was that the reputation of OR would be tarnished if operations researchers, working *as* operations researchers, strayed beyond the bounds of their special expertise to flail at extremely complex problems that other kinds of policy advisers might be better equipped to handle. To drive his point home, he pounced on several reckless statements in Ackoff's talk on the nature of the development problem.

In his response to Hitch's criticism, Ackoff more or less accepted Hitch's specific criticisms, but repeated his claim that OR should not be defined by its techniques, but by its "method," which he did not define except insofar as it revealed a "logic" of "procedure." While he acknowledged that he, personally, did not have the knowledge necessary to address India's problems, he argued that it if he were to engage himself with these problems, "I would surround myself with such knowledge." Hearkening back to the broader wartime notion of OR, he pointed out, "The concept of a *team* of *mixed disciplines* is an essential part of OR, is it not?"[4] Ackoff's response marked the end of the dispute, but the problems that Hitch had exposed remained. Because Churchman and Ackoff did not see OR as a science *in aid* of management, but as an all-encompassing science *of* management, they understood it as an abdication of scientific responsibility to recuse themselves from aspects of policymaking in which they had no special experience. Recall here not only their feeling that one of the most pressing problems facing the sciences was specialization and compartmentalization, but also their commitment to including ethics as a part of the problem of decision.

Churchman and Ackoff would both repeatedly insist over the following years that improvements enacted through mainstream OR techniques could lead to bad results out of a failure to consider their strategic, and particularly their ethical implications. But the OR profession continued to be unwilling to follow their lead into these domains. For this reason, Churchman and Ackoff became increasingly disillusioned with the fields of OR and management science. Churchman left the Case Institute of Technology for the School of Business Administration at the University of California, Berkeley, in 1957, where he helped establish the Center for Research in Management. He remained there for the rest of his career. In 1964 Ackoff took the entire Case OR Group to the Wharton School at the University of Pennsylvania, where it joined with the statistics department.[5] However, facing faculty resistance to his efforts to expand beyond OR into broader, less mathematical areas of "systems thinking," Ackoff would leave the department to set up his own academic program in "Social Systems Science." Finally, in the late 1970s, citing a diminished status of OR at universities and in the organizational hierarchies of companies, he took it upon himself to pronounce OR scientifically dead and to sever his relationship with it.[6]

Patrick Blackett versus Charles Hitch

Inasmuch as the RAND Corporation embraced Charles Hitch's ideas about the limits of useful analysis, toward the end of the 1950s RAND also became a frequent target of criticisms that its analytical reach surpassed the bounds of legitimacy. One important source of these criticisms was the dean of World War II-era OR, Patrick Blackett. As we have already seen, although Blackett did not take part in the development of any of the postwar manifestations of OR, he and others of like mind had eagerly invoked the wartime OR experience to suggest how scientists could be better integrated into the work of the British state. Such rhetorical uses of the wartime experience waned after 1948, but Blackett continued to take a personal interest in protecting the legacy of his wartime work for many years after. In April 1948 he republished two of his wartime memoranda on OR in the British Association journal *The Advancement of Science*. In 1950 he wrote a short piece reflecting on the lessons of the wartime experience for the first issue of the new *Operational Research Quarterly*. In 1953 he published another piece in the military affairs yearbook *Brassey's Annual* entitled "Recollections of Problems Studied, 1940–45." He also spoke about wartime OR in a number of addresses that he gave throughout this period.[7]

Initially, Blackett continued to discuss wartime OR as an advance in the use of science in the military, and thus as a path toward greater involvement in the state. In a 1950 speech at the Royal Statistical Society, he recalled, "We just did the obvious thing that seemed sensible at the time, but by the end of the war the scientists had, to a great extent, gate-crashed into the Holy of Holies, higher staff direction of the war, and this was a unique development."[8] However, within a few years he became less facile in his recollections about what had made wartime OR valuable. In a 1953 lecture at the Institute of Physics, he explained, "Operational research is not, as some people think, a case of the bright scientist suddenly intervening and telling the experts what to do. It is very much more the slow and careful enquiry into extremely complicated matters by scientists who have soaked themselves in the atmosphere of an operational command."[9] Blackett was thus prepared to deliver a new polemic later in the decade castigating a bastardized version of OR, which he supposed illegitimately leveraged scientific prestige to intervene in policy.

In December 1958 Sidney Dell, a member of Blackett's wartime group at the Admiralty then working as an economist for the United Nations in New York, wrote to Blackett to inform him of a new article in the *Review of Economics and Statistics* entitled "Economics and Military Operations Research," written by Charles Hitch. Dell highlighted Hitch's discussion of Philip Morse and George Kimball's treatment of the problem of convoy size, which, as we have seen, Hitch had been using since 1952 as a case example of the importance of selecting proper criteria for suboptimizations. However, Blackett's group had originally handled the convoy problem, and he and Dell took Hitch to be making an assault on the quality of their work. They were particularly incensed by Hitch's apparent "smugness" in remarking that the wartime choice of an "exchange rate" of U-boats to merchant vessels sunk, and the selection of a ratio as an optimization criterion in general, "would not have been made by a person with good economic intuition." No doubt they regarded as condescending his allowance that the wartime "participants" were able, due to their "good sense" and "intuitive restraint," to make a good recommendation on the convoy problem, in spite of their failure to make a "sophisticated choice of criterion."[10]

Even before obtaining and reading Hitch's paper, Blackett also took immediate umbrage at what he (mistakenly) understood to be an argument that factors supporting smaller convoys should have been included in their wartime study in order to define an optimum convoy size precisely. For him, such an argument confirmed a diagnosis that Hitch was fatally detached from the realities of policymaking. He wrote to Dell that to estimate an optimum size it would have been necessary to take into account "second partial differentials, which was out of the question." Unaware that Hitch was actually an important critic of overly elaborate formal analysis, he went on, "I fear that a great deal of Operational Research done by clever but rather conceited young men, who know very little about the subject about which they are talking, could easily discredit Operational Research seriously." After Dell had crafted a response, which he sent to Blackett for comment, Blackett again fumed, "Hitch must be super academic if he thinks an urgent operational paper in wartime should go into all the arguments. No one would read it if it did."[11]

Dell's response and Hitch's counter-response only appeared in the *Review* in the spring of 1960. Per his correspondence with Blackett, Dell emphasized the wartime Admiralty group's commitment to delivering useful and timely advice, as well as the impracticality of using a priori optimization calculations in the difficult and ongoing process of forming proper policy. For his part, Hitch was caught unawares by how his didactic point about criterion selection had been transformed into a slight against Blackett's wartime work. He pointed out that he, too, valued practicality in analysis, and that he shared "Mr. Dell's objections to 'a useless search for refinement.'" Nevertheless, he maintained that some estimate of the difficulties associated with larger convoys had to be considered for the recommendation in favor of them to stand.[12]

Had Blackett and Dell not been so predisposed to disagree with Hitch, they might well have agreed with him. The Admiralty's wartime report had in fact contained just such a discussion of the "handling" difficulties associated with larger convoys, which pointed to the lack of difficulties experienced where larger convoys had already been used. On the question of the optimal size of a convoy, while Blackett and his group had not advocated a specific upper limit, suggesting merely that increasing average size from forty to sixty ships would be unproblematic, they actually did include a discussion of questions that the Admiralty would face in fixing one.[13]

Nuclear Strategy and Intellectual Authority

If the intellectual predispositions of Blackett and Hitch and his RAND colleagues were not actually far removed from each other, by the early 1960s Blackett had little incentive to acknowledge it. Following the war, Blackett had become a pundit in the growing public debates over nuclear policy, making a name for himself through his books *Military and Political Consequences of Atomic Energy*, published in 1948, and *Atomic Weapons and East–West Relations*, published in 1956, as well as through a number of articles he wrote in this period.[14] In the late 1950s, RAND also became publicly associated with nuclear policy, particularly following the 1959 publication of Albert Wohlstetter's article "The Delicate Balance of Terror" in the elite journal *Foreign Affairs*. In it Wohlstetter objected to the commonly voiced idea that the mere existence of a nuclear arsenal of any size and configuration would be sufficient to deter war, and reprised his 1953 basing study's argument for the need to maintain an arsenal that could survive a sneak attack.[15] For his part, Blackett rejected as fatally unrealistic any policy calculation that supposed that nuclear weapons could be depended upon to achieve any strategic objectives, including crippling another nation's nuclear arsenal, thus invalidating the logic of first strike and second strike as a legitimate intellectual framework for formulating nuclear policy.

Blackett publicly blasted Wohlstetter and other nuclear strategists in a 1961 speech at the Royal United Service Institution (RUSI) entitled

"Operational Research and Nuclear Weapons," which was quickly revised and reprinted in the widely read journal *Encounter* as "A Critique of Some Contemporary Defence Thinking."[16] The polemics that Blackett had honed against Charles Hitch became the centerpiece of his new attack. He argued that strategists' work was premised on academic ideas, including the abstract theory of games, and therefore ignored inscrutable operational and political realities. He contrasted their work to his wartime work, which had been "an addition to, and not a substitute for, the exercise by the trained staffs of conventional military wisdom." In cases of uncertainty, he argued, scientists had an obligation to "keep silent: never should they fall into the trap of decking out what is essentially only a hunch with a pseudo-scientific backing."[17] He also explained why these principles did not preclude his own punditry on nuclear affairs, remarking, "I do not think that my experience of four years of active operational analysis during the War has given me any reason to suppose that my views on the present [strategic] situation are any more likely to be reliable than those of any other academic who has studied the subject deeply, except in one respect: my experience has given me some personal acquaintance with how major decisions in war are in fact taken." In critiquing nuclear strategists like Wohlstetter, he stressed the importance of professional identity: he spoke "not as an ex-operational researcher, or even as an ex-professional fighting man," referring to his World War I Royal Navy experience, "but as one who is deeply concerned with making the right choice of policy—right, that is, from the broad considerations which govern world history."[18]

The Legacy of Henry Tizard

Circa 1960, Blackett was walking a critical tightrope. On the one hand, he was, at that time, once again an active proponent for new technocratic planning in Britain, and served as an adviser on scientific issues to the out-of-power Labour Party.[19] On the other hand, he also felt he had to stand against the dangers of *inappropriate* technocratic influence. His diagnosis of the intellectual failures of American nuclear strategists was part of this stand, but so, too, was his rehearsal of wartime grievances occasioned by the deaths of Frederick Lindemann, Viscount Cherwell in 1957, and Henry Tizard in 1959. As we have seen, Blackett, among others, had held Tizard up since the war's end as an exemplary adviser. The memorial lecture he gave for Tizard at the Institute for Strategic Studies in London in February 1960, reprinted in *Nature* that August, simply provided him with a new opportunity to lionize Tizard's qualities. Now, however, he was also free to contrast those qualities with how a bad adviser might act, as illustrated by Tizard's wartime conflicts with Lindemann, including Tizard's still-topical criticisms of Lindemann's arguments for area bombing.[20]

Shortly thereafter, Blackett's tales found a much larger audience when they were taken up by his close acquaintance, the scientist-turned-novelist C. P. Snow. During the war, Snow had participated in the organization of science for the war effort, helping to administer the Ministry of Labour's "central register" of Britain's scientific and technical personnel. Thereafter, the proper exploitation of science remained a question close to his heart. In 1959 he made it the subject of his Rede Lecture at Cambridge University, "The Two Cultures and the Scientific Revolution," which was quickly published as a widely discussed book. Among his claims, he asserted that the British administrative class was dominated by people trained in literature and the classics, who understood tradition-based ideals to constitute a sound basis of governance. Snow lamented that these people were dangerously removed from the nation's scientists, who were committed to scrutinizing, revising, and overturning ideas.[21]

Snow's identification of the two-cultures problem was, in many ways, a recapitulation of wartime discussions about the importance of integrating science properly into the war effort. It is therefore no surprise that, to express ideas about how to integrate science into policymaking *successfully*, he reached back to that same discourse. In 1960 Snow delivered the three-part Godkin Lectures at Harvard University, which he entitled "Science and Government" and published as a book in 1961. These lectures expanded on Blackett's account of the conflicts between Tizard and Lindemann. Snow used this history to illustrate the operation of what he called the "closed politics" of science in government, which he defined as policymaking in which there could be no appeal to the consensus of the broader scientific community. He supposed that closed politics could be subdivided into "committee politics," "hierarchical politics," and "court politics." Tizard, he argued, prevailed over Lindemann in the early disputes of the Committee for the Scientific Survey of Air Defence, because the other members of his committee were sympathetic to his perspective. Further, as a scientific adviser, Tizard had proven himself a master of building effective bureaucratic hierarchies, and of marshaling the knowledge within them into sound policy recommendations. In contrast, by Snow's account, Lindemann was a master of court politics, and so was able to have a major influence on policies, albeit very often in cases far beyond what his personal command of available facts should have allowed.[22]

Snow's taxonomy might have led the way to new conversations about how scientific expertise operates in bureaucratic structures, but his ideas on this score were overshadowed by the ensuing controversies over his handling of the Tizard-Lindemann disputes. No one denied that Lindemann could be difficult and that Tizard was a talented coordinator. But many of Snow's critics—including Robert Watson-Watt, Lindemann's former student R. V. Jones, and Lindemann's official biographer the Earl of Birkenhead—felt that Lindemann's positions, such as his supposed opposition to radar research, were portrayed unfairly.[23] And these complaints were not without justice. For the sake of using Lindemann's work as an exemplar of a kind of politics, Snow had misrepresented some of Lindemann's views, and excluded some of his most important contributions to the war effort, including the creation of statistical groups at the Admiralty, in the War Cabinet, and in Churchill's Prime Ministerial office. These groups, not unlike OR groups, had used research and new bureaucratic arrangements to bring additional factual rigor to war planning and policymaking. For those who supported Snow's account, most notably Blackett, it was not only the bitter memories of their encounters with Lindemann, but also Lindemann's ability to serve as an icon of the inappropriate conduct of scientific advisers that made it unconscionable to defend him.[24]

Snow and his supporters were concerned over the proper behavior of advisers not only because the similarly detached nuclear strategists were supposedly tilting the world toward nuclear oblivion, but—just as important—because it had potential consequences for bureaucratic reform at a moment when science policy was coming back into Britain's national spotlight.[25] In 1963, when Blackett and R. V. Jones privately clashed over a new account of the wartime conflicts that Jones was to publish in *The Oxford Magazine*, the debate swung effortlessly to current affairs. In response to Snow's warning in *Science and Government*, "We must never tolerate a scientific overlord again," Jones emphasized the need for a strong scientific voice in government, or else "a non-scientific administration may be able to act as it pleases, exploiting the differences between the scientists, or at best making up its mind on what it thinks of the individual scientists as persons, in which event we are nearly back to the overlord concept."[26] Blackett, however, was nonplussed by what he viewed as the statement's recklessness. He wrote to Jones:

You give no hint as to where in the Government machine these men should be placed, or what authority they should have, and over what departments or organisations. You mention no formal relation with other scientists but do make the surprising remark "Provided that the scientist concerned does not isolate himself from the experience and opinions of his fellow scientists. ..." You mention no obligation to consult, to do or not to do seems to be left to the overlord's initiative.

These details were pressing matters for him, because he was at that time in consultation with a committee meeting under Cabinet Secretary Burke Trend to consider a new round of civil service reforms. For Blackett, only statutory specificity could help ensure that any new bureaucratic machinery would not be dependent on the choice of appointees to ensure that it embodied the qualities of Tizard and not Lindemann.[27] Two decades following the war, Blackett's thinking on the subject of the relationship between science and the state continued to be deeply influenced by that experience.

Policy Analysis and the Public Sphere

All proponents of the sciences of policy agreed that the legitimacy of advisory work was tightly tied to advisers' ability to engage with the deliberations of traditional authorities. There was, then, little sense among those working in these sciences that their work could wield any final authority, particularly in the public sphere. In fact, because authorities had to be free to reject advisers' conclusions, and because those conclusions had the potential to embarrass decision makers, it was generally understood that the relationship between advisers and decision makers was to be kept private, and that advisers should generally be responsible to a single authority within an organization. In his 1957 "Ten Common Pitfalls" report on guidelines for quality analysis, Herman Kahn was very explicit about this point. "One of the trickier questions," he observed near the end of the report, "involves under what circumstances one is justified in jumping over [military] staff members to their superiors." He figured it was permissible, provided one understood the lower-level objections. But, he cautioned, "One is, of course, almost never justified in jumping any channels clandestinely," adding that "the only really unforgivable sin is going to the public press or Congress." Following this statement, Kahn inserted a footnote, indicating that RAND strategist Bernard Brodie had "pointed out that public journals and books are sometimes legitimate and even advisable channels of communication." Articles and books might "be read in high quarters" and attract comment, where a "report may just gather dust."[28]

Kahn would take Brodie's advice in spectacular fashion. The year after Wohlstetter published his "Delicate Balance of Terror" article in *Foreign Affairs*, Kahn published *On Thermonuclear War*, an enormously controversial book on nuclear policy. In it, he swept quickly, and often glibly through scenarios in which nuclear wars might begin, how they might unfold, and what the survivors might face.[29] The book was supposed to initiate new public discussions about nuclear arms and civil defense, but, whether or not it did, its sensational handling of nuclear war created a public relations debacle for RAND. In 1961 Kahn departed to start his own policy analysis organization in New York, the Hudson Institute. There he became increasingly engaged in extrapolating political and social trends into preliminary sketches of possible future states of affairs—a line of thought soon labeled "futurology."[30]

The conclusiveness and authority of these sorts of publications were unclear. Critics feared that figures such as Wohlstetter and Kahn were exploiting a scientific authority that was grounded in suspect academic ideas to further suspect policy recommendations. The critics also feared that analysts' association with the RAND Corporation, a prominent Air Force contractor, gave their ideas undue influence in high circles. In February 1961, for example, Solly Zuckerman, who was then serving as the Chief Scientific Adviser to the Ministry of Defence, wrote to Blackett praising his RUSI lecture:

One can only hope that the derivative nonsense from Johnny von Neumann's games theory—a trend of thought which could only have been developed by someone utterly naive about human affairs—will dissipate itself quickly. The interesting thing is that more and more intelligent people in America are now kicking against the products of the Rand and [Oskar] "Morgenstern" views. ... [Thomas] Schelling and [American defense hawk and foreign policy insider Paul] Nitze unfortunately have their followers in high places here. Slow and tactful work will be necessary within official circles if the kind of thinking they are supposed to represent is revealed for what it is. Unfortunately Nitze firmly believes in the silly sums which have produced concepts like "counter force," "first strike," "second strike," etc.

Zuckerman subsequently packaged these accusations into a January 1962 article he wrote for *Foreign Affairs* entitled "Judgment and Control in Modern Warfare."[31]

No doubt the prospect that America's defense policy was being set through the application of academic ideas seemed quite credible at that moment. Publicly hawkish scientists, in particular Edward Teller, were known to exercise an influence from the government's defense laboratories. In his 1961 farewell address warning of the influence of the "military-industrial complex," President Dwight Eisenhower specifically warned that "public policy could itself become the captive of a scientific-technological elite." President John Kennedy, in turn, appointed Robert McNamara Secretary of Defense with the promise of rationalizing defense spending. While McNamara's aim was to curb the influence of the arms industry and the ambitions of individual services in setting national defense policy, it was easy to interpret his intervention as a deepening of scientific influence. This was particularly the case since he drew much of his staff, including Charles Hitch, from the RAND Corporation, which was rapidly gaining its reputation as a center for the work of defense intellectuals. When Stanley Kubrick took the concept of a Doomsday Machine from Kahn's *On Thermonuclear War* and portrayed it as an initiative that government leaders might consider, he was drawing on the plausibility of the ideas that Kahn's public musings were representative of RAND's private advice, and that this advice could lead directly to government policy.[32]

Needless to say, this was not how people like Wohlstetter and Kahn viewed their public writing. Wohlstetter exhibited a particular talent for criticism in his appreciation of how the polemics used against him and his colleagues in strategic theory and policy analysis exploited ambiguities haunting discussions of "science" and its relationship to politics, so as to privilege arbitrarily some scientists' arguments over others'. In a 1963 article in *Foreign Affairs* entitled "Scientists, Seers and Strategy," he acidly observed how figures such as Blackett and Snow looked to scientists' special "foresight" to enlighten policymaking, while simply dismissing the arguments of scientists with whom they disagreed. He noted how in *Science and Government*, "Lindemann is the dark angel, sadistic and violent, without the gift of foresight," and how "Blackett, a passionate battler against the forces of darkness, uses the story in his innumerable present feuds." In a 1964 article entitled "Sin and Games in America," published in an esoteric collection of articles on game theory, he further observed how associating analysts' work with the theory functioned as a means of identifying their ideas as unrealistic and irresponsible, even though few of these analysts actually made any use of the theory. He argued that not only did these accusations skirt the need to grapple with the substance of the arguments they were criticizing, but they also routinely misrepresented the work that game theorists actually did.[33]

To Wohlstetter, the polemics his opponents used only served, as he put it in "Scientists, Seers and Strategy," to increase the "supply of blunt weapons for the factional quarrels and feuds among scientists."[34] In "Sin and Games in America," he observed,

Many of these comments exhale a faint but distinct odor of invidious comparison: the British against the Yanks, the older against the "new" military writers, World War II operational researchers against contemporary strategists and systems analysts, military against civilians, practical men against professors, followers of the heart against those relying on their heads. There are differences, of course, between the members of each of these pairs. However, they are hardly the differences between Truth and Error, or Virtue and Sin.[35]

He felt that the tragedy of such comparisons was that they increased the distance between disputants, making pointed policy debate impossible. By releasing them from the necessity of engaging each others' arguments in their detail, it led to "bias" in scientists' advocacy for one or another technology or policy. Thus, ironically, the debates over the legitimacy of a particular scientist's contributions to a debate tended "to discourage the use of the patient and tentative method of science, as distinct from the *authority* of science, in assisting the cardinal choices [in policy] of which Snow speaks." For him this constituted "a rather surprising anti-rationalism" from someone who spoke of scientists' special foresight.[36]

Wohlstetter's response to the criticisms leveled against him and his analyst colleagues reflected their sense of their role in the public sphere. They were, generally, very cognizant of the fact that their contributions were speculative. They justified their participation on the grounds that existing arguments implicitly harbored speculative assumptions, which they felt compelled to question. Accordingly, they portrayed themselves as lending subtlety, intellectual diversity, and rigor to complacent discourses, and prided themselves on their own willingness to answer objections. Although they often did take strong positions on issues, as a rule they did not portray their positions as definitively correct. They also emphasized their willingness to question views from across the political spectrum. In "Scientists, Seers and Strategy," for instance, Wohlstetter took pains to criticize both Edward Teller's arguments against a nuclear test ban, as well as physicist Hans Bethe's arguments in favor of it.[37] At the same time, though, they did not suppose themselves to be apolitical. At the Hudson Institute, Herman Kahn became a particularly trenchant critic of what he viewed as the moralistic certitude underlying New Left intellectuals' diagnosis of, and recommended remedies for, various political and social ills.[38]

In a sense, when policy analysts entered the public sphere, they viewed themselves as abiding by the rules that applied to theorists. Because public arguments would never result directly in policy decisions, they understood their main role to be to question the presuppositions, logic, and completeness of others' arguments. The implication of arguing in this way is that the superior contribution is the one that is more thoroughly worked out and defended. This, however, upsets the ordinary order of much public political discourse, wherein the objective is to make a compelling, but not necessarily a fully rigorous argument. It is thus not surprising that analysts were frequently accused of partaking in a kind of rationalist sophistry, wherein sophisticated arguments are made, but ultimately with the effect of concealing rather than revealing the truth. Much as a theorist would object to the imposition of an axiom or a logical argument that could not be amended into a body of theory, so the analysts objected to suggestions that their detailed arguments were somehow illegitimate, while their opponents evidently felt they were exempted from the burden of answering objections. Beginning with his 1962 volume, *Thinking about the Unthinkable*, and especially its essay "In Defense of Thinking," Kahn proved particularly keen to defend his speculative work by reference to the benefits to be reaped by not arbitrarily limiting the acceptable content of public policy debate.[39]

Ultimately, the traditional rules governing advising and theorization did not apply in the realm of public discourse. There, argumentation may well persuade, but arguments coexist with polemics, which themselves often take the form of arguments that an opponent's arguments are simply not worth entertaining. It is perfectly possible to accuse opponents of being ignorant Colonel Blimps, but if one appeals to the sophistication of one's own arguments, there is nothing to prevent one from being accused of being a rationalizing Dr. Strangelove—and it is even possible that one's accusers would have a point. If rigorous academic standards can help arguments to engage with and build on each other, and if legally appointed authorities are the final judge of whether or not their advisers add something to their deliberations, argumentation in the public sphere devolves more easily into a state of nature: nasty, brutish, and often all too prolonged.

30 Conclusion

Some days we are convinced no one is listening; on others, we are, like the ants riding downstream on a log, convinced we are steering. We are doubtless wrong both times; for our prestige usually ensures us a hearing, but our views do not dominate in decision-making. However, it does seem that a lot of things go, eventually, the way we pull, whether it is a field of mathematics, of economics, or of operations analysis; a flood of computers, of ICBMs, or of satellites; a command-and-control philosophy, a civil defense concept, or some other facet of a war-fighting capability. We cannot prove that anything happens that would not happen anyway, but things may at least happen sooner when we start the discussion early, at a level of discourse that requires attention and thought by others. Moreover, it is inevitable that we do influence the future—when Merlin tells Guinevere that a dark, handsome stranger will enter her life, that changes the probabilities; he does not have to invent Lancelot too.

John Williams, "An Overview of RAND," 1962[1]

There are, in effect, two incompatible histories of the sciences of policy. The first casts the history as one of a momentous encounter between science and policy. Within this history the ability of science to exert an appropriate influence is contingent on the ability of inhabitants of the science domain and the policy domain to overcome their cultural and intellectual differences. The second history is one of more incremental change, in which a number of distinct but interrelated activities found, or failed to find, places within an institutional and intellectual environment committed to improvement. In this environment, distinctions between things that were "science" and things that were not had little practical significance. This book aims to move our understanding of the sciences of policy away from the first picture and toward the second. At the same time, it is important to realize that if the first picture is misleading, it is not completely misplaced. The intricate intertwining of the various strands of the sciences of policy owed a great deal to various proponents' common, if vague, interest in the application of "science" to policy. Moreover, these proponents' advocacy drew upon, and did much to fuel, an influential discourse in which the *proper* application of "science" to policy was viewed as a problem of profound historical importance with systematic implications for the quality of future governance.

The importance of deftly navigating the promises and false promises of "science," "knowledge," "reason," or "rationality" occupies a strikingly important place within our rhetorical inheritance. The idea found early expressions in the stories of Prometheus, Icarus, the oracles, and, later, Faust and Frankenstein, and it informs what we take to be most significant about the historical periods of the Renaissance and the Enlightenment. In the twentieth century, the idea that scientists introduced something peculiar, even mystical-seeming, into others' affairs was still very common. General Frederick Pile referred to Patrick Blackett as his "Magician," and, in his history of World War II, Winston Churchill famously referred to its more technical aspects as the "Wizard War."[2] In 1962 the RAND Corporation's John Williams referred to RAND as playing the Merlin to the military's Guinevere. Around that time, related sobriquets began to be used routinely to convey skepticism and hostility, as in reference to Robert McNamara's "Whiz Kids," or, in the wake of the horrors of Vietnam, journalist David Halberstam's sardonic label, "the best and the brightest."[3] It has become a common trope used in historical accounts, such as in the titles of journalist Fred Kaplan's book *The Wizards of Armageddon*, Thomas Hughes's *Rescuing Prometheus*, and Bruce Kuklick's *Blind Oracles*. In the conclusion to this book, I would like to look a little closer at why it is that, in a world replete with ideas and schemes of various stripes, the attribution of epistemological peculiarity to certain ideas and schemes has proven so rhetorically influential. I will then end by discussing how this rhetorical tradition was reinforced rather than overturned by subsequent historians of the sciences of policy, and of science and technology more generally.

"The Author Is the Real Hero": Narrative and Historiography

In 1972 Henry Tizard's son, Peter Tizard, wrote to an aging A. V. Hill, thanking him for sending him a copy of one of his father's speeches. He took the opportunity to complain that nobody had "done real justice to my father in writing about him," and that all attempts to do so had "been heavily and unfortunately influenced by that deplorable ... book by Snow in which the author is the real hero!"[4] The complaint might seem odd, given the extent to which C. P. Snow had lionized Henry Tizard in *Science and Government*. In fact, it was an astute point. By portraying Tizard's conflicts with Lindemann as an important tragedy, the book established a powerful narrative of Tizard's life as one of stifled potential. Tizard had, of course, played a heroic role in the development and implementation of radar prior to the Battle of Britain. Yet, Snow clearly implied that the truly heroic role remained to be played by those who could effect a more stable and enduring partnership between science and government.

It is important to recognize that the cogency of Snow's narrative rested completely on the assumptions, spelled out in *The Two Cultures*, that Britain was a fundamentally troubled nation, and that much of its trouble could be traced to its inability to master the "scientific revolution." The point of *Science and Government* was to emphasize that simply embracing science was not a sufficient remedy. Snow's exegesis on Tizard's successes in integrating science into the government machinery implied that the principles underlying those successes remained poorly understood and had to be recovered from history. This narrative ploy swiftly catalyzed the growth of an incipient historiography of twentieth-century British science and technology policy. In 1962 Patrick Blackett published his Tizard memorial lecture, his writings on OR, and much of his nuclear punditry into a volume entitled *Studies of War, Nuclear and Conventional*. The same year, the writer Ronald Clark published *The Rise of the Boffins*, narrating the development of a science-oriented bureaucracy in the British government before and during World War II. In 1965 he published a full biography of Tizard. Also in 1965 the radical science journalist J. G. Crowther included Tizard and Lindemann as the most recent figures in his *Statesmen of Science*. In 1967 he published *Science in Modern Society*, which began with a discussion of wartime developments, including OR, bringing the story up to the mid-1960s. Sociologist Hilary Rose and biologist Steven Rose's 1969 book *Science and Society* covered similar ground.[5]

As we have seen, J. D. Bernal had argued as early as 1941 that the wartime organization of science held important lessons for the future, and further developed this point in his 1945 "Lessons of the War for Science" lecture. When Snow adopted this idea in *Science and Government*, he transformed the exhilaration and optimism of Bernal's story of the war into one of frustrated potential, which created plenty of room for the ambitions of 1960s-era reformers. But it was left to Crowther and the Roses to use Snow's narrative structure to develop a fully realized Bernalist account of the social relations of science in mid-twentieth-century Britain. Just as Snow portrayed Tizard's accomplishments as hard-won, tenuous, and frustrated, thus did Crowther and the Roses depict the progress of British scientists and policymakers in understanding and taking advantage of Bernal's ideas about the social relations of science.

Beyond their contributions to the historiography of twentieth-century science-state relations in Britain, Crowther and the Roses also implicated the broader historiography of science and technology in society's ongoing failure to establish a sound relationship with those forces. They were particularly concerned that historians would inappropriately privilege the academic perspective of science as a "pure" form of intellectual pursuit. Crowther, for instance, worried that the focus of a new generation of historians on a "disembodied history of scientific ideas" would license the idea that "science exists without any obligations to society." He believed such history reflected "a long-range natural protective action, by dominant interests that do not wish to have the social and political implications of their scientific policy comprehensively investigated."[6] The Roses similarly resisted histories depicting science as proceeding "in a more or less ordered manner irrespective of the prevailing social environment in which it is performed." But, where Crowther was anxious about future historiography, the Roses looked forward to new portraits that would dispel the sense of "inevitability" and "neutrality" surrounding science and technology, which would, in turn, encourage researchers, engineers, and policymakers to work together more responsibly than they previously had.[7]

In the long run, Crowther's and the Roses' critique of the historiography of science would exercise an indirect but crucial influence on the historiography of the sciences of policy. The path of this influence led through a generation of historians appearing in the 1970s and 1980s who saw themselves as inaugurating an era of a newly sophisticated history of science and technology founded in large part on sociological insights descended from Bernal's work. The 1985 book *Leviathan and the Air-Pump*, written by Steven Shapin and Simon Schaffer, is often cited as the cornerstone of this new historiography. And, indeed, we find within the book not simply an insightful reading of the practical and intellectual contours of seventeenth-century experimental and political philosophy, but also a bold argument that a naive understanding of science and its relationship to society dates all the way back to that moment. According to Shapin and Schaffer, the rise of experimental science was made possible by neglecting the fact that experimental conclusions, like political ones, depend on agreements concerning the validity of the evidence used to justify them. By "the late twentieth century," they claimed, the failure to confront that reality had led to a crisis of confidence in science, paving the way for a reconsideration of the nature of science and its relationship to politics and society. They wrote, "Neither our scientific knowledge, nor the constitution of our society, nor traditional statements about the connections between our society and our knowledge are taken for granted any longer. As we come to recognize the conventional and artifactual status of our forms of knowing, we put ourselves in a position to realize that it is ourselves and not reality that is responsible for what we know." The implication, of course, was that their rather esoteric study of certain aspects of Robert Boyle's and Thomas Hobbes's work had broad importance because those who read it would find themselves able to discuss contemporary science with a maturity and poise that others still lacked.[8]

Shapin and Schaffer's rather audacious claim might have been taken as mere flourish if not for the fact that that claim's implied history of a centuries-long naiveté concerning science-society relations was becoming a convenient way of justifying the expansion of a new genre of critical studies of science. In 1988 Bruno Latour, one of the key authors in this genre, specifically cited Shapin and Schaffer's book as the history of how "nature" and "culture" had come to be regarded as separate, thus marking the ascendancy of "modern" thought, which he aimed to replace with a new metaphysics of knowledge.[9] In her landmark 1990 book *The Fifth Branch: Science Advisers as Policymakers*, Sheila Jasanoff similarly invoked a history of naiveté to lend her work a sense of cogency. According to her, in the twentieth century, as science became more prominent in society, society had become increasingly trapped between "democratic" and "technocratic" models of political authority, leading to a widespread "oscillation between deference and skepticism toward experts." She argued that, fortunately, the "sociology of science" had "in recent years" exposed "the contingent and relativistic character of knowledge." Thus, she could claim, the "notion that the scientific component of decisionmaking can be separated from the political and entrusted to independent experts has effectively been dismantled."[10]

The genre of work that built up around figures such as Shapin, Schaffer, Latour, and Jasanoff was, in many ways, a continuation of the critique of Bernal, Snow, Crowther, and the Roses. Yet it had also changed in important ways. In the 1960s it had been common to use a widespread failure to understand science-society relations as a way of explaining apparent political failures to take full advantage of science and technology. From the 1990s, it was much more common to use a similar failure of understanding to explain an apparent widespread overenthusiasm for science and technology as means of solving social problems.[11] Historians' treatments of the sciences of policy changed accordingly. The crucial problem was no longer to diagnose why operational research and figures such as Tizard struggled to find influence; it was to establish what sort of political and intellectual culture could possibly hope to produce a "science" of policy. To locate such aspirations, attention almost instinctually turned from postwar Britain to Cold War America. This new literature's analytical approach borrowed heavily, albeit largely unknowingly, from the criticisms of Patrick Blackett, Solly Zuckerman, and others by casting American developments in the sciences of policy as the product of a culture that did not so much understand science as fetishize it. This portrait was congenial to historians' sense of their own heroic role as diagnosticians of persistent flaws in science-society relations. Unfortunately, it also created a forceful and misleading picture of the sciences of policy on the western side of the Atlantic Ocean.

Historiography and the Ideology of Science in Policy

Our problem must be stated as if it were closed, so that it can be solved, and yet its elements must contain within themselves the possibility of fitting into a larger model. The objective function must reflect the fact that the problem will have implications for the future. ... We must, in short, build the possibility of learning into our decision models. ... This kind of view of the world may seem somewhat disturbing to some. One likes the idea that there is a finite solution to problems— that they all can be solved with expenditure of sufficient mental energy. But the whole history of mathematics and science is a disproof of any such simple notion. And personally, I now find it rather attractive that the world is an open one, and that there will always be new problems to solve. At no stage can we rest content. As Goethe says: "Who ever strives, him can we save."

Kenneth Arrow, "Decision Theory and Operations Research," 1957

The Rand thinkers inhabited a closed world of their own making, one in which calculations and abstractions mattered more than experiences and observations, since so few of the latter even existed to be applied. Jonathan Schell notes that the phrase "think tank," of which Rand is indisputably the paradigmatic case, evokes "a hermetic world of thought" that "exactly reflects the circumstances of those thinkers whose job it is to deduce from pure theory, without the lessons of experience, what might happen if nuclear hostilities broke out."

Paul Edwards, *The Closed World*, 1996

If historians and sociologists of science and technology are to be our guide, scientists and engineers who involved themselves in political affairs were psychologically characterized by a deep confidence that their skills could be applied to social and political questions as easily as to natural ones. Sociologically, policymakers and the broader public trusted scientists and engineers to solve pressing problems because they were convinced that those professions' methods were reliable and objective.[12] Because, the story goes, this ideology put such great faith in the integrity and capability of a peculiarly scientific methodology, those who purported to employ scientific methods worked feverishly to demarcate their perspective from other, less authoritative ones.[13] This demarcation, in turn, insulated scientific activities from political accountability, leading to serial controversies and policy failures, and, ultimately, to destructive crises of confidence in science and technology. For this reason, historians and sociologists have been eager to remind their audiences that attempts to give epistemological priority to certain perspectives are socially unstable, and mainly serve to conceal the artificiality, tenuousness, and cultural biases of expert claims to authority.

This basic narrative and analytical strategy informs the great majority of professional historians' recent treatments of the sciences of policy. For the historian Paul Edwards, in his 1996 book *The Closed World*, epistemological and technical authority in the Cold War era rested in instruments, including the sciences of policy and technological systems, that could enclose their subject in intellectual and engineering frameworks, thus rendering them knowable and controllable. But, Edwards claimed, this "closed world discourse" proved vulnerable to the encroaching open-endedness of a "green world."[14] Similarly, the eminent historian of technology Thomas Hughes argued in his 1998 book *Rescuing Prometheus* that the sciences of policy were paired with large-scale systems engineering in the Cold War era as part of an influential ideology of technology. Ultimately, he supposed, the shortcomings of this ideology had to be solved by introducing "postmodern" social and political concerns into otherwise purely technocratic managerial and engineering orthodoxies.[15]

Since the year 2000, a scientistic or rationalistic ideology has been systematically read into a variety of specific historical episodes associated with the sciences of policy. In Martin Collins's generally very useful history of the early years of the RAND Corporation, *Cold War Laboratory*, he argues that RAND initially adhered to a pre-New Deal "associationalist" ideology of voluntary state-industrial cooperation. However, in adopting systems analysis, it shifted quickly to a mistaken belief that "the authority of knowledge, legitimated through scientific method" could provide "the means to overcome the pull and counterpull of American pluralism."[16] Historian of economics Philip Mirowski traces agent-based neoclassical economics models, which he disdains, to formalistic models of decision favored by American operations researchers. He claims, "The crusading aura of the operations researcher was based on the assumption that he would bring the tonic of quantification to fields and questions which had previously been sorely wanting in that dimension; and as such, he helped to foster the mistaken belief that quantification and precision measurement were methodologically stabilizing influences bestowing scientific credibility in their own right, entirely removed from the larger social dimension of science."[17] Cowles Commission economists, he argues, were "bedazzled and in awe of the nascent computer," and, therefore, unthinkingly adopted OR-type optimization tools into their theories.[18] In fact, Mirowski takes this ideological mode of explanation to be so powerful that OR could even be regarded as "the anvil upon which the postwar relationship between scientists and the American state was hammered out," and that "the blade was then turned to carve out a new model of society that could be amenable to the rapprochement of science and the military."[19]

This sort of attribution of extraordinary sociological authority to formalism and computers has continued to exercise a strong grip on historians. In his preface to the 2012 volume *Cold War Social Science*, the prominent historian of science Theodore Porter claims that for social scientists of the Cold War era, it was "almost axiomatic that the more abstract, theoretically or quantitatively rigorous form of science is inherently the more powerful." Echoing Crowther and the Roses almost a half century earlier, he supposes that social scientists' "preoccupation with neutral objectivity," exhibited in their emphasis on the "technical tools of science," constituted an "ideological obfuscation," which made it plausible to claim that social science could "stand outside the value-laden character of the processes and interests that shaped the production and uses of social knowledge."[20] Lorraine Daston, Michael Gordin, and the other authors of the 2013 essay collection *How Reason Almost Lost Its Mind* claim that people associated with the sciences of policy adhered to a strict rule-based concept of "rationality," which, fueled by Cold War largesse, somehow failed to grasp a centuries-old concept of "reason" connoting the judgment that governs how rules are to be applied.[21]

While popular as bugbears, formalism, precision, and digital computation have not been absolutely indispensable to the ideology that historians have constructed. In her 2005 book, *The Worlds of Herman Kahn*, Sharon Ghamari-Tabrizi does much to illuminate RAND analysts', and particularly Kahn's, full awareness of the uncertainties, informalities, and inconclusiveness of policy analysis. Yet, she still requires a scientistic ideology in order to explain how such analyses could exist, and, of course, to distance herself from them. "For myself," she writes, "I cannot place my hope optimistically in the sciences underwriting war, whether this is the science that extrapolates the weapons effects of possible wars, the social sciences that shape foreign policy and invent, play, and analyze war games and forecasts, the fantastic imaginary of threat assessment, or science as cosmic metaphysics—all of which found a merry devotee in Herman Kahn."[22] She traces the rise of this scientism to the war, and to Patrick Blackett, whose "smugness," she claims, "was intolerable to the services." But, according to her narrative, "Even in an age of wizards' war, martial prowess still presided over the military, however scientifically enhanced it might be."[23] It was only in the milieu of Cold War America that the ideology of scientism could take full root.

A historiography that regards itself as a harmonizing force in science-society relations is heavily incentivized to depict the history of those relations as plagued by ideologies that have brought them systematically into discord. This was as true of the historiography of British science-state relations in the 1960s as it is of today's historiography of "Cold War" science. What now needs to be recovered from the historical record are the ideas that actually did govern those relations, through their successes, their failures, and their day-to-day activities. The task will be an ongoing one. In this book, I have emphasized the intellectual distinctions that historical actors drew between the theoretical investigation of concepts and logical structures, the creation of practical tools, and the analysis of policies and engineering designs. I have emphasized the patterns of responsibilities that people doing different kinds of work understood themselves to have toward each other, such as theorists' responsibility to other theorists, engineers' responsibility to the users of their creations, and practical analysts' responsibility to further the goals of decision makers. I have also emphasized practical analysts' recognition of experienced decision makers' tacit knowledge, and their appreciation that rational actions did not have to flow from explicit, rational thoughts. Finally, I have emphasized various individuals' attempts to reinforce these important contours of their thought by building appropriate institutional structures.

This reconstituted picture of historical ideas upends some fundamental tenets of historians' prevailing one. Most notably, it places comparatively little emphasis on historical ascriptions of significance to the concept of "science." In my estimation, scientific methodology, particularly highly formalized methodology, was neither strenuously resisted nor unthinkingly trusted by established decision makers or the general public. Although scientific figures sometimes invoked the broad concept of science, they also, in many cases, possessed a more refined sense of how their work functioned intellectually within organizational deliberations. In historians' accounts, the intellectual relationship between decision makers and scientific figures is one of conflict, negotiation, and blind trust rather than collaboration. However, I would urge that decision makers' main difficulties involved intelligently assessing various kinds of expertise to locate useful expertise, and incorporating that expertise constructively into their deliberations. While decision makers typically could not comprehend expert work in all its detail, they could develop some sense of its validity and its ability to aid them in their own work. Of course, this was as true of using expertise in, for instance, geographically localized problems or legally technical areas as it was in scientifically technical areas.[24]

While it is true that outside scientific experts were sometimes described as "objective," this objectivity was not taken to arise from any transcendental neutrality in the methodology they employed, but from their lack of an institutional stake in one or another outcome of a deliberation. It was well understood that the viability of an analytical organization's place within an institution was contingent upon the institution's leadership's commitment to accepting honest advice. Moreover, it was well accepted that analytical organizations working within or for an institution could not be absolutely relied upon to give objective information concerning that institution or its competitors to outsiders. Higher-level institutions, such as the U.S. Department of Defense, that were charged with mediating between lower-level institutions, such as the individual U.S. military services, were understood to require their own analytical organizations. Similarly, no analytical organization within an institution had the intellectual authority to justify that institution's actions as somehow transcendentally correct, or of optimal value to the general public, at least without risking serious criticism if journalistic or public perception could convincingly claim otherwise. Of course, this was not usually an issue, since organizations as a rule—but certainly not always—kept internal expert advice private, so as not to embarrass the institution if it acted contrary to that advice.[25] The burden was on the institutions to act in the public interest, and to seek out and employ expert advice to help them do so. The burden was on journalists and the public to judge whether institutions were, in fact, acting as they claimed, and whether they were employing the best available and most appropriate experts to meet their expressed goals.

Of course, as historian Brian Balogh has emphasized, as the twentieth century progressed, public groups increasingly began to employ their own grassroots experts to aid their assessments and advocacy. As time has passed, previously marginalized groups have gained access to additional journalistic and expert resources, though perhaps not all that they require. But the overarching point of my argument is that the employment of appropriate expertise has always been regarded as important to furthering the public interest, and it has never been regarded as a straightforward matter to locate and constructively exploit such expertise. There was never any moment in history when an idealistic conception of science, technology, and their authority clearly prevailed among scientists, decision makers, or the public at large.[26] The crises in confidence in science and expertise identified by so many authors in science studies are, I would suggest, misleadingly broad readings of more specific disruptions in institutional legitimacy.

The title of this book, *Rational Action*, is not an endorsement of the rationality of the sciences of policy, but neither is it a sardonic condemnation. It is a description of a project that developed an astounding array of intellectual and institutional facets, which leaned both on each other and on existing ideas and institutions, in an attempt to find ways to act more rationally. At least some of the paths that people working on this project traveled, such as the more abstruse arms of decision theory and economics, may seem very foreign to us. We may not agree with the actions that some of them recommended or furthered, particularly where military policies were concerned. We might find comfort in supposing that such work derived from some altogether different rationality from the one by which we suppose we ourselves abide. I hope that this book has shown that, in fact, we share many of our ideas with those who pursued this project; that, for all the extraordinary changes that the sciences of policy wrought on the landscape of expertise, the basic idea of what it means to act rationally has remained essentially constant through them. Any impulse we might have to find the missing element in their ideas about rationality, rendering them fatally flawed or archaic, is the same impulse that produced many of their most successful accomplishments. If we hope to surpass or supplant their ideas, we must first honestly come to terms with those ideas' diversity and reach, and then, in one way or another, continue the project on which they worked. This is the striving that Kenneth Arrow recognized in Goethe's Faust. And all those who believe that learning and argument can lead to a better world should concede that they, too, hope to find in that striving a chance for salvation.

Notes

Introduction

1. Bernal 1945/1975.

2. Weaver 1945; on the Society for Freedom in Science, see McGucken 1984.

3. Warren Weaver, "Comments on a General Theory of Air Warfare," AMP Note No. 27, January 1946, LOC ELB, Box 43, Folder 5; also available in TNA: PRO AIR 52/106.

4. On operational/operations research, readers should also consult Kirby 2003b; Gass and Assad 2005; and Assad and Gass 2011 for a general picture encompassing Britain, the United States, and, to a very limited extent, the broader world. Works on more specific topics will be cited at appropriate points in the text.

5. On the discourse surrounding the idea of "science" at that time, see especially Hollinger 1995; Edgerton 1996a; Ortolano 2009; and Desmarais 2010.

6. On wartime activities as representing contributions of "science," see especially Crowther and Whiddington 1947; and Thiesmeyer and Burchard 1947.

7. Edgerton 1996b, 1996c, and 2006.

8. Rosenhead 1989 argues that OR proponents missed an opportunity to realize a more radical accomplishment in the immediate postwar period. Kirby 2000 suggests that OR did not spread in Britain due to the undeveloped state of British management. Kirby 2003b envelops the history of OR into a broader history of science and the British state, which includes the establishment of the Ministry of Technology. Kirby 2006 and Kirby 2007 pay close attention to related British debates over whether OR should be delimited to a particular specialty.

9. "An Uneasy Alliance" was the title of Solly Zuckerman's Lees Knowles Lecture, presented in 1965 and published in Zuckerman 1967.

10. Mumford 1934, 1967/1970; Horkheimer 1947; Horkheimer and Adorno 1947; Ellul 1954/1964; Marcuse 1964; Habermas 1968, which was translated and

incorporated into Habermas 1970. For a sympathetic overview of this discourse, see Winner 1977.

11. On America's enthusiasm for technology, see especially Noble 1977 and Hughes 1989; on liberalism, see Matusow 1984.

12. President Dwight Eisenhower's 1961 farewell speech highlighting the influence of the "military-industrial complex" on research, policy, and economic life in America, and of a "scientific elite" on national policy was extremely influential. An early exposition on the theme was Cook 1962. The prominence of the "scientific elite" was, however, challenged, as in Greenberg 1965. Halberstam 1972 highlighted intellectuals' influence on the Vietnam War, and was influential in placing a focus on intellectualism as a source of faulty American defense and foreign policy. A historiography of intellectuals and nuclear arms policy emerged in the 1980s, which includes Kaplan 1983; Herken 1985; Waring 1995; May 1998; Ghamari-Tabrizi 2005; and Kuklick 2007. However, an early overview, Freedman 1981 did not single out intellectuals, and Rosenberg 1986 argued explicitly against overemphasizing their influence. Some intellectuals have captured particular attention. The role of Walt Rostow is addressed in Milne 2008. The budgeting reforms of Secretary of Defense Robert McNamara and his direction of the Vietnam War are frequently cited as part of a general intellectualist influence, but there are few works directly on the subject. For an insightful history and historiographical overview of McNamara's reforms, see Young 2009. A general American enthusiasm for technology has been implicated in American war making in Adas 2006.

13. In the 1960s, there was a major debate about the relations between social scientific research and foreign policy, with a particular emphasis on Project Camelot, an abortive project to study insurgency movements. On Project Camelot, see Lowe 1966 and Horowitz 1967, and, for a broader view of relations between social science and the state, see Lyons 1969. For more recent historiography, see Simpson 1994; Herman 1995, 1998; Solovey 2001. Relations between social science and war have been traced in Robin 2001 and Rohde 2013. Social scientists' theories of "modernization" have recently been recounted and critiqued in Latham 2000, 2011; Gilman 2003; Engerman et al. 2003; Ekbladh 2010; and Shah 2011. Rostow (see note 12) usually features prominently in this history because of the links he drew between economic development and anti-socialism in Rostow 1960. On American intellectuals' interest in Russian modernization, see Engerman 2003. Engerman 2009 is an examination of the rise of Russian and Soviet studies, which emphasizes the pre–World War II origins of area studies. Krige 2006 connects American science policy, including its support for operations research, to a hegemonic political strategy in Western Europe.

14. On links between defense and domestic policy studies, see Jardini 1996 and Light 2003.

15. The instrumentality of modern American middle-class life was a theme developed in influential works of the time, notably Mills 1951 and Whyte 1956. Waring 1991; Amadae 2003; and Adam Curtis's documentary film, *The Trap* 2007, all take the sciences of policy to be a product of, and a crucial support to, this culture of instrumentalism. Similar stories concerning the relationships between technology, rationalizing science (manifested particularly in "cybernetics"), and political and cultural ideology have been told about communist and socialist nations; see Kotkin 1995; Josephson 1996; Gerovitch 2002; Caldwell 2003; Augustine 2007; and Medina 2011. Also see Spufford 2010, a novel. For leavening on this theme, see Kojevnikov 2008.

16. In discussions of United States science policy, anxiety over Soviet progress, rather than scientific enthusiasm and triumphalism, is often cited as the crucial driver of government action; for recent discussions, see Kaiser 2002; Wang 2008; and Wolfe 2013. On the impacts of ideology on the content of science, Forman 1987 is the classic argument, focusing on the case of physics. Edwards 1996 influentially linked the engineering of Cold War defense systems, the sciences of policy, artificial intelligence, and cognitive science together as participating in a "closed world" discourse. The literature on nuclear arms policy, cited in note 12, typically links social science with strategic theory, but, for additional links between social science, the sciences of policy, and the Cold War more generally, see Poundstone 1992; Pickering 1995; Mirowski 2002; Amadae 2003; chapters in Solovey and Cravens 2012; most chapters in Erickson et al. 2013; and Cohen-Cole 2014. Other literature on strategic theory, game theory, economics, and finance, which does not prioritize the Cold War context, includes essays in Weintraub 1992 (particularly O'Rand 1992, Riker 1992, and Rider 1992); Crowther-Heyck 2005; Mehrling 2005; Erickson 2006; MacKenzie 2006; and Leonard 2010. The problem of evaluating the role of the "Cold War" as a crucial context has recently come into open debate; see especially Isaac 2007; Engerman 2010; and essays in Isaac and Bell 2012 (especially Mandler 2012). For counter-argument, see Mirowski 2012 in the same volume.

17. Edgerton 1991, xv–xvi, suggests the idea of "inverted Whig" history, which emphasizes serial failure and absence of progress.

18. Collini 2006 notes that British intellectuals have habitually presented themselves as absent or inefficacious in the history of their country, which rhetorically creates a space for future contributions. Similarly, Edgerton 2005 and 2006 develop the concept of "anti-history." Edgerton argues that historians of British science and technology have been anti-historians in that they have made the history of their subject one of absence and failure in order to establish a need for their own (usually rather mundane) calls for reform.

19. Gieryn 1983 influentially suggested that demarcating a boundary of what is science, and what is not, is an important ideological means of commanding intellectual authority. While I would not deny that such demarcationist rhetoric has

some power, and has often been employed, I believe that it is important not to overstate its historical importance.

20. I have previously outlined the importance of not just institutional, but also intellectual relations between scientists and military officers in Thomas 2007.

21. Bureau of Labor Statistics, U.S. Department of Labor, *Occupational Outlook Handbook*, 2014–2015 Edition, available at http://www.bls.gov/ooh/.

22. I have previously outlined this argument in Thomas 2012.

Chapter 1

1. On advances in gunnery before and during World War I, see Sumida 1989, 1996; Mindell 2002; Brooks 2005; and David 2009.

2. On Darwin and his company see Cattermole and Wolfe 1987.

3. On Darwin, Hill, the Mirror Position Finder, and the state of gunnery in World War I, see David 2009, chapters 4 and 5; and, for context, Hogg 1978. Also see Hill's unpublished history of his group, "The Anti-Aircraft Experimental Section of the Munitions Inventions Dep't (1916–1918)," CAC AVHL I 1/37, 1918.

4. War Office 1924–1925, 252–255.

5. Hill, "Anti-Aircraft Experimental Section"; War Office 1924–1925.

6. War Office 1924–1925.

7. Hill, "Anti-Aircraft Experimental Section."

8. See, for instance, Munitions Inventions Department, Gunsights and Rangefinders Committee, Meeting No. 40 minutes, 12/11/1918, CAC AVHL I 1/24; and Cattermole and Wolfe 1987, 127–129.

Chapter 2

1. See Mandler 2006, chapter 5, on interwar anxieties over and portrayals of "English national character."

2. Low 1937/1991.

Chapter 3

1. For an overview of scientists' criticisms, see Crook 1994. The most recent entry in a vast literature on the bombing campaign in Europe is Overy 2013.

2. Sherry 1987, to take a prominent example, regards strategic bombing as the product of "technological fanaticism."

3. Blackett 1960/1962, followed by Snow 1961, focused on disputes between long-standing government adviser Henry Tizard and Winston Churchill's friend and adviser, Oxford physicist Frederick Lindemann, Viscount Cherwell. The issue quickly made its way into the lore of British science-state relations, and thus its historiography; see Clark 1965, and the broader overview Rose and Rose 1969. Blackett 1960/1962 also implicated the Bomber Command Operational Research Section in disputes over bombing strategy, a point later echoed in Zuckerman 1978. Zuckerman had clashed with the section in his efforts to aid Air Marshal Arthur Tedder in diverting planes from Bomber Command to tactical bombing in support of the invasion of Normandy. Prominent physicist Freeman Dyson was a junior scientist in the section, and argued in Dyson 1979 and Dyson 2006 that working on bombing tactics was dispiriting and often futile. Wakelam 2009 challenges such claims of its futility.

4. Harris 1947, 35–36.

5. Minute from Squadron Leader, Tr. 2., to Senior Air Staff Officer, Bomber Command, 11/24/1936, TNA: PRO AIR 14/1.

6. Minute from Commander-in-Chief to Senior Air Staff Officer, Bomber Command, 3/10/1938, and subsequent minutes, TNA: PRO AIR 14/1.

7. Minute from Ops.2 to G/C Air, 7/21/1939; "Programme of Trials for the Bomber Development Unit," 8/19/1939; and "Bomber Development Unit" from Senior Air Staff Officer, Bomber Command, to the Officer Commanding Bomber Development Unit, 11/14/1940; all from TNA: PRO AIR 14/190. Commander-in-Chief, Bomber Command to the Under Secretary of State, Air Ministry, 3/26/1941, TNA: PRO AIR 14/682.

8. Ludlow-Hewitt to Group Leaders, 1/5/1940; and Senior Air Staff Officer to Group Headquarters, "Ventilation of Ideas," 1/8/1940; both TNA: PRO AIR 14/101.

9. Harris 1947, 35–36.

10. Deputy Chief of Air Staff to Commander-in-Chief, Bomber Command, 2/15/1942, TNA: PRO AIR 14/682.

11. "Submarine and Anti-submarine," October 1942, TNA: PRO AIR 15/155.

12. On LeMay, see LeMay 1965; Coffey 1986; and Kozak 2009.

13. Zimmerman 1988 discusses LeMay's command style from the perspective of his chief operations analyst at the Strategic Air Command. During World War II, Carroll Zimmerman led the operations analysis section at the Ninth Air Force from 1943 to 1945; see United States Army Air Forces 1948.

14. LOC, CEL, Box B6, Critique, 11/8/1944.

15. LOC, CEL, Box B6, Critique, 11/26/1944. On LeMay's copy of the critique transcript the last line is underlined in pencil.

16. XXI Bomber Command Tactical Doctrine, 3/12/1945, LOC CEL, Box B13, emphasis added.

17. LOC, CEL, Box B6, Critique, 11/26/1944. On heaters, see Critique, 2/12/1945.

18. LOC, CEL, Box B6, Critique 10/22/1944.

19. On statistical staff, see Hay 1994. Operations analysts, at least occasionally, sat in on LeMay's critiques; see LOC, CEL, Box B6, Critique, 11/8/1944. Sherry 1987 properly casts LeMay as a kind of technical expert alongside his staff of statistical analysts and operations analysts.

20. LOC, CEL, Box B6, Critique 11/8/1944.

Chapter 4

1. Holley 1953 and 2004 offer a military perspective on doctrine development.

2. For a brief biography, see Cornford 1946. Additional fragments of information, including the facts that he was "a bachelor, peculiar, [and] used to drink heavily," can be found in Warren Weaver, Diary of Chairman, Section 2 of Division D, 11/27/1942, NARA RG 227, Entry 153, Box 13, "Project No. 7—Columbia University SRG-C (2)" folder. Cunningham's single publication was Cunningham and Hynd 1946, a discussion of some of his wartime work.

3. Director of Armament Development to Chief Superintendent, Royal Aircraft Establishment, 11/9/1937, TNA: PRO AIR 13/879. The original paper is unavailable, but see a subsequent, longer draft of the paper: Cunningham, "The Theory of Machine Gun Combat: An Amplified Introduction" dated 2/3/1938, TNA: PRO AIR 13/879. For the differences between Cunningham's theory and prior theories, see Cunningham, "The Mathematical Theory of Combat Applied to Tanks," AWA Report No. 24, May 1941, TNA: PRO AIR 20/12944.

4. In World War I, Lockspeiser had been affiliated with the Anti-Aircraft Experimental Section; see David 2009, appendix 4. He would go on to have a distinguished career in the civil service, and be elected a Fellow of the Royal Society in 1949; see Edwards 1994.

5. The problem of "duels" would fuel much of the early postwar research in game theory; see Erickson 2006 and Leonard 2010.

6. Memorandum from Lockspeiser to Dr. Roxbee Cox, 11/19/1937, TNA: PRO AIR 13/879.

7. Cunningham, "Theory of Machine Gun Combat," 1 and 13.

8. Deputy Director of Research and Development (Armament) to Superintendent of Scientific Research, Royal Aircraft Establishment, 5/2/1938, TNA: PRO AIR 13/879.

9. B. G. Dickins to Chief Superintendent, Royal Aircraft Establishment, 6/10/1938, TNA: PRO AIR 13/879.

10. D. L. Webb to Air Officer Commanding No. 25 (Armament) Group, 6/13/1938; and attached "The Statement of Statistical Problems," TNA: PRO AIR 13/879.

11. Lockspeiser to Under Secretary of State, 6/22/1938; and W. J. Richards to Under Secretary of State, 6/27/1938; both TNA: PRO AIR 13/879.

12. Cunningham to Lockspeiser, 11/27/1938, TNA: PRO AIR 13/879.

13. G. W. H. Stevens to Cunningham, 12/6/1938, TNA: PRO AIR 13/879.

14. Cunningham, "The Mathematical Theory of Combat," AWAS Note No. 1, TNA: PRO AIR 20/12820; Cunningham, "Report on the Influence of Numerical Superiority in Air Combat as Distinct from the Total Fire Power on Either Side," TNA: PRO AIR 20/371.

15. See Cunningham, "Theorems of Multiple Combat," draft, TNA: PRO AIR 13/879; and Cunningham, "Mathematical Theory of Combat Applied to Tanks."

16. Cunningham to Stevens, 9/13/1939, AIR 13/879; Air Ministry 1963, 39.

17. AWA Report No. BC/13, "Bomb Census Analysis, Periodical Report No. 12," October 1941, TNA: PRO AIR 14/3639.

18. Cunningham, "Note on the Correlation of Bombing Probability Calculations with Plans for Bombing Operations," AWA Paper No. 3, n.d. [1940], TNA: PRO AIR 20/371.

19. See, for instance, "Analysis of Anti-Aircraft Gunfire from Data Contained in the A. A. Intelligence Proforma," AWAS Report No. 16, 12/12/1940, TNA: PRO AIR 20/12936.

20. Cunningham, "Note on the Theory of Aerial Gunnery," AWAS Report No. 48, May 1943, TNA: PRO AIR 20/12712.

21. Cornford 1946.

Chapter 5

1. On fire-control technology through World War II, see Mindell 2002.

2. "Notes on Conference to Discuss A. A. Gunnery," 8/14/1940, CAC AVHL I 2/3 Pt. I.

3. On Hill, see Katz 1978.

4. "Notes on Conference"; on Blackett's work at the RAE, see letter from Patrick Blackett to Henry Tizard, 9/16/1939, IWM HTT 32. On Blackett and his work, see Lovell 1975 and Nye 2004.

5. The best account of this work is L. E. Bayliss, "The Origins of Operational Research in the Army," Army Operational Research Group Memorandum No. 615, 10/11/1945, TNA: PRO WO 291/887 (copies are also available in RS PMSB, PB/4/7/2/8; and in CT HPR 14.9). For additional details, see I. Evans, "The Beginnings of Operational Research," RS PMSB, PB/4/7/2/16.

6. Bayliss, "Origins."

7. See ADRDE (ORG), "Memorandum on Operational Research Sections, Ministry of Supply," June 1942, IWM HTT 303.

8. In addition to Bayliss, "Origins," see "AORG Radar Section 1941–5," c. 1946, TNA: PRO WO 291/1288; and D. W. Ewer, "The History of O.R.S. 1(b). A Summary of the Analysis made of the Operational Performance of the H.A.A. Defences. 1940–1946," Operational Research Group (Weapons & Equipment), Report No. 328, 12/6/1946, TNA: PRO WO 2901/304.

Chapter 6

1. The secretary of DSIR was its head; the position of rector is equivalent to a university president.

2. On Tizard, see Clark 1965. The other initial members of the CSSAD were the Air Ministry's Director of Scientific Research Harry Wimperis; and the committee's secretary, A. P. Rowe, who was Wimperis's assistant.

3. See memo from HQ Bomber Command, 11/12/1939, TNA: PRO AIR 14/98.

4. On Tizard's wartime advisory career, see Clark 1965; on his mission to North America, see especially Zimmerman 1996.

5. Watson-Watt began hyphenating his name when he received a knighthood in 1942. Throughout the text, I will be referring to him with and without a hyphenated name depending on whether it is historically accurate.

6. See material in TNA: PRO AIR 2/3181 on Watson-Watt's appointment.

7. Minute by Joubert to Permanent Undersecretary, 10/4/1940, TNA: PRO AIR 2/8587.

8. See minutes and correspondence in TNA: PRO AIR 2/3181 and TNA: PRO AIR 2/8587. A. F. Wilkins, Watson Watt's longstanding assistant, would continue to serve that function as the ORO at Air Ministry HQ.

9. Stanmore Research Section continued to be administered by the TRE; see minute from Watson Watt, 1/7/1941, TNA: PRO AIR 2/8587.

10. On this committee, see especially McGucken 1984, chapters 7 and 8, and chapter 5 for more on Hill's advocacy.

11. Letter from Lord Hankey to Sir Archibald Sinclair, 1/6/1940 [sic, 1941], TNA: PRO AIR 19/148.

12. Letter from Lord Hankey to Sir Archibald Sinclair, 1/21/1941, TNA: PRO AIR 19/148.

13. Unsigned memorandum to Private Secretary to the Secretary of State, 1/24/1941; a reference to "my" operational research officers reveals that Watson Watt wrote it.

14. Memorandum from Robert Watson Watt to Private Secretary to the Secretary of State, 1/24/1941, TNA: PRO AIR 19/148.

15. Ibid.

16. When Blackett arrived at Coastal Command the Air Ministry did encourage the Admiralty to make use of his services; letter from Sir Archibald Sinclair to First Lord A. V. Alexander, 2/15/1941, TNA: PRO AIR 19/148.

17. See "Memorandum on Operational Research" (unsigned and undated) and Watson Watt to VCAS 5/23/1941, attached to Letter from Watson Watt to DCAS, 10/3/1941, TNA: PRO AIR 2/5352. Freeman had been an early supporter of Watson Watt's radar work, when, before the establishment of MAP, he was the Air Member [of the Air Council] for Research and Development. The Air Council was the main policymaking body of the Air Ministry.

18. I paraphrase liberally. Watson Watt's description of OR's "extended" uses in this memorandum is nearly incomprehensible—a characteristic trait of his writing.

19. On Tizard's and Freeman's approval, see Memorandum from Tizard to VCAS and CAS, "Operational Research," 7/17/1941, IWM HTT 302. On Jones's position, see Jones 1978. The Permanent Undersecretary was Arthur Street.

20. There is an intriguing annotation in pencil on Watson Watt's proposal to VCAS (see note 17) that we cannot connect with a specific author or time, but is worth quoting: "This seems to overlook entirely matters of operational practice, tactics and intelligence; confine the field to technical research. These sections are essentially OPERATIONAL research sections."

21. ORS Coastal Command (CC), Report No. 125, "Activity of Enemy Aircraft in the Western Approaches in May, 1941," TNA: PRO AIR 15/731. It is unclear why numbering began with 125.

22. Letter from Holt Smith to Prof. Williams, 7/7/1942, TNA: PRO AVIA 7/1005. Holt Smith was complaining that ORS CC Report No. 185, "Air Escort for Convoys" did not address questions related to ASV radar.

23. Tizard, "Operational Research" (see note 18). All quotes not otherwise cited in the subsequent four paragraphs are from this document.

24. Minute from Watson Watt to ACAS(I), 9/1/1941, TNA: PRO AIR 2/5352; and minute from Tizard to Freeman, 9/2/1941, HTT 302.

25. Letter from Rankine to Tizard, 7/21/1941, HTT 352.

26. Letter from Peirse to Under Secretary of State, 8/8/1941, TNA: PRO AIR 2/5352; and Minute from DCAS to Tizard, 8/12/1941, TNA: PRO AIR 2/5352.

27. "Dr. Roberts," from MAP's Director of Communication Development's office was culling information from wireless interceptions and radio activities. "Miss Goggin" of David Pye's staff at MAP was working under Bomber Command's Chief Armaments Officer; see minute from DCAS to Tizard, 8/12/1941, and minute from DSR to DCAS, 8/27/1941; TNA: PRO AIR 2/5352.

28. Minute from Tizard to DCAS, 8/21/1941, TNA: PRO AIR 2/5352.

29. Letter from J. D Bernal to C. P. Snow, 4/11/1961, CUL JDB, J.217. See also Crook 1994, 81–89.

30. A copy of Butt's report, dated 8/18/1941, is available in TNA: PRO AIR 14/1218.

31. Tizard apparently received this information from Bernal, who was performing a survey of German bombing for the Ministry of Home Security. See Letter from Tizard to Medhurst, 9/4/1941 (second of two), TNA: PRO AIR 2/5352.

32. Minutes of the 22nd Meeting of the Air Fighting Committee, 8/28/1941, TNA: PRO AIR 2/8653.

33. See responses by the Senior Air Staff Officer (SASO) at Bomber Command Air Vice-Marshal Saundby, dated 8/21 and 8/29/1941; and memorandum from John Baker to SASO, 9/5/1941; all from TNA: PRO AIR 14/1218.

34. Letter from ACAS(I) to Tizard, 8/30/1941. The ACAS(I) was Air Commodore C. E. H. Medhurst.

35. On the OR Centre and Committee, see Memoranda from DCAS, 9/20/1941; from DCAS to Joubert, 10/2/1941; and from David Pye to DCAS, 10/9/1941; all from TNA: PRO AIR 2/5352. The OR Committee held its first meeting on October 31, 1941 and typically met once every one or two months between then and April 1943. Minutes of the meetings can be found in TNA: PRO AIR 20/6172.

36. See note from Watson Watt to DCAS, 10/3/1941; and DCAS to Watson Watt, 10/6/1941; TNA: PRO AIR 2/5352.

37. The DCAS was Air Vice-Marshal Norman Bottomley. Watson Watt did not mention the episode in his memoirs, but claimed to have coined the terms "operational research" and "Operational Research Section"; see Watson-Watt 1957, 203. A. P. Rowe is generally accepted, following Williams 1968, as having coined "operational research" to describe work involved in implementing radar.

Chapter 7

1. Babbage 1830.

2. Morrell and Thackray 1982.

3. See Cardwell 1957; Haines 1969; MacLeod 1971; Turner 1980; and Alter 1987. On Imperial College, see Gay 2007. The "red brick" universities were the Victoria University of Manchester, and the Universities of Birmingham, Bristol, Leeds, Liverpool, and Sheffield.

4. See Clark 1962 and McGucken 1984. On the Development Commission, see Olby 1991; on the Agricultural Research Council, see Cooke 1981 and DeJager 1993.

5. See Werskey 1978 and McGucken 1984. On the British Science Guild, see MacLeod 1994. For a recollection of Tots and Quots, see Zuckerman 1978.

6. See Hall et al. 1935, and especially the chapter by Patrick Blackett, also entitled "The Frustration of Science." See also Crowther 1941 for a long-term historical view of these relations.

7. Bernal 1939, 97. On the Marxist critique in general, see Werskey 1978.

8. *Science in War* 1940; on its production, see Zuckerman 1978, 111–112 and 398–399.

9. Letter from Lovell to Blackett, 10/14/1939, IWM HTT 32. The book was Lovell 1939. On Blackett's politics, see Lovell 1975; Hore 2000; and Nye 2004. Blackett was a member of Tots and Quots.

10. For the best summary of Britain's radar effort, see Zimmerman 2001; on Tizard and his work on the committee and in the radar research effort, see Farren and Jones 1961; and Clark 1965.

11. The friction between Lindemann and the rest of the committee became a well-rehearsed episode following the publication of Snow 1961. Accounts appear in Birkenhead 1961; Clark 1965; Wilson 1995; and Fort 2003. See Zimmerman 2001, chapters 5, 7, and 9, for a recent account of prewar radar work.

12. On Tizard's proposal, see McGucken 1984, 168–169, and material in TNA: PRO CAB 21/711; quote from Edward Bridges, memorandum, 1/14/1939, in that file.

13. McGucken 1984, chapters 7 and 8, describes the history of Hill's proposal in exhaustive detail. Also see material in TNA: PRO CAB 21/711, 21/712, 21/829, and 21/1163. Quotes from Wing Commander William Elliot memorandum to Secretary, 10/2/1939; and Elliot to Secretary, 10/30/1939; both in TNA: PRO CAB 21/1163.

14. Tizard submitted his resignation to the Secretary of State for Air, Archibald Sinclair. Letter from Tizard to Sinclair, 6/19/1940, IWM HTT 58, reproduced in Clark 1965, 233–234.

15. Letter from Hill to Blackett, 8/12/1940, RS PB/9/1/52.

16. Letter from Hill to Henry Dale, 9/17/1941, IWM HTT 58.

17. Letter from Hill to Neville Chamberlain, 9/21/1940, TNA: PRO CAB 21/829.

18. When Tizard resigned as chair of the CSSAD, Hill resigned as well, and explicitly threatened to use this position to make his complaints public; letter from Hill to the Secretary of State for Air, 6/21/1940, HTT 58.

19. On the committee's activities, see McGucken 1984, chapter 8.

20. Alan Barlow to Chancellor of the Duchy of Lancaster, 9/18/1840, TNA: PRO CAB 21/829. Barlow was the son-in-law of Horace Darwin. Hankey agreed to chair the committee in good faith.

21. Letter from Hill to Tizard, 3/1/1942, IWM HTT 57.

22. Hill 1941 (*Science*), 582. The speech was to the Parliamentary and Scientific Committee.

23. Hill, Letter to the Editor of *The Times*, published 7/1/1942, copy in TNA: PRO CAB 127/218.

24. Hill 1942, 7, wrote, "The spirit is likely to be better where contact with outside science is the rule, where publication is normally permitted, where criticism and discussion are possible, where something more like the atmosphere of a university exists, and where a guiding and inspiring influence can be exercised, sometimes behind the scenes, by advisory bodies of experienced independent scientists." Hill apparently did not relish his brash parliamentary persona, but thought it necessary in order to achieve progress; see Katz 1978, 116.

25. The memorandum is reproduced in Blackett 1948b/1962.

26. Bernal 1942.

27. Hill 1942. Despite Bernal's claim, Hill did not mention OR specifically. He had mentioned it in passing as early as a March 1941 speech in Parliament; see Hansard HC Deb 11 March 1941, vol. 369, cols. 1226–1230. He would address the subject again in Parliament in February 1942; see "The War Situation," reprinted in Hill 1960, 288–295; see also Hansard HC Deb 24 February 1942, vol. 378, cols. 125–132.

28. Speech by Sir Henry Tizard at Parliamentary and Scientific Committee luncheon, 2/3/1942, IWM HTT 572.

Chapter 8

1. Minutes of the first meeting of the Operational Research Committee, 10/31/1941, TNA: PRO AIR 2/5352.

2. Minute from D of E, 12/14/1941; and Office Acquaint for Operational Research Staff from Henry Markham, 12/18/1941; and D of E to T. Padmore, 3/11/1942, TNA: PRO ADM 1/20113. Fowler had been a member of Hill's group in World War I; on Fowler, see Milne 1945.

3. On Schonland, see Austin 2001. On the appointment of Darwin and the succession of Ellis, see materials in TNA: PRO WO 32/10330. Darwin was the grandson of the naturalist Charles Darwin, and nephew of Horace Darwin.

4. Minutes of these meetings can be found in IWM HTT 298. For context, see Edgerton 2006, chapter 4, especially 163–164.

5. Minutes of the seventh informal meeting of "independent scientific advisers," 9/8/1942, IWM HTT 298.

6. Minute from CAOR, 10/5/1943, TNA: PRO ADM 1/20113.

7. Minute from D of E, 1/10/1944; Memorandum from CAOR to D of E, 4/20/1944; Minute from D of E, 5/30/1944; TNA: PRO ADM 1/20113. Blackett became Director of Naval Operational Research (DNOR).

8. See "Reconstitution of the Operational Research Group, Ministry of Supply," 1/26/1943, TNA: PRO WO 291/1286; "The Department of the Scientific Adviser to the Army Council," 9/11/1946, TNA: PRO WO 291/1300; and "Operational Research in the British Army, 1939–1945," October 1947, WO TNA: PRO 291/1301.

9. See Memorandum from Tizard to Air Chief Marshal Sir Charles Portal, 7/15/1943, TNA: PRO AIR 20/3411; Letter from Tizard to VCAS, 8/7/1943, TNA: PRO AIR 20/3145.

10. On Thomson's departure see M. J. Dean to VCAS, 10/19/1944, TNA: PRO AIR 30/3411.

11. Rau 1999; Morse 1977, chapter 7; and Tidman 1984, chapter 1.

12. Rau 2000.

13. McArthur 1990; Rau 1999; and Shrader 2006.

14. On the circumstances surrounding his appointment, see interview with Bowles, 7/14/1987, NASM, RAND Oral History Project, 41 and 45–50. On Bowles's office, see Kevles 1977, 309–312; Guerlac 1987, 699–704; Collins 2002, 17–26; and Allen V. Hazeltine, "A Summary of Activities: Office of Dr. Edward L. Bowles," 11/1/1945, LOC ELB, Box 43, Folder 3.

15. Hazeltine, "Summary," 6–18.

16. Ibid., 2–5 and 19–27.

17. Interview with Bowles, 7/14/1987, 59–60; Hazeltine, "Summary," 19, 60–68, and 155–161.

18. See especially a memorandum from Bowles to Henry Stimson, 8/24/1943, LOC ELB, Box 32, Folder 3; and Louis Ridenour, "Trip Report," 1/20/1943, LOC ELB, Box 39, Folder 6.

19. Diary of Warren Weaver, 4/20/1944, NARA RG 227, Applied Mathematics Panel: General Records, 1942–1946, Box 4, "Operational Research" folder.

20. See Guerlac 1987, 683–687 on the crash development program, and 697 on the utility of field liaison; on BBRL and the Advanced Service Base, see chapters 37, 39, and 40.

21. Letter from Bowles to Ridenour, 9/27/1944, LOC ELB, Box 30, Folder 8.

22. See Guerlac 1987, 870–873; Hazeltine, "Summary," 172; and John G. Trump, "A War Diary, 1944–5," entry for 11/14/1944, LOC ELB, Box 42, Folder 1.

23. See Thiesmeyer and Burchard 1947, esp. 47, 50–51, and 300–317; Rau 1999, chapter 6; Rau 2000; and "Final Report of Activities, Operational Research Section, Headquarters AFMIDPAC, 31 May 1944 to 2 September 1945," NARA RG 227, Office of Field Service: Records of the Operations Research Group, Pacific Ocean Area, 1944–1945, Box 331.

24. Trump, "War Diary," note 22, entry for 4/20/1945.

Chapter 9

1. Letter from Tizard to Zuckerman, 3/30/1943, IWM HTT 360. On Zuckerman's wartime activities, see Zuckerman 1978.

2. Letter from Zuckerman to Tizard, 5/22/1943, IWM HTT 311.

3. Blackett, "Scientists at the Operational Level" [1941], published in Blackett 1948b/1962.

4. "Planned Flying and Planned Maintenance in Coastal Command, Part V: The Approach to a System," 8/22/1945, TNA: PRO AIR 15/154; see also Rosenhead 1989.

5. "Instructions for ASWORG Members," 12/1/1942, NARA RG 38, Records of the Anti-Submarine Warfare Analysis and Statistical Section, Tenth Fleet, Series II, Administrative Files, Box 47.

6. Morse and Kimball 1946, 10. The statement reappeared in the published version, Morse and Kimball 1951, 10b.

7. Morse 1977, 183. The Chief of Naval Operations is the military head of the U.S. Navy.

Chapter 10

1. "Notes for an Operational Research Study of the Planning of Landings," 11/11/1943, TNA: PRO ADM 219/630.

2. Schoolteachers with degrees in mathematics and physics were sought out by RAF Fighter Command to replace personnel who had left to form newer OR groups; Air Ministry 1963, 11. Future Nobel Prize winners included Patrick Blackett, John Kendrew, William Shockley, and Maurice Wilkins.

3. "Summary of work of Coastal Command O.R.S.," n.d. [c. 1942], TNA: PRO AIR 14/763. Merton was the son of physicist Thomas Merton, who had done studies on camouflage paint for the Royal Aircraft Establishment; see Hartley and Gabor 1970 and Fort 2003. On Gordon, see Rosenhead 1989, and his original reports, "The Efficient Utilisation of Manpower Resources, With Special Reference to Maintenance Manpower," ORS Coastal Command Report No. 206, 11/15/1942; and "Final Report on the Experiment on Planned Flying and Maintenance with No. 502 Squadron: August–December 1942," ORS Coastal Command Report No. 223, 3/16/1943, both in TNA: PRO AIR 15/732.

4. "Operational Research Section Bomber Command, Present Allocation of Staff to Items under Investigation," 9/26/1942, TNA: PRO AIR 14/763.

5. "Study of Incendiary Attack on Dock and Storage Area, Hankow, China," XX Bomber Command, Operations Analysis Report No. 19, LOC CEL, Box B37, Folder 3.

6. "Street Fighting," Army Operational Research Group, Report No. 167, 1/2/1945, TNA: PRO WO 291/156.

7. "Carrier Aircraft in Battle for Leyte Gulf, 24–26 October 1944," Air Operations Research Group, Air Research Report #18, 4/10/1945, quote on 50; report courtesy of the CNA Corporation library.

8. See G. Research, "Problems for Operational Research Section," September 1944, TNA: PRO WO 203/716; and "Problems for No. 10 O.R.S.," October 1944, TNA: PRO WO 203/716, Appendix A. "Operational Research Reports from SEAC 1944–47," Army Operational Research Group Memorandum No. B5, January 1951, TNA: PRO WO 291/1199, reveals just as much variety.

9. "A Review of A. S. V. Performance," ORS Coastal Command Report No. 201, 10/14/1942, TNA: PRO AVIA 7/1005.

10. "Disappearing Contacts with Mk. II. A.S.V.," ORS Coastal Command Report No. 226, 4/10/1943, TNA: PRO AIR 15/732.

11. MacDougall 1951, 1978; Wilson 1995; and Fort 2003.

12. Hay 1994.

13. The first ZZ report not compiled by AWAS was "Report on the Analysis of ZZ forms for the period August 1st–November 30th, 1941," ADRDE (ORG) Report No. 43, December 1941, TNA: PRO WO 291/41.

14. Many such files are to be found in collections of reports from the ORS at RAF Coastal Command (TNA: PRO AIR 15/731–734), the Admiralty's OR reports (TNA: PRO ADM 219), and among files produced by the wartime U.S. Navy's Operations Research Section, provided to the author courtesy of the CNA Corporation library.

15. The most frequently cited wartime OR report is ORS Coastal Command Report No. 142, "Analysis of Attacks on U-boats by Aircraft," 9/11/1941, TNA: PRO AVIA 7/1004. See, for example, Goodeve 1948. Among other things, the report observed that although depth charges were less effective at shallow depths, their chances of hitting a target at those depths were much higher. But the report noted that others in the military had also made this suggestion. Blackett 1953/1962, 215–216, pointed out it was a restatement of military wisdom to aim at good targets and ignore poor ones.

16. See "Air Offensive against U-boats in Transit," ORS Coastal Command Report No. 204, 10/12/1942, Appendix I, TNA: PRO AIR 15/732. See also "Diary of Dr. Philip M. Morse and Dr. William Shockley," LOC ELB Box 39, Folder 5, entry for 11/24/1942.

17. "Disappearing Contacts with Mk. II. A.S.V.," ORS Coastal Command Report No. 226, 4/10/1943, TNA: PRO AIR 15/732. On the importance of scrutinizing data, see also section 8.3.2 of Morse and Kimball 1946. See also Morse 1977, 179–181, for similar problems encountered in the course of developing search theory.

18. See Blackett 1953, 211; unlabeled, unsigned note, 9/16/1940, TNA: PRO AIR 16/166; and "The Trend of Air Defence at Night," ORS Fighter Command Report No. 234, 8/24/1941, TNA: PRO AIR 16/1043.

19. "Intruders: A Statistical Report comparing the Relative Success of British and German Intrusive Efforts, with reference to British Defensive Success," ORS Fighter Command, Report No. 267, 11/8/1941, TNA: PRO AIR 16/1043.

20. "Intrusion. The Value of Intruders compared with Defensive Fighters for Night Defence," ORS Fighter Command, Report No. 326, 4/9/1942, TNA: PRO AIR 16/1043.

21. Llewellyn-Jones 2000 concludes that a celebrated Admiralty OR study on convoy size was not decisive; it merely lent "confidence" to arguments already in circulation. Waddington 1973 and Wakelam 2009 offer a similar sense of how the intellectual economy in which OR scientists participated worked.

22. Physicist Freeman Dyson's prevailing memory of working at RAF Bomber Command ORS as a young mathematician was one of futility; see Dyson 2006. Wakelam 2009 offers an alternative perspective.

Chapter 11

1. Blackett, "Methods of Operational Research" [1943], published in Blackett 1948b/1962.

2. Minutes of the first meeting of the Operational Research Committee, 10/31/1941, TNA: PRO AIR 2/5352.

3. "Preliminary Report on the Submarine Search Problem by the A/S/W Operations Research Group, Section C-4, N.D.R.C.," 5/1/1942, NARA, RG 38, Records of the Anti-Submarine Measures Division, Tenth Fleet, Administrative Files, Box 3.

4. On the early development of search theory, see Morse 1977, 177–181. See Thomas 2007 for discussion of the relevance of Morse's acoustical methodology.

5. "Digest of minutes, conference on anti-submarine warfare," Enclosure C, 8/10/1943, NARA, RG 38, Records of the Anti-Submarine Warfare Analysis and Statistical Section, Tenth Fleet, Series II, Administrative Files, Box 47, "ASM Conferences (2)" folder. See also Morse 1953, 163–164.

6. Leonard 2010, 270–275; and Morse and Kimball 1946, section 5.4.

7. "Digest of minutes."

8. Morse 1953, 163.

9. "Digest of minutes."

10. See Weaver 1970 and Rees 1987; on molecular biology, see Kohler 1976.

11. Mindell 2002, 276–282, quote on 280.

12. On the importance of bibliography, see, for instance, Applied Mathematics Panel Study No. 25 (AMG-C No. 172), NARA, RG 227, Applied Mathematics Panel: Studies and Notes 1943–1946, Box 6, Study 25, especially letter from Merrill Flood to Warren Weaver, 3/16/1943; letter from Warren Weaver to D. C. Lewis, 8/11/1943; and D. C. Lewis, "A Bibliography of AA Artillery," 5/19/1944.

13. Rau 1999, 238–239.

14. Warren Weaver, "Comments on a General Theory of Air Warfare," January 1946, 35, LOC ELB Box 43, Folder 5.

15. On the recruitment process, see NARA RG 227, Applied Mathematics Panel: General Records 1942–1946, Box 13, "Project No. 7—Columbia University SRG-C" folders 1 and 2; on Wolfowitz, see Zacks 2003.

16. Weaver, "General Theory," 35–37; and Statistical Research Group, Columbia University, "The Mathematical Theory of Combat," 8/26/1942, TNA: PRO AIR 52/97.

17. On the Applied Mathematics Panel, see Rees 1980; Wallis 1980; Rosser 1982; MacLane 1989; and Owens 1989.

18. See Wallis 1980, 329.

19. See Klein 2000.

20. Wallis 1980 and Klein 2000.

21. Diary of Warren Weaver, 12/7/1942; and letter from John Tate to Warren Weaver, 1/12/1943; both NARA RG 227, Applied Mathematics Panel: General Records 1942–1946, Box 10, "Division 6" folder.

22. Diary of Warren Weaver, 12/16/1942, NARA RG 227, Applied Mathematics Panel: General Records 1942–1946, Box 9, "Wilks, Samuel S." folder.

23. Diary of Warren Weaver, 12/21/1942, NARA RG 227, Applied Mathematics Panel: General Records, 1942–1946, Box 10, "Division 2" folder. The head of Division 2 was the MIT architect John Burchard.

24. See letter from Warren Weaver to Vannevar Bush, 2/8/1943, NARA RG 227, Applied Mathematics Panel: General Records 1942–1946, Box 4, "Operational Research" folder.

25. Letter from Warren Weaver to Vannevar Bush, 2/25/1943. Weaver was aware of Bush's opinions on OR at least as early as November 1942; see memorandum from MHT [M. H. Trytton] to WW, 11/6/1942. Both letters are located in NARA RG 227, Applied Mathematics Panel: General Records 1942–1946, Box 4, "Operational Research" folder.

26. Ibid.

27. "Diary of Executive Committee Meeting," 3/8/1943, NARA RG 227, Applied Mathematics Panel: General Records 1942–1946, Box 1, "AMP Meetings, 1943–1946" folder.

28. Letter from Warren Weaver to John Tate, 3/15/1943, NARA RG 227, Applied Mathematics Panel: General Records 1942–1946, Box 10, "Division 6" folder.

29. B. O. Koopman, "A Quantitative Aspect of Combat," 6/23/1943, NARA RG 227, Applied Mathematics Panel: General Records 1942–1946, Box 7, "Koopman, B. O." folder.

30. Letter from Philip Morse to Warren Weaver, 10/1/1943, NARA RG 227, Applied Mathematics Panel: General Records 1942–1946, Box 10, "Division 6" folder; letter from E. J. Moulton to Philip Morse, 10/6/1943, and letter from E. J. Moulton to J. F. Ritt, 12/3/1943, both in NARA RG 227, Applied Mathematics Panel: General Records 1942–1946, Box 7, "Koopman, B. O." folder. Lanchester 1916 was regarded as a predecessor to OR; see Blackett 1948b/1962, 198.

31. Operations Evaluation Group 1946 encapsulated wartime work on search; it was later published as Koopman 1980.

32. On these devices, see Green 1993, 2001.

33. Especially useful on the ongoing relationship between AMP and OR is "Excerpt from WW's AMP diary," 12/17/1943, NARA RG 227, Applied Mathematics Panel: General Records 1942–1946, Box 4, "Operational Research" folder.

Chapter 12

1. Diary of J. D. Williams, 2/13/1945, NARA RG 227, Box 1, "Williams, J. D. Diary" folder.

2. Letter from Edwin Paxson to Oswald Veblen, 2/15/1945, NARA RG 227, Applied Mathematics Panel: General Records 1942–1946, Box 8, "Paxson, E. W. Correspondence" folder.

3. Vaughn D. Bornet, "John Williams: A Personal Reminiscence (August, 1962)," 8/12/1969, RAND Document D-19036, 29, RAND, John Williams papers, Box 1.

4. Ibid.

5. Diary of J. D. Williams, 11/20/1943, NARA RG 227, Applied Mathematics Panel: General Records 1942–1946, Box 1, "Williams, J. D. Diary" folder.

6. Diary of J. D. Williams, 10/6/1943, NARA RG 227, Applied Mathematics Panel: General Records 1942–1946, Box 1, "Williams, J. D. Diary" folder.

7. Diary of J. D. Williams, 1/30/45 (Second Report) and 2/13/1945, NARA RG 227, Applied Mathematics Panel: General Records 1942–1946, Box 1, "Williams, J. D. Diary" folder. On Youden, see Eisenhart and Rosenblatt 1972, and Miser 1992.

8. Diary of J. D. Williams, 4/26/1944, NARA RG 227, Applied Mathematics Panel: General Records 1942–1946, Box 1, "Williams, J. D. Diary" folder.

9. See Paxson biography in RAND, Box 13 "Bob Specht's RAND biographies, M–R." On Paxson's activities at the Jam Handy Organization, see Rosser 1982; and Thiesmeyer and Burchard 1947, 241–242.

10. See memorandum from Bowles for General Arnold, 1/3/1944, LOC ELB, Box 30, Folder 5; Diary of Warren Weaver, 2/24/1944, NARA RG 227, Applied Mathematics Panel: General Records 1942–1946, Box 8, "Paxson, E. W. Correspondence" folder.

11. Diary of Warren Weaver, 2/24/1944, NARA RG 227, Applied Mathematics Panel: General Records 1942–1946, Box 8, "Paxson, E. W. Correspondence" folder.

12. Letter from Weaver to Paxson, 7/28/1944, NARA RG 227, Applied Mathematics Panel: General Records 1942–1946, Box 8, "Paxson, E. W. Correspondence" folder.

13. See memorandum from Paxson, 7/23/1944, "Paxson, E. W. Diary" folder; and letter from Paxson to Weaver, 8/15/1944, NARA RG 227, Applied Mathematics Panel: General Records 1942–1946, Box 8, "Paxson, E. W. Correspondence" folder. It is unclear whether, in the end, Gardner received any new personnel to replace Paxson.

14. Letter from Weaver to Edward W. Paxson [sic], 8/21/1944, NARA RG 227, Applied Mathematics Panel: General Records 1942–1946, Box 8, "Paxson, E. W. Correspondence" folder.

Chapter 13

1. On the B-29, see Wolf 2005.

2. Edward Bowles's note, 12/15/1971, LOC ELB, Box 42, Folder 4; interview with Bowles, 7/15/1987, NASM, RAND Oral History Project, 66; and Morse 1977, 187.

3. Letter from Bowles to Shockley, 1/4/1944, LOC ELB, Box 32, Folder 6. On Shockley's activities in Bowles's office, see Hazeltine, "Summary of Activities: Office of Dr. Edward L. Bowles," 11/1/1945, 84–87, LOC ELB, Box 43, Folder 3.

4. Memorandum from Shockley to Bowles, 2/3/1944, LOC ELB, Box 32, Folder 6; and Hazeltine, "Summary," 85–86.

5. Hazeltine, "Summary," 86. See also interview with Edward Bowles, NASM RAND Oral History, 65–66, 70–71; from a second interview on 8/20/1987, 92–93 and 112.

6. On Shockley and the transistor, see Riordan and Hoddeson 1997.

7. Memorandum from Shockley to Bowles, "Discussion of a Proposed Program on the Quantitative Aspects of Modern Warfare," 2/15/1945, LOC ELB, Box 31, Folder 1.

8. See Hazeltine, "Summary," 115–119, quote on 116.

9. The best concise history of the study is Warren Weaver, "Comments on a General Theory of Air Warfare," AMP Note No. 27, January 1946, 37–53. Copies are available at TNA: PRO AIR 52/106, and LOC ELB, Box 43, Folder 5. Also see Hazeltine, "Summary of Activities: Office of Dr. Edward L. Bowles," 11/1/1945, 107–109, LOC ELB, Box 43, Folder 3, for a perspective on the project from outside AMP. Owens 1989, 292–293, offers a brief overview.

10. Weaver, "Comments," 37.

11. Ibid., 40.

12. Ibid., 41–42.

13. "Discussion of Bombing Effectiveness at High and at Very High Altitudes," SRG-P No. 70, 10/28/1944, NARA RG 227, Box 19, "Study No. 128—Papers by J. D. Williams" folder. Muroc Army Air Field in California was later renamed Edwards Air Force Base.

14. Ibid., 41–42.

15. Ibid., 46.

16. Weaver listed numerous important results achieved under the AC-92 contract; see ibid., 48–49, 52–53.

Chapter 14

1. See Warren Weaver, "Comments on a General Theory of Air Warfare," AMP Note No. 27, January 1946, Appendix B, LOC ELB, Box 43, Folder 5. By April the plan appeared to be moving forward; see Diary of Warren Weaver, 4/10/1945, NARA RG 227, Applied Mathematics Panel: General Records, 1942–1946, Box 1, "WW—Diaries—General" folder.

2. Letter from Warren Weaver to members of AMP, 12/14/1945, NARA RG 227, Applied Mathematics Panel: Records of Certain Consultants and Aids to the Division, 1942–1946, Box 8, "Raisins in the Oatmeal" folder.

3. Discussions of the report have appeared in Collins 2002, 112 and 116–119; Ghamari-Tabrizi 2005, 174–175; and Erickson 2006, 111–116.

4. Weaver, "Comments." Weaver pointed out that his notion of military worth was akin to "economic utility," which, he observed, had recently been redefined in Von Neumann and Morgenstern 1944.

5. Ibid., 5.

6. Ibid., 12–16, quote on 15.

7. Ibid., 15.

8. Vaughn D. Bornet, "John Williams: A Personal Reminiscence (August, 1962)," 8/12/1969, RAND Document D-19036, 29, RAND, John Williams papers, Box 1.

9. Weaver, "Comments," 17–18, Weaver's emphasis.

10. Ibid., 16 and 33–34.

11. Letter from Paxson to Weaver, 12/19/1945, Paxson's emphasis, NARA RG 227, Applied Mathematics Panel: Records of Certain Consultants and Aids to the Division, 1942–1946, Box 8, "Raisins in the Oatmeal" folder.

Chapter 15

1. Official histories exist for OR in the U.S. Navy and Army; see Tidman 1984, and Shrader 2006, 2008, 2009. On "operations analysis" in the postwar Air Force, see Holley 1971 and LeRoy Brothers, "The Office of the Assistant for Operations Analysis, US Air Force," *The Second Tripartite Conference on Army Operations Research, Volume II: Technical Proceedings*, 23–27 October 1950, NARA RG 319, General Decimal File (Security Classified Correspondence, 1950–1955), 1950–1951, Box 40, (020, ORO). The only history for the British services is Pratt 1981, also available at TNA: PRO ADM 219/731. See also Kirby and Godwin 2010. A number of brief overviews of military OR work were also published in the 1950s and 1960s; see: Brothers 1954; Pennycuick 1954; Morton 1956; Engel 1960; Lowe 1960; Pugh 1960; Whitson 1960; and Gadsby 1965. On OR in the British Army, see also A. W. Ross, "An Introduction to the Army Operational Research Group," AORG Memorandum No. F. 10, August 1955, TNA: PRO WO 291/1437. On the Weapons Systems Evaluation Group and the Institute for Defense Analyses, see Ponturo 1979. In the 1950s, military OR began to be adopted in other nations' armed services, as well as in the forces assigned to the North Atlantic Treaty Organization; see Ross 1980 and Krige 2006, chapter 8.

2. "The Causes of Bombing Errors as determined from analysis of Eighth Air Force Combat Operations," Operations Analysis Report No. 3, 7/15/1947, quote on 13; NARA RG 341, Technical Operations Analysis Reports 1945–1957, Box 12.

3. "Aircraft losses due to U-boat flak 1943–1945," Research Branch Coastal Command Report No. 389, August 1951, TNA: PRO AIR 15/734.

4. "Effects on U-Boat Performance of Intelligence from Decryption of Allied Communication," OEG Study 533, 3/22/1954, copy provided by the CNA Corporation library.

5. Pratt 1981.

6. Wohlstetter 1964b, 208, made this point in response to criticisms that defense analysts' work was purely speculative.

7. J. E. Henderson, "Training in Air to Air Firing with the G.M.II. Fixed Ring Gun Sight and the Gyro Gun Sight, Film Assessment as the Method of Recording the Results from the Training Exercises," RAF Fighter Command Research Branch Report No. 680, 12/20/1946, TNA: PRO AIR 16/1045.

8. W. F. Druckenbrod et al., "An Operational Evaluation of the LACROSSE Guided Missile, the 155-mm Gun, and the 8-in. How (C)," Technical Memorandum ORO-T-312, September 1955, NARA RG 319, General Decimal File (Security Classified Correspondence, 1955), 1955, Box 18, (040, ORO).

9. "A Study of Defence in a Mining War against the United Kingdom," Admiralty DOR Report No. 15, August 1950, TNA: PRO ADM 219/406.

10. "The Choice of the Magnetic Safe Depths for Coastal and Inshore Minesweepers," DOR Report No. 18, August 1951, TNA: PRO ADM 219/409.

11. Norman C. Dahl, "A Method for Estimating the Radius of Blast Damage from High Explosive or Atomic Bombs," Technical Memorandum No. 12, 3/7/1949, NARA RG 341, Box 13; J. C. Mouzon, "The Effect of Thermal Radiation from the Super Bomb on the Bombing Aircraft," Working Paper No. 32, 4/15/1952, NARA RG 341, Box 11.

12. W. A. Taylor et al., "Vulnerability of the Infantry Rifle Company to the Effects of Atomic Weapons," Technical Memorandum ORO-T-1 (CONARC), CORG-FER-1, August 1955, NARA RG 319, 1955, Box 18, (040, ORO); Robert J. Best, "The Feasibility of Chemical Warfare in the Defense of a Perimeter in the Naktong Valley Basin," Technical Memorandum ORO-T-5 (FEC), 1/31/1951, NARA RG 319, ORO decimal files; "Redstone Missiles as Atomic Warhead Carriers," Technical Memorandum ORO-T-311 has been removed from its box—only the title is available. On ORO projects, see Shrader 2006, 98.

13. An overview of British and U.S. Army OR problems can be found in *Second Tripartite Conference*.

14. Martha A. Olson and Richard T. Sandborn, "Aircraft Attrition in Korea: An Analysis of MiG-15 Effectiveness," Operations Analysis Technical Memorandum No. 31, 2/11/1952, NARA RG 341, Box 13.

15. Discussions of the use of OR personnel in Korea can be found in Shrader 2006, 2008; Tidman 1984; and Pennycuick 1954. On the use of the social sciences in Korea and elsewhere, see Robin 2001 and Rohde 2013.

16. Alfred H. Hausrath et al., "The Utilization of Negro Manpower in the Army," Report ORO-R-11, August 1955, NARA, RG 319, General Decimal File {Security Classified Correspondence, 1955}, Box 24, (040, ORO). Hausrath 1954 summarized the conclusions. The study was later published as Johns Hopkins University Operations Research Office 1967. The subcontractors were the American Institute for Research, Inc.; International Public Opinion Research, Inc.; and the Bureau of Applied Social Science Research of Columbia University.

17. See, for example, Joseph A. Joseph, "The Application of Linear Programming to Weapons Selection and Target Analysis," Operations Analysis Technical Memorandum No. 42, 1/5/1954, NARA RG 341, Technical Operations Analysis Reports 1945–1957, Box 13.

18. See Hausrath 1971 and Ghamari-Tabrizi 2000. On gaming in Britain, see "Preliminary Note on the Use of Gaming in Operational Research," Admiralty DOR Memorandum No. 181, June 1959, TNA: PRO ADM 219/533.

19. The point is well illustrated by a memorandum that LeRoy Brothers wrote to the USAF's Assistant Chief of Staff for Operations concerning an Air University report urging a militant stance against the Soviet Union. Although Brothers found the report compelling, he argued against its distribution beyond the Air Force in part because it offered "strong conclusions allowing very little doubt," and he feared that unless the conclusions were represented as judgments and not facts, the report would "be subjected to very damaging criticism." Memorandum from LeRoy A. Brothers to Major General James E. Briggs, 6/16/1954, NARA RG 341, Vice Chief of Staff; Operations Analysis, 1949–1961, Box 3, "TS-1011" folder.

20. These statements are based on the examination of a large sample of British and American postwar OR documents contained in the U.S. and UK National Archives.

21. The U.S. Army kept files containing commentary on OR studies, from which this discussion is largely drawn. See, for example, a memorandum from John G. Hill to Deputy Assistant Chief of Staff (G-3), re: ORO Final Report ORO-R-5, 2/27/1953, NARA RG 319, General Decimal File {Security Classified Correspondence, 1950–1955}, 1953, Box 20, (040, ORO).

22. Letter from Ellis Johnson to Brigadier General Robert A. McClure, Chief of Psychological Warfare, 12/26/1951, NARA RG 319, General Decimal File {Security Classified Correspondence, 1950–1955}, 1950–1951, Box 40, (020, ORO).

23. See "The Professional Evaluation of ORO Publications," an ORO booklet, in JHU 02.001, Papers of the Office of the President, Series I, Box 34, "Jan. 1956–Dec 1956" folder; and L. D. Flory, "Analysis of the ORO Work Program with Respect to Timeliness," 11/1/1960, and Helen S. Milton, "ORO Publications: An Army-Wide Review, Summary of Questionnaire Responses of 85 Army Agencies," 11/21/1960, both in JHU 03.001, Papers of the Office of the Provost, Series I, Subseries 4, Box 28, "Associated Universities/ORO, ORO Publications, 1960–1962" folder. See Whitson 1960 for some published discussion of ORO practices.

24. Johnson, "African Study Project," nd. [c. 2/1958], JHU 02.001, Papers of the Office of the President, Series I, Box 34, "Jan–Dec 1958" folder; on Johnson's broader ambitions for OR, see Johnson 1958 and Johnson 1960.

25. Jean G. Taylor, "Report on the Operations Research Office Summer Training Program for High School Students," 11/2/1956, JHU 02.001, Papers of the Office of the President, Series I, Box 34, "Jan. 1956–Dec 1956" folder; "A-Attack to Kill 80%, 11 Boy Scientists Predict," *Washington Post* 12/28/1956, B1; the students' publication was Bentz et al. 1957.

26. See especially Shrader 2006, 120–122. On "Relax And Cooperate," see Smith 1966, 272. In an email to the author, Mark Eisner, a former ORO/RAC researcher recalls the joke as "Relax and Comply." Also see Johns Hopkins University's internal assessment of the situation, memorandum from P. Stewart Macaulay, provost, to Milton Eisenhower, president, 11/17/1960, JHU 03.001, Papers of the Office of the

Provost, Series I, Subseries 4, "Associated Universities/ORO Review of ORO-Army Relationships, 1960–1961" folder, Box 28. Johnson moved to the OR program at the Case Institute of Technology.

27. Steinhardt, "Providing for the Navy's long-term needs for operations research and naval warfare studies," memorandum for Deputy Chief of Naval Operations, 5/8/1959 MIT MC 75, Box 10, "Operations Evaluation Group, Applied Science Division."

28. See Tidman 1984, 204–205 and 220–221; and Shrader 2008, chapters 2 and 3.

Chapter 16

1. Edgerton 2006, chapter 6. On the importance of the Ministry of Supply in particular, see Edgerton 1992.

2. On Tizard's postwar career, see Clark 1965; on the DRPC, see Agar and Balmer 1998. On Zuckerman's experiences in postwar defense, see Zuckerman 1988. For additional detail on government decision making concerning defense priorities, see Oikonomou 2010.

3. The Office of Naval Research became the preeminent sponsor of military and academic research in the postwar period; see Schweber 1988 and Sapolsky 1990.

4. On the organization of R&D in the Air Force, see Komons 1966.

5. See autobiographical note attached much later to a letter from Dwight Eisenhower to Bowles, 8/27/1947, LOC ELB, Box 35, Folder 5. Bowles became a full-time consultant, and was also accepted back at MIT as a "consulting professor."

6. See Collins 2002, chapter 2; and letters from Bowles to Karl Compton, 12/20/1945 and 1/2/1946, MIT AC 4, Box 34, Folders 6 and 7.

7. Interview with Bowles, 7/15/1987, 70–71; 8/20/1987, 92–93, NASM, RAND Oral History Project.

8. Collins 2002, 40–54.

9. Smith 1966, 30.

10. For instance, to study the question of air defense, he urged that an organization similar to RAND be set up at MIT; see letter from Bowles to James Killian and attached memorandum to Major-General F. L. Ankenbrandt, 4/26/1948, MIT AC 4, Box 34, Folder 7.

11. Collins 2002, chapter 3 is the best source on this transition.

12. RAND was unofficially an acronym for "Research ANd Development," but the joke at the time was that was really an acronym for "Research And No Development."

13. For "blue sky," see Vaughn D. Bornet, "John Williams: A Personal Reminiscence (August, 1962)," 8/12/1969, RAND Document D-19036, RAND, John Williams papers, Box 1, 1.

14. On these figures, see Shrader 2006, 61. On the RDB in general, see Needell 2000, chapter 4.

15. On the policy council and the proposal to form WSEG, see also Needell 2000, 114–122.

16. "Fifth Meeting of the Scientific Advisors to the Policy Council," 10/20/1947; and see Lt. Col. Edwin Black to I. I. Rabi, 11/26/1947, and attached letter and memorandum drafts; all in NARA RG 330, Records Concerning Organization, Budget, and the Allocation of Research and Development, 1946–1953, Box 521, Folder 3.

17. Memorandum from James Forrestal to the Joint Chiefs of Staff, 2/4/1948, NARA RG 218, Central Decimal File, 1948–1950, Box 129, "334 (WSEG) (2-4-48) Establishment of Weapons Systems Evaluation Group Sec. 1" folder.

18. In the spring the RDB set up an ad hoc committee consisting of Loomis, Shockley, Purdue University president Frederick Hovde, with Lloyd Berkner as chair; see their final report, "Weapons Systems Evaluation: Report of the Ad Hoc Committee Appointed to Study the Problem of Weapons Systems Evaluation," 5/6/1948, LOC ELB, Box 45, Folder 5.

19. The ORO would later move to other locations in Maryland, and ultimately Virginia; see Whitson 1960, Shrader 2006, and Ceruzzi 2008.

20. Ponturo 1979 provides a general history; Morse 1977, 244–261, provides a personal recollection of WSEG's first year and a half.

21. Smith 1966; Jardini 1996; Hounshell 1997; and Collins 2002. Ghamari-Tabrizi 2005, chapter 2 is especially vivid concerning the intellectual atmosphere at RAND.

22. Ponturo 1979, 43, and "National Military Establishment Weapons Systems Evaluation Group, Preliminary Budget Estimate, Fiscal Year 1950," 10/1/1948, NARA RG 330, "Records Concerning Organization, Budget, and the Allocation of Research and Development, 1946–1953," Box 521, Folder 3. In initial planning, total expenditure for 1950 was projected to be $280,415. RAND was founded on a $10 million grant from the Air Force. When that grant ran out, RAND began receiving annual appropriations. In 1950, the second year RAND received appropriations, its expenditures were projected to be at $4 million; see sources in the previous note.

23. Morse 1977 and letter from Morse to Karl Compton, 2/7/1949, NARA RG 330, "Records Concerning Organization, Budget, and the Allocation of Research and Development, 1946–1953," Box 521, Folder 1. Compton was Bush's successor on the RDB.

24. Ponturo 1979, 118–119 and A-4; H. P. Robertson's motivations for taking the post are discussed in a letter to John von Neumann, 5/23/1950, LOC JvN, Box 19, Folder 8.

25. Declassified reports located in NARA RG 330 (some of which were only available in part) were unusually large in comparison to those of the other civilian analysis organizations.

26. Ponturo 1979, 120–127.

27. RAND Management Committee meeting minutes, 4/15/1952 and 4/22/1953, RAND, Box 32.

28. On IDA and related science policy developments, see Ponturo 1979, chapter 4; and Aaserud 1995.

29. MIT, the Case Institute of Technology, the California Institute of Technology, Tulane University, and Stanford University. This list soon expanded.

30. See Aaserud 1995 and Finkbeiner 2006.

31. The best information on IDA's activities is to be found in its annual reports, held by some libraries.

32. On Ramo-Wooldridge, see Hughes 1998, chapter 3; and Dyer 1998. On Stanford Research Institute's relationship with the sciences of policy, see Platt 1954; Gibson 1980 and 1986 constitute the official history of the organization.

33. Robin 2003 and Engerman 2009.

34. MITRE Corporation 1979; Baum 1981; Everett 1983; Klingaman 1993; Dyer and Dennis 1998; and Redmond and Smith 2000. See Ceruzzi 2008 on assorted military contactors in Northern Virginia.

35. Westwick 2003. On the postwar development of the atomic weapons laboratories, see Hewlett and Duncan 1972; and Hewlett and Holl 1989.

36. See Herken 1992; Greene 2007; and Wang 2008.

37. On McNamara, see Shapley 1993; on PPBS, see Young 2009.

Chapter 17

1. Bernal 1945/1975, 562–563.

2. Ibid., quotes on 557, 559, and 563–564.

3. Ibid., quotes on 565 and 570–574.

4. Easterfield 1983; Rosenhead 1989; and Kirby 2003b, chapter 6. Easterfield suggests that Patrick Blackett persuaded Labour politician Stafford Cripps to create the new body.

5. Association of Scientific Workers 1947, chapter 11.

6. Blackett, Presidential Address, 5/24/1947, RS PB/5/1/4/19. The winter of 1947 was unusually harsh, and created severe problems in Britain.

7. Ibid.

8. Zuckerman 1948; Watson-Watt 1948; and Wansbrough-Jones 1948.

9. On the planning of science controversies, see McGucken 1984, chapter 9.

10. "The Deployment of Scientific Effort in Britain" 1948.

11. Waddington 1948.

12. Rosenhead 1989, quotes on 19 and 20.

13. Brundrett, "The Place of Science in the Machinery of Government," 2/21/1951, TNA: PRO DEFE 19/24; see also lectures from 1955 and 1957 in TNA: PRO DEFE 19/2.

14. Tizard, "Science and the Machinery of Government," 11/24/1951, IWM HTT 611.

Chapter 18

1. Kirby 2003b contains extensive and complementary discussion of these events.

2. "The Organisation of the Scientific Department of the National Coal Board: Memorandum by the Scientific Member" n.d. (probably 3/18/1946), TNA: PRO COAL 33/21.

3. Tomlinson 1971 and Kirby 2003b, chapters 7 and 8. On Ellis, see Hutchison, Gray, and Massey 1981.

4. See Goodeve 1948. On Goodeve, see Richardson 1981 and Kirby 2011. On OR in the iron and steel industry, see Kirby 2003b, chapter 6.

5. Memorandum from G. Essame to Ellis, 1/28/1947, TNA: PRO COAL 33/21.

6. Charles Goodeve, "The Future of Co-operative Research in the Iron and Steel Industry," 2/18/1946, CAC GOEV 5/1.

7. See Goodeve 1948, quotes from 377 and 383; also see "Operational Research: A paper based on recent lectures given by Sir Charles Goodeve, F.R.S.," December 1947, CAC GOEV 5/1.

8. See Yates 1948. During the war Yates had been involved in assessing bombing damage in collaboration with Solly Zuckerman; see Zuckerman 1978, 209–210 and 242–243.

9. On the OR Club and its journal, also see Kirby 2003b, chapter 11. On the Club's view of its publication practices, see "Editorial Notes [1]" 1950. On the Club's view of its activities, see "Operational Research Club" 1950.

10. See Kirby 2003b, chapters 7–10, for many further details on these and other individuals and groups. On Smeed, see Dyson 2006 and Wakelam 2009. On Wansbrough-Jones, see Hicks and Smith 1983. Rivett was the head of the NCB's Field Investigation Group through the 1950s; see Mercer 2005 and Rand 2011. Stafford Beer was best known as a proponent of cybernetics; for his early efforts to link his broader interests to OR, see Beer 1954, 1959; on the scope of his vision, see Pickering 2009, chapter 6, and Medina 2011; for biography, see Rosenhead 2011. On Tippett, see Tippett 1950 and Daniels 1982. Smeed, Tippett, and Wansbrough-Jones were all editorial advisers to *OR Quarterly*.

11. Bell 1949a and b.

12. Smeed 1952.

13. "Editorial Notes [2]" 1950.

14. Blackett 1950/1962.

15. Kirby 2000 argues against the claims of Trefethen 1954 and Ackoff 1957a, that Britain experienced a rapid adoption of OR in industry relative to the United States. However, where he posits that Britain's adoption was slow because the country was not sufficiently professionalized for OR, I would urge that such explanations are unnecessary since British OR proponents limited their ambitions in view of their sense that OR described an array of existing professional practices in that country.

16. Williams 1951.

17. "Editorial Notes" 1953.

18. See Kirby 2003b, 382.

19. On Slater, see Kay 1971; on Kendall, see Stuart 1984; on Giffard, see Lindsay 2001.

20. Whether OR should be a profession would remain controversial; see Kirby 2006.

Chapter 19

1. Morse and Kimball 1946.

2. It was, for instance, used at Arthur D. Little consulting group as they were setting up their OR group; see Magee 2002. It was also a textbook used for an early OR course at MIT; see Operations Evaluation Group memorandum, "A Course on Operations Research," MIT MC 75, Box 9, "NRC—Comm on OR" folder. See also Morse 1977, 288–290.

3. See materials in LOC ELB, Box 52, folder 4, especially a letter from Wilks to Edward Bowles, 11/24/1947. Participants were Bowles, Jacinto Steinhardt, LeRoy Brothers, Merrill Flood, Roger Wilkerson, A. E. Brandt, E. S. Lamar, George Dantzig, Arthur A. Brown, Thomas I. Edwards, W. J. Youden, and G. Baley Price.

4. Letter from Wilks to Bowles, 3/15/1948, and attached proposal, LOC ELB, Box 35, Folder 8.

5. A letter from R. C. Gibbs to Detlev Bronk, then head of the NRC, 10/17/1949, refers to "two exploratory conferences" indicating that the 1948 meeting proposed by Wilks did take place; NAA COR, "PS: Com on Operations Research, 1949–1949, Beginning of Program" folder.

6. Biographical details from letter from Gibbs to Bronk, 10/17/1949. On Levinson's career in commercial research, see Levinson 1953.

7. "Minutes of the Meeting on Operations Research held at the National Research Council," June 22, 1949, MIT MC 75, Box 9, "National Research Council" folder.

8. See NAA COR, "PS: Com on Operations Research 1950–1951, Operations Research Center: Massachusetts Institute of Technology: Proposed" folder.

9. Morse 1950.

10. Morse and Kimball 1951. On extent of early sales and international translations, see Morse 1977, 290–291.

11. Levinson and Brown 1951.

12. "Operations Research for the Guidance of Management: Report of a Symposium," 9/24/1951; and "Proceedings of the Conference on Operations Research held at the University of Illinois," 9/27/1951, NAA COR, in folders of the same title. See also other conferences listed in memorandum from Levinson to members of the executive committee, 8/13/1951, NAA COR, "PS: Com on Operations Research 1950–1954, General" folder.

13. Committee on Operations Research, "Operations Research with Special Reference to Non-Military Applications," April 1951, NAA COR, "PS: Com on Operations Research 1951, Operations Research with Special Reference to Non-Military Applications" folder.

14. "Minutes of the Committee on Operations Research, March 23, 1951," 6/1/1951, NAA COR, "PS: Com on Operations Research 1951–1953, Meetings: Minutes."

15. "Proceedings of the Conference on Operations Research held at the University of Illinois, 27 September 1951."

16. Solow 1951; on the use of the article, see Memorandum from Levinson to members of the executive committee, 8/13/1951, NAA COR, "PS: Com on Operations Research 1950–1954, General" folder.

17. Camp et al. 1953.

18. Johnson 1954 contains an institutional overview of OR circa 1953. According to Johnson, "A reasonable estimate would appear to be that there are about 1,500 analysts now in the United States, about one-half of whom are engaged in industry. About one-third of the total are invested with the title of operations research workers and the remainder are engaged in it under different names." Of the 168 ORSA members who worked in industry in July 1953, 36 were employed by aircraft producers across 11 different companies.

19. The NRC committee had appointed a subcommittee consisting of Philip Morse and Jacinto Steinhardt to establish a professional society, and steps apparently were also taken outside of NRC auspices; see letter from Horace C. Levinson to R. C. Gibbs, 5/6/1953, NAA COR, "PS: Com on Operations Research 1950–1954, General" folder. The NRC committee was formally disbanded in 1955.

20. Morse 1953, 159.

Chapter 20

1. Wald 1947. On Wald, see Wolfowitz 1952.

2. Flood 1956b, 61; emphasis added. The actual number is $(n-1)!/2$, since backward routes are equivalent to forward routes.

3. Dantzig, Fulkerson, and Johnson 1954, 393; for further detail, see Hoffman and Wolfe 1985.

4. On the history of linear programming, see Dantzig 1982; Albers 1990; and Kjeldsen 2002.

5. This system, alternatively known as "program control," "program planning," and "program monitoring" is described in depth in Hay 1994, 99–112.

6. SCOOP stood for "Scientific Computation of Optimal Programs."

7. Koopmans's wartime problem was later published as Koopmans 1970. He had emigrated to the United States in 1940 with the aid of Merrill Flood, who also spurred his interest in the TSP; see "The Princeton Mathematics Community in the 1930s," transcript 11, 5/14/1984. http://www.princeton.edu/~mudd/finding_aids/mathoral/pmc11.htm.

8. Stigler 1945.

9. In addition to previously cited works on the history of linear programing, see Erickson 2006 and Leonard 2010.

10. Koopmans 1951.

11. See Kjeldsen 2000.

12. See Kjeldsen 2002; on the relationship between linear programming and game theory and the economic framework of Walrasian equilibrium, see Düppe and Weintraub 2014; for a more radical interpretation, see Mirowski 2002. Dorfman, Samuelson, and Solow 1958 provided an early explication of the links. On Kantorovich, see Gass and Rosenhead 2011; Kantorovich and his work are fictionalized in Spufford 2010.

13. On early practical applications of linear programming, see Gass 2002 and Cooper 2002.

14. Dantzig, Fulkerson, and Johnson 1954; and Flood 1956b, who cites Robinson's unpublished work.

15. Leonard 2010 offers the most up-to-date scholarship.

16. Dantzig 1963, 13.

Chapter 21

1. Whitin 1953, chapter 1.

2. Trundle 1936, cited by Whitin 1953.

3. Whitin 1953, 208.

4. Ibid., 110–114. See Keynes 1930, 116.

5. Whitin 1953, 115.

6. Ibid., 5.

7. Ibid., v.

8. See Cherrier 2010 for exposition on Marschak.

9. Arrow 2002, 2; the bibliographers are identified in the references section of Arrow, Harris, and Marschak 1951.

10. Arrow, Harris, and Marschak 1951, quotes on 252 and 255.

11. Ibid., quote on 257.

12. Arrow 2002, 2.

13. Arrow, Harris, and Marschak 1951.

14. Ibid.

15. Arrow 2002, 2.

16. Arrow dealt with a similar issue in Arrow, Blackwell, and Girshick 1949.

17. Dvoretzky, Kiefer, and Wolowitz 1952a, 1952b.

18. For discussion of dynamic programming, see Klein, forthcoming.

19. Bellman 1958. Bellman also suggested a third fork in which "the functional equation can be utilized to determine the structure of optimal policies as a function of the structure of the cost and penalty functions, and also of the probability distribution occurring." In 1960, Scarf would prove the optimality of the (s,S) form for a wide array of assumptions; Scarf 1960.

20. Arrow, Harris, and Marschak 1951, 251–252.

Chapter 22

1. Koopman 1952.

2. "Proceedings of the Conference on Operations Research held at the University of Illinois," 9/27/1951," NAA COR, in folder of the same title.

3. Karchere 1953.

4. Morse 1955.

5. Abbott 1988, 237–238, holds OR became inward looking and succumbed to a "professional regression." Mirowski 2002 argues that a theory-enamored "American" version of OR displaced a more prosaic "British" version. Hughes 1998 presumes OR was a technocratic enterprise, which only later was compelled to adopt "postmodern" concerns and become less theoretical. For "authority and prestige," see Fortun and Schweber 1993, 629.

6. Jardini 1996, 106.

7. Mood 1953.

8. I develop this point further in Thomas 2012.

Chapter 23

1. Warren Weaver, "Opening Remarks, Plenary Session, RAND Conference," Conference of Social Scientists, September 14–19, 1947, LOC ELB, Box 44, Folder 4.

2. See Kaplan 1983; Jardini 1996; Hounshell 1997; Collins 2002; Mirowski 2002, chapter 6; Amadae 2003, chapter 1; and Ghamari-Tabrizi 2005.

3. Kaplan 1983, 86–87; Collins 2002 quotes Kaplan on 164.

4. Collins 2002, 208, correctly identifies systems analysis with design engineering, but also takes it to be a means of transcending policy dispute, as on 163.

5. Merrill M. Flood, Memorandum No. 8, "Aerial Bombing Tactics, I. General Considerations," 10/26/1944, NARA RG 227, Entry 152, AMP Notes and Studies, Study 128, Box 19, "Princeton FCR Memorandas" folder; see also discussion in Leonard 2010, 281–284.

6. Paxson biography in RAND, Box 13, Bob Specht's RAND biographies, M–R.

7. NOTS Working Paper No. 10, Memorandum from Paxson to Dr. L. T. E. Thompson, "The Role of Mathematical Research in the Airborne Fire Control Development Program," 5/13/1946, RAND, Ed Paxson papers, Box 4.

8. Ibid.

9. Jardini 1996, 50–52, which quotes a letter from Paxson to von Neumann, 10/6/1946, RAND D-63.

10. Williams joined RAND in June 1946. See Vaughn D. Bornet, "John Williams: A Personal Reminiscence (August, 1962)," RAND Document D-19036, 8/12/1969, RAND, John D. Williams collection, Box 1, 1–3, 12, and 15, quote on 3.

11. Collins 2002, chapter 4, discusses this history in some depth.

12. Paxson became known as "The Systems Analyst"; see Kaplan 1983, 87.

13. See Collins 2002, chapter 5, and esp. 179–183 for a discussion of Paxson's systems analysis in light of ongoing disputes over the design of the B-52.

14. "Trip Report of E. W. Paxson, May 3 to 17, 1947," RAD-127, 7, RAND Corporation library.

15. E. W. Paxson, "Sixteen Hours of Discussion at Fort Leavenworth," D(L) 1115-PR, 12/15/1951, RAND Corporation library.

16. Ibid. This was a suggestion for the use of the "Delphi Method," which was developed by RAND analysts, particularly Olaf Helmer.

17. See "Trip Report of E. W. Paxson, May 3 to 17, 1947," RAD-127; and E. W. Paxson, "Second Order Theory of Bomber Attrition," RAD-150, 7/3/1947, RAND Corporation library.

18. See Edward Quade interview, 2/18/1988, NASM, RAND Oral History Project, 9–10; and Collins 2002, 19.

19. RAND would become an important player in the development of the postwar social sciences; see Robin 2001; Amadae 2003; Crowther-Heyck 2006; Engerman 2009; and Solovey 2013.

20. Bornet, "John Williams," 21.

21. Weaver, "Opening Remarks."

22. Herbert Goldhamer, "Human Factors in Systems Analysis," RM-388, 4/15/1950, RAND Corporation library.

23. Kaplan 1983, 88.

24. Paxson, "Second Order Theory."

25. Collins 2002, 200–201. Even with the new costs taken into account, the systems analysis still yielded a preference for turboprop engines.

26. Quade interview, 15. Contrast to Kaplan 1983, 89: "Air Force officers, almost all of whom were pilots, hated the study. They didn't care about systems analysis. They liked to fly airplanes. They wanted a bomber that could go highest, farthest, fastest." Collins 2002, chapter 5, provides a more nuanced view of Air Force perspectives; see esp. 204 and 210.

27. E. S. Quade, "The Proposed RAND Course in Systems Analysis," 12/15/1953, D-1991, 5–6, RAND Corporation library.

28. Albert Wohlstetter, who was instrumental in turning systems analysis away from Paxson's style, gave him substantial credit, noting this fact; interview with Albert Wohlstetter, 7/29/1987, 5, NASM, RAND Oral History Project.

Chapter 24

1. Armatte and Dahan Dalmedico 2004 is a preliminary attempt to come to terms with the varieties of theorization that flourished in the postwar era.

2. Dvoretzky, Kiefer, and Wolfowitz 1952a, 190.

3. On Girshick, see Blackwell and Bowker 1955; on the invitation to Arrow, see Arrow 2002. On Blackwell, see Albers 1985 and DeGroot 1986.

4. Arrow, Blackwell, and Girshick 1949; Wald and Wolfowitz 1948.

5. Albers 1985.

6. Arrow, Blackwell, and Girshick 1949.

7. Wald 1950.

8. Savage 1954; for "Bible," see Gass and Assad 2005, 100.

9. Blackwell and Girshick 1954; on Blackwell and Bayesian statistics, see DeGroot 1986; on Blackwell's historic election to the NAS, see Grimes 2010.

10. Arrow 1951; Arrow 2002 notes the connections between military worth and social choice; Amadae 2003 discusses the origins and influences of Arrow's work in some depth, while arguing that it constituted an intellectualized defense of the ideology of capitalist self-interest.

11. Arrow 1957, quotes on 766–767 and 769.

12. See Shubik 1987 and Shubik 2002 for a discussion of what he called the "high church," "low church," and "conversational" varieties of game theory.

13. See Mirowski 1992 and Leonard 2010, especially chapter 10.

14. See Erickson 2006, esp. 125–147; and Leonard 2010, 308–327.

15. Nash 1950, 1951.

16. See Nassar 1998; Giocoli 2004; and Erickson 2006, 154–160, 163–164, and 183–190. Mirowski 2002, 331–349, criticizes Nash's work as a harbinger of what he views as overly rationalistic economic models.

17. Williams 1954; Luce and Raiffa 1957; Schelling 1960. Poundstone 1992 provides the background history of the Prisoner's Dilemma. For perspectives on the absorption of game theory into the behavioral and social sciences, see O'Rand 1992; Riker 1992; Amadae 2003; and Erickson 2006, chapter 3.

18. In the context of finance theory, derived from OR and economics, MacKenzie 2006 describes this tension between description and prescription in terms of "performativity."

19. This issue surfaced with the "socialist calculation" debate of the 1930s; see Chaloupek 1990. On the epistemological evolution of economic modeling, see Morgan 2012.

20. Arrow and Debreu 1954; on the links between Morgenstern and the Austrian School, see Leonard 2010; the reformulation of Walrasian economics plays a crucial role in the critique of neoclassical economic theory found in Mirowski 2002; the moment is treated with new sensitivity in Düppe and Weintraub 2014.

21. Backhouse 2012; on MIT economics and the style of Samuelson, Solow, and their colleagues, see Weintraub 2014, particularly essays by Backhouse, Halsmayer, and Thomas.

22. Samuelson 1951, and other essays by Arrow, Koopmans, and Georgescu-Roegen in Koopmans 1951.

23. Arrow and Debreu 1954, 265.

24. This emphasis on heuristics might be contrasted with the supposition of Heyck 2012 that postwar theorization transposed an ontology of rationality from the individual to the group.

25. Samuelson 1938, 1948; for critical analysis of the history of the concept, see Wong 1978.

26. Alchian 1950, 211.

27. Simon 1947, quotes on xiii and 1. For further discussion on this phase of Simon's career, see Crowther-Heyck 2006, chapter 5.

28. On the Graduate School of Industrial Administration, see Fourcade and Khurana 2011.

29. Simon 1951, 1952.

30. For further detail on Simon and his approach, see Sent 2000; Augier 2000; Mirowski 2002, 452–479; and Crowther-Heyck 2005. Newell worked at the RAND Corporation before being lured by Simon to the Carnegie Institute of Technology.

31. Simon 1969, especially 13–22.

32. Dantzig 1982, 46, Dantzig's emphasis.

Chapter 25

1. See Johnson and Kaplan 1987; Miranti 1990; and McKenna 2006 on prior management-oriented professions and their relationship to the well-known "scientific management" of F. W. Taylor.

2. This chapter is an abridged and reorganized version of Thomas 2012.

3. See a memorandum from James R. Killian to George Harrison and Karl Compton, 9/5/1944, and a letter from Henry Loomis to Philip Morse, 10/24/1947, both in MIT AC 4, Box 150, Folder 6; see also a series of mostly undated memoranda in MIT MC 75, Box 2, "Institute Committee on O/R" folder (one item is dated 5/7/1946).

4. Kittel 1947.

5. Letter from Karl Compton to Patrick Blackett, 11/21/1947; letter from Blackett to Compton, 2/6/1948; both in MIT AC 4, Box 32, Folder 2. See also Blackett's hand-written notes from the meeting in RS PB/4/7/2/4/4.

6. Wadsworth was the overseer of MIT's contract with the OEG. The course was taught out of the mathematics department. It was also repeated in subsequent years.

7. Operations Evaluation Group memorandum, "A Course on Operations Research," MIT MC 75, Box 9, "NRC—Comm on OR" folder.

8. Ibid., Appendix D and Appendix F.

9. Ibid., Appendix D.

10. See Kahn 1986 for a general history of the firm. On the prewar boom in management consulting, see McKenna 2006.

11. On Old and the Navy, see Sapolsky 1990. On King, see "Appendix: Partial List of Consultants to the Weapons Systems Evaluation Group and to the Institute for

Defense Analysis," JCS Info Memo 1084, 3/15/1957, NARA RG 218, Central Decimal File, 1958: 334 (2-4-48) Weapons Systems Evaluation Group Sec. 8, Box 53.

12. John Magee, interview with the author, April 11, 2005, Concord, Massachusetts.

13. Magee 2002.

14. Ibid.

15. Ibid.; Magee 1953.

16. Herrmann and Magee 1953.

17. Ibid., quotes on 100, 109, and 112. Articles in *HBR* were often used as means of promoting consulting services; see McKenna 2006, 72. Pocock 1953 also discussed likely tensions between management and OR. J. W. Pocock was an operations researcher at the Booz, Allen & Hamilton consulting firm.

18. Herrmann and Magee 1953, 102 and 110.

19. Magee 2002, 151.

20. Magee 1956a, 1956b, 1956c.

21. Magee 1956a, 50.

22. Magee 1956a, 50–57, quotes on 52 and 56; see also Magee 1956b, 105.

23. Magee 1956a, 58–60; Magee 1956c, 60–63; and Magee 1958, 136–138.

24. Magee 2002, 150; see also Magee and Ernst 1961.

25. On Ernst, see "Announcements" 1959.

26. George Harrison, "Memorandum of Meeting of the Advisory Committee on the Operations Evaluation Group," 1/9/1951, MIT AC 4, Box 165, Folder 4.

27. On proposals at the foundation of SIM, see memoranda in MIT AC 132, Box 10, "Industrial Management, School of," folder; also see MIT MC 75, Box 10, "Committee on Operations Research (NRC)" folder, including diary note, "Meeting of Levinson and Waterman with Stratton and Advisory Council of MIT to discuss Operations Research," 11/20/1950, and a letter from Julius Stratton to Horace Levinson, 12/1/1950. On subsequent development of the plan, see letter from Levinson to Stratton, 3/1/1951; and Levinson, "Memorandum of Telephone Conversation with Dr. Stratton," 4/27/1951, both in NRC COR, "PS: Com on Operations Research 1950–1951, Operations Research Center: Massachusetts Institute of Technology: Proposed" folder.

28. Brooks evidently had not been privy to Arthur D. Little's work for Sears.

29. Memorandum from T. M. Hill to Dean Brooks, "Operations Research," 1/9/1952, MIT AC 132, Box 12, "Operations Research—P. M. Morse" folder.

30. Ibid.

31. Hill appears on the membership list printed in the fourth issue of the *Journal of the Operations Research Society of America*.

32. See memorandum from Morse to Brooks, 7/10/1952; memorandum from Brooks to Morse, 9/16/1952; memorandum from Morse to Stratton, 10/1/1952, emphasis in original; and memorandum from Stratton to Morse 12/15/1952, formally authorizing the establishment of the committee; all in MIT AC 132, Box 12, "Operations Research—P. M. Morse" folder.

33. Philip Morse's reply to ORSA Education Committee survey, 1/5/1953, MIT AC 4, Box 165, Folder 5; OR committee membership taken from the committee's meeting minutes, in MIT MC 75, Box 2, "Committee on OR Minutes" folder; Philip Morse also discusses the OR program at MIT in Morse 1977, chapter 9; on work at Arthur D. Little, I rely on Magee, interview with the author, April 11, 2005.

34. Morse 1977, 296; on Little's interest in computation, see Institute Committee on Operations Research meeting minutes, 5/27/1953, MIT MC 75, Box 2, "Committee on OR Minutes" folder. For biographical detail, see Hauser and Urban 2011.

35. Little 1954; he published the results in Little 1955.

36. "Minutes of the Planning Committee for the 1953 Summer Session in O/R," 4/22/1953, MIT MC 75, Box 2, "Committee on OR Minutes" folder.

37. Morse 1977, 292.

38. Memorandum from Hill to Brooks, 5/10/1954, MIT MC 75, Box 10, "ORSA Corres." folder.

39. Memorandum from Morse to Hill, 5/14/1954, MIT MC 75, Box 10, "ORSA Corres." folder.

40. See Morse 1977, 296 and 355.

Chapter 26

1. Churchman 1994, 99.

2. Ibid., 107. For biographical background on Churchman, see Assad 2011; on Ackoff, see Kirby and Rosenhead 2011.

3. Churchman 1948, 58–59; see also Ackoff 1962.

4. Churchman 1948, chapter 1. On Churchman's wartime experience, see Churchman 1994, 99–101.

5. Ibid., 22–23, 247–251.

6. Ibid., chapter 9.

7. See Churchman and Ackoff 1947; Churchman 1948, 236–247; and Ackoff 1949; see also Churchman 1961.

8. See, for instance, Churchman 1948, chapter 13.

9. See Churchman and Ackoff 1946. On the common mid-twentieth-century portrayal of the personal equation as a contribution of "psychology" to astronomy, see Schaffer 1988.

10. Churchman, Ackoff, and Wax 1947.

11. Ibid., 3.

12. Wilks 1947.

13. Ackoff 1988 discusses his and Churchman's migration in more detail.

14. The precise sequence of events remains somewhat unclear; see Ackoff 1988; Churchman 1994; and Dean 1994. On the railroad project, see Churchman 1955.

15. See Churchman, Ackoff, and Arnoff 1957, 4–7; and Churchman 1961, 314.

16. Ackoff 1953.

17. Churchman and Ackoff 1954 responded to Smith et al. 1953; quotes are from the counter-response, Smith 1954, 181–182.

18. Churchman, Ackoff, and Arnoff 1957.

19. Arnoff 1957 and Dean 1994.

20. Ackoff 1952 was, with Koopman 1952, one of the first papers advocating the development of mathematical methods in OR.

21. See Kirby 2003a and Kirby 2007.

Chapter 27

1. Morse 1977, 291.

2. Vazsonyi 2002, chapters 5–11, quote on 132. For biographical background, see Weida 2011.

3. Salveson 1997, quote on 77.

4. Ibid., 79.

5. William W. Cooper, "The Founding of TIMS," n.d., written for the Online History Section of the *Interfaces* journal, accessed October 26, 2014, http://pubsonline .informs.org/page/inte/stats-history. On Cooper's work, see Cooper 2002.

6. See Salveson 1997, and Cooper, "Founding of TIMS"; on Hurd's work at IBM, see Akera 2007, 227–246.

7. Churchman 1994, 107.

8. Ibid., 107–109.

9. Vazsonyi 1954, quotes on 70 and 71. See also Vazsonyi 1955 and Vazsonyi 1956. On "Gozinto," see Vazsonyi 2002, 205–211.

10. Churchman 1994, 107.

11. Flood 1956a.

12. Lathrop 1957.

13. Hertz 1958. For biographical information, see Baker and Plant 2011. The other letter was Katcher 1957.

Chapter 28

1. "Program for the Evaluation Section of Project RAND," 1/20/1947, RAD-26, RAND Corporation library.

2. Ibid.

3. NOTS Working Paper No. 10, "The Role of Mathematical Research in the Airborne Fire Control Development Program," 5/13/1946, RAND, Ed Paxson papers, Box 4.

4. "Air Defense Study," RAND Document R-227, 10/15/1951, RAND Corporation library. See Smith 1966, 90; and Jardini 1996, 64–69, for further discussion.

5. "Air Defense Study," chapter 3, sections IV and V on "The Defense Systems Analysis: The Study's Numerical Phase," and "Some Limitations of the Defense Systems Analysis," respectively.

6. "Air Defense Study," chapter 2, section VI.

7. RE stands for Research and Experiments. RE-8 is where Solly Zuckerman and J. D. Bernal had worked earlier in the war.

8. See interview with Charles Hitch, 2/9/1988, NASM RAND Oral History Project, 1–13; and Vaughn D. Bornet, "John Williams: A Personal Reminiscence (August, 1962)," RAND Document D-19036, 8/12/1969, RAND, John D. Williams collection, Box 1, 3. After leaving the Pentagon, Hitch would become president of the University of California system.

9. Hitch 1950, 193.

10. Ibid., 195.

11. Hitch and McKean 1960, quotes on 130–131, except for *"optimum optimorum,"* which is from Hitch 1953, 98.

12. Hitch 1953 and Hitch and McKean 1960.

13. E. S. Quade, "The Proposed RAND Course in Systems Analysis," RAND document D-1991, 12/15/1953, RAND Corporation library. For a perspective on this problem from Lockheed, see Bailey 1953.

14. Quade, "Proposed RAND Course in Systems Analysis."

15. Ibid.

16. Ibid.

17. Alchian received his PhD in economics from Stanford in 1944, and was an associate of RAND from December 1949 to February 1960. Kessel received his PhD in economics from the University of Chicago in 1954, and was an employee of RAND from September 1952 to September 1956. See Hounshell 2000, 260–268, for further context; for biographical information, see 293–295.

18. Armen A. Alchian and Reuben A. Kessel, "A Proper Role for Systems Analysis," RAND document D-2057, 1/27/1954, RAND Corporation library.

19. Ibid.

20. Malcolm Hoag received his PhD in economics from the University of Chicago in 1950, and was an Assistant Professor of Economics at the University of Illinois from 1948 to 1950. He lectured on international finance at the University of California, Los Angeles, in 1951, and was a program officer for the Marshall Plan mission to the UK from 1950 to 1952. He joined the Economics Department at RAND in 1952.

21. E. S. Quade and M. W. Moag, "An Outline for the Proposed Course in the Appreciation of Systems Analysis," RAND document D-2132, 3/15/1954, RAND Corporation library.

22. NASM, RAND Oral History Project, Interview with Albert Wohlstetter, 7/29/1987, 7–8.

23. A. J. Wohlstetter, F. S. Hoffman, R. J. Lutz, and H. S. Rowen, "Selection and Use of Strategic Air Bases," RAND document R-266, April 1954, http://www.rand.org/pubs/reports/R0266.html. An overview of the study can be found in Smith 1966, chapter 6; see also Quade 1964b, and Jardini 1996, 123–125.

24. See Wohlstetter 1964a; see also Albert Wohlstetter, "Systems Analysis Versus Systems Design," RAND document P-1530, 10/29/1958, RAND Corporation library, http://www.rand.org/about/history/wohlstetter/P1530/P1530.html.

25. Hitch 1955, 480. Erickson 2006, 167–168, traces these changes at RAND, and also correctly notes, at 102, that the historiography of game theory has traditionally conflated its earlier and later uses, implying that deterministic uses of game theory were accepted at RAND as a legitimate means of drawing conclusions concerning high-level strategic problems. Examples of this assumption can be found in Kaplan 1983, 63–68; and Jardini 1996, chapter 3.

26. On gaming at RAND, see Ghamari-Tabrizi 2000, 2005, especially chapter 6 in the latter on "faith and insight in war-gaming." For further details, see Hausrath 1971.

27. Brodie became one of the first civilian nuclear strategists upon the publication of Brodie 1946. The literature on nuclear strategy at RAND is substantial; see especially Kaplan 1983; Herken 1985; May 1998; and James Digby, "Strategic Thought at RAND, 1948–1963: The Ideas, Their Origins, Their Fates," RAND document N-3096-RC, June 1990, http://www.rand.org/pubs/notes/N3096.html. On game theory in strategic thought, see Erickson 2006, chapter 3.

28. Herman Kahn and Irwin Mann, "Ten Common Pitfalls," vii, RAND document RM-1937, 7/17/1957, RAND Corporation library, http://www.rand.org/pubs/research_memoranda/RM1937.html. Mann was Kahn's assistant; see Bruce-Biggs 2000, 51–52. See also Ghamari-Tabrizi 2005, for discussion of Kahn's live presentations of this document at briefings.

29. Koopman 1956 and Hitch 1956.

30. Quade 1964a, and Quade and Boucher 1968.

Chapter 29

1. Contrast Kubrick's Strangelove, whose fascist logic is only ultimately revealed, with the obvious arrogance, sadism, over certainty, and occasional nihilism of Prof. Groeteschele, played by Walter Matthau in the Sidney Lumet film *Fail-Safe*, which was filmed at the same time as *Dr. Strangelove* and appeared shortly after it.

2. Ackoff 1957b, quote on 464.

3. Hitch 1957, Hitch's emphasis.

4. Ackoff 1958, Ackoff's emphasis. Two decades later Ackoff unapologetically revisited the issue in Ackoff 1977.

5. The Wharton School is the University of Pennsylvania's business school. Ackoff's OR program also took over the existing Management Science Center at the school.

6. Churchman 1994 exhibits his keen sense of disappointment. Ackoff 2001 recounts both his frustrations and his departmental moves. His declarations of the

scientific death of OR can be found in Ackoff 1979, and again in Ackoff 1987. Historical discussion of Ackoff's (and others') disaffection with OR can be found in Kirby 2003a, 2007. Ackoff's criticisms have been instrumental in some subsequent assessments of the OR profession; in addition to Kirby's works, see Waring 1991, 38–42, and Hughes 1998, 192–194.

7. Blackett 1948b/1962, 1950/1962, 1953/1962. While visiting Stockholm to collect his Nobel Prize in Physics in 1948, Blackett was apparently successful in sparking interest in OR with lectures he gave there; see Kaijser and Tiberg 2000.

8. P. M. S. Blackett, "Operational Research," ca. 1950, "Transcript of informal talk, including discussion following lecture, given to a Royal Statistical Society group," RS PB/4/7/2/7. Related speeches on OR from the same period are also available in the PB/4/7/2 group.

9. "Operational Research, Notes on a lecture given by PMSB to Institute of Physics," 3/17/1953, RS PB/4/7/2/10.

10. Letter from Sidney Dell to Patrick Blackett, 12/31/1958, RS PB/4/7/3/8; Hitch 1958.

11. Letters from Patrick Blackett to Sidney Dell, 1/20/1959 and 3/25/1959, RS PB/4/7/3/8.

12. Dell 1960 and Hitch 1960. Hitch, 8/18/1959, RS PB/4/7/3/8, wrote to Blackett directly attempting to reconcile their points of view when it appeared as though Dell's response would not be published.

13. CAOR, "The Case for Large Convoys," TNA: PRO ADM 219/19. The document is not dated, but Llewellyn-Jones 2000 indicates that it was completed in May 1943.

14. Blackett 1948a, published in America the following year as *Fear, War, and the Bomb*; and Blackett 1956. On Blackett's punditry, see Howard 1984/1991.

15. Wohlstetter 1959.

16. Blackett 1961c. The lecture was delivered on March 22, 1961. The *Encounter* version, Blackett 1961a/1962, was significantly altered. A seventeen-page draft of the introduction to the original speech, containing material not included in the published version, can be found in RS PB/4/7/2/13. The finding aid to Blackett's papers claims that this is the "introduction to lecture as delivered"; it is cited here as an "unpublished introduction," as I can find no further confirmation that it was, in fact, delivered. For specific quotes, I cite the most publicly available version.

17. Blackett 1961a/1962, 129–130.

18. Blackett, unpublished introduction to Blackett 1961c, 15.

19. See Nye 2004, chapter 3.

20. Blackett 1960/1962.

21. Snow 1959. On the crucial intellectual and institutional context of Snow's punditry in this period, see Ortolano 2009. On the intellectual legacy of "The Two Cultures," see Ortolano 2008.

22. Snow 1961; on his conception of "closed politics," see chapters 9, 10, and 11.

23. On the controversy surrounding the book, see Lear 1961, and Snow's letter to the editor in the 3/7/1961 issue of *New Scientist*; Watson-Watt 1961; Jones 1961, a response to a letter to the editor of *The Times* from Snow on 4/8/1961, and a counter-response from Jones on 4/12/1961; Philip Joubert 1961 in the *Daily Telegraph*, and A. P. Rowe's response to Joubert in a letter to the editor on 4/12/1961; and Birkenhead 1961. Joubert and Rowe both supported Snow's account. Lutyens 1961, the *Daily Telegraph* review of the book, made the connection between Lindemann and RAND strategists' nuclear strategy explicit.

24. See Blackett 1961b/1962, his review of Snow's book for *Scientific American*.

25. Labour Party leader Harold Wilson made science and technology a major part of the party's platform, culminating in his promise to marshal the "white heat" of the "revolution" in science and technology for Britain's future; see Edgerton 1996c and 2006, chapter 6.

26. Jones 1963.

27. Letter from Blackett to Jones 1/4/1963 (from which the quote is taken); from Jones to Blackett, 1/9/1963; from Blackett to Jones, 1/14/1963; and from Jones to Blackett, 2/13/1963; all in RS PB/9/1/64.

28. Herman Kahn and Irwin Mann, "Ten Common Pitfalls," RAND document RM-1937, 7/17/1957, RAND Corporation library, 46, http://www.rand.org/pubs/research_memoranda/RM1937.html.

29. Kahn 1960.

30. See Bruce-Biggs 2000 and Ghamari-Tabrizi 2005.

31. Zuckerman 1962.

32. On the prevalence of the concept of a "scientific establishment" at the time, see Greenberg 1965. Kahn intended the Doomsday Machine as a parody of a government defense policy that relied on the threat of a full-scale nuclear war as a deterrent to a conventional war or the use of nuclear weapons. He argued that such a policy could only be fully credible if leaders could be believed to be truly willing to escalate a conflict to that level.

33. Wohlstetter 1963, which was expanded on in Wohlstetter 1964b; Wohlstetter 1964c, in which he tartly observed, 214, "It is not at all clear that the critics who

have made these attacks have read very much either of the 'strategists' who, they suppose, all use game theory, or of game theory itself. However, it is very clear that they have read each other."

34. Wohlstetter 1963, 466.

35. Wohlstetter 1964c, 221.

36. Wohlstetter 1963, 472, Wohlstetter's emphasis.

37. On the various scientific perspectives in the test-ban debate and their influence on policy, see Greene 2007.

38. See Aligica and Weinstein 2009 for essays by Kahn on these points; see also Bruce-Biggs 2000 for insight into Kahn's critiques of the "New Class" and various leftwing positions.

39. Kahn 1962.

Chapter 30

1. J. D. Williams, "An Overview of RAND," RAND document D-10053, 5/14/1962, RAND, John Williams papers, Box 1.

2. Churchill 1949, chapter 19.

3. Halberstam 1972.

4. Letter from Peter Tizard to A. V. Hill, 8/2/1972, CAC AVHL II 4/80.

5. Blackett 1962; Clark 1962, 1965; Crowther 1965, 1967, and Rose and Rose 1969.

6. Crowther 1967, 142 and 290.

7. Rose and Rose 1969, 240–241.

8. Shapin and Schaffer 1985, 344. Iconoclasm and maturity with respect to science and society is a recurring theme in Shapin's writing; see Shapin 2010.

9. Latour 1988, 5; and Latour 1993. For a critique of Latour's historical narrative, see Kusch 1995.

10. Jasanoff 1990; see also Ezrahi 1990. For an early criticism of sociologists' claims about historical actors' naiveté concerning science and technology, see Edgerton 1993.

11. This transition was already under way in the 1960s; the Roses were as concerned with overenthusiasm as they were with neglect.

12. The theme of "trust" runs throughout the literature, but see especially Porter 1995.

13. The touchstone work on the "ideological" function of "boundaries" is Gieryn 1983.

14. Edwards 1996; see also Pickering 1995 on what he calls the "World War II regime."

15. Hughes 1998.

16. Collins 2002, 163.

17. Mirowski 2004a, 16.

18. Mirowski 2002, 259.

19. Mirowski 2004b, 301.

20. Porter 2012, ix and xii.

21. Erickson et al. 2013.

22. Ghamari-Tabrizi 2005, 9.

23. Ibid., 46–47.

24. This point is in rough accord with the treatment of expertise in the "Sociology of Expertise and Experience" championed by Harry Collins and Robert Evans, which they frame as a "third wave" of science studies; see Collins and Evans 2002, 2007.

25. See Evans 2007 on the private uses of expertise.

26. Balogh 1991a, 1991b, discuss the tandem development of institutional and grassroots experts, but supposes that the development of the latter group followed on a loss of confidence in the former. I would suggest that grassroots expertise grew in the wake of a loss of confidence in institutions as acting in the public interest, rather than their loss of confidence in expertise as a neutral arbiter of what constituted action in the public interest.

Bibliography

Archives

CAC Churchill Archives Centre, Cambridge University, Cambridge, UK

AVHL Papers of Archibald Vivian Hill
GOEV Papers of Charles Goodeve

CT California Institute of Technology Archives, Pasadena, CA, USA

HPR Papers of Howard Percy Robertson

CUL Cambridge University Library, Manuscripts and University Archives, Cambridge, UK

JDB Papers of John Desmond Bernal

IWM Imperial War Museum, Documents Collection, London, UK

HTT Papers of Henry T. Tizard

JHU Johns Hopkins University Archives, Baltimore, MD, USA

02.001 Papers of the Office of the President
03.001 Papers of the Office of the Provost

LOC Library of Congress, Manuscript Collections, Washington, DC, USA

CEL Papers of Curtis E. LeMay
ELB Papers of Edward L. Bowles

MIT Massachusetts Institute of Technology, Institute Archives and Special Collections, Cambridge, MA, USA

AC 4 Records of the Office of the President, 1930–1959
AC 132 Records of the Office of the Chancellor, 1949–1957
MC 75 Papers of Philip McCord Morse

NAA National Academies Archives, Washington, DC, USA

COR Records of the National Research Council Committee on Operations Research

NARA National Archives and Records Administration, Archives II, College Park, MD, USA

RG 38 Records of the Office of the Chief of Naval Operations
RG 218 Records of the U.S. Joint Chiefs of Staff
RG 227 Records of the Office of Scientific Research and Development
RG 319 Records of the Army Staff
RG 330 Records of the Office of the Secretary of Defense
RG 341 Records of Headquarters, U.S. Air Force (Air Staff)

NASM National Air and Space Museum Archives, Washington, DC, USA

RAND RAND Corporation archives, Santa Monica, CA, USA

RS Royal Society Archives, London, UK

PB Papers of Patrick Maynard Stuart Blackett

TNA: PRO The National Archives of the United Kingdom, Public Record Office, Kew, London, UK

ADM Records of the Admiralty
AIR Records of the Air Ministry
AVIA Records of the Ministry of Aircraft Production
CAB Records of the Cabinet
COAL Records of the National Coal Board
DEFE Records of the Ministry of Defence
WO Records of the War Office

Published Sources

Aaserud, Finn. "Sputnik and the 'Princeton Three': The National Security Laboratory That Was Not to Be." *Historical Studies in the Physical and Biological Sciences* 25 (1995): 185–239.

Abbott, Andrew. *The System of Professions: An Essay on the Division of Expert Labor.* Chicago: University of Chicago Press, 1988.

Ackoff, Russell L. "On a Science of Ethics." *Philosophy and Phenomenological Research* 9 (1949): 663–672.

Ackoff, Russell L. "Some New Statistical Techniques Applicable to Operations Research." *Journal of the Operations Research Society of America* 1 (1952): 10–17.

Ackoff, Russell L. *The Design of Social Research*. Chicago: University of Chicago Press, 1953.

Ackoff, Russell L. "A Comparison of Operational Research in the USA and Great Britain." *Operational Research Quarterly* 8 (1957a): 88–100.

Ackoff, Russell L. "Operations Research and National Planning." *Operations Research* 5 (1957b): 457–468.

Ackoff, Russell L. "On Hitch's Dissent on 'Operations Research and National Planning.'" *Operations Research* 6 (1958): 121–124.

Ackoff, Russell L. *Scientific Method: Optimizing Applied Research Decisions*. New York: John Wiley & Sons, 1962.

Ackoff, Russell L. "National Development Planning Revisited." *Operations Research* 25 (1977): 207–218.

Ackoff, Russell L. "The Future of Operational Research Is Past." *Journal of the Operational Research Society* 30 (1979): 93–104.

Ackoff, Russell L. "OR, A Post Mortem." *Operations Research* 35 (1987): 471–474.

Ackoff, Russell L. "C. West Churchman." *Systems Practice* 1 (1988): 351–355.

Ackoff, Russell L. "OR: After the Post Mortem." *System Dynamics Review* 17 (2001): 341–346.

Adas, Michael. *Dominance by Design: Technological Imperatives and America's Civilizing Mission*. Cambridge, MA: The Belknap Press of Harvard University Press, 2006.

Agar, Jon. *The Government Machine: A Revolutionary History of the Computer*. Cambridge, MA: The MIT Press, 2003.

Agar, Jon, and Brian Balmer. "British Scientists and the Cold War: The Defence Research Policy Committee and Information Networks, 1947–1963." *Historical Studies in the Physical and Biological Sciences* 28 (1998): 209–252.

Air Ministry. *The Origins and Development of Operational Research in the Royal Air Force*. London: HMSO, 1963.

Akera, Atsushi. *Calculating a Natural World: Scientists, Engineers, and Computers During the Rise of U.S. Cold War Research*. Cambridge, MA: MIT Press, 2007.

Albers, Donald J. "David Blackwell." In *Mathematical People: Profiles and Interviews*, ed. Donald J. Albers and G. L. Alexanderson, 17–32. Boston: Birkhäuser, 1985.

Albers, Donald J. "George B. Dantzig." In *More Mathematical People: Contemporary Conversations*, ed. Donald J. Albers, Gerald L. Alexanderson, and Constance Reid, 61–79. San Diego: Academic Press, 1990.

Alchian, Armen A. "Uncertainty, Evolution, and Economic Theory." *Journal of Political Economy* 58 (1950): 211–221.

Aligica, Paul Dragos, and Kenneth R. Weinstein, eds. *The Essential Herman Kahn: In Defense of Thinking*. Lanham: Lexington Books, 2009.

Alter, Peter. *The Reluctant Patron: Science and the State in Britain, 1850–1920*. New York: Berg, 1987.

Amadae, S. M. *Rationalizing Capitalist Democracy: The Cold War Origins of Rational Choice Liberalism*. Chicago: University of Chicago Press, 2003.

"Announcements." *Operations Research* 7 (1959): 420–422.

Armatte, Michel, and Amy Dahan Dalmedico. "Modèles et Modélisations, 1950–2000: Nouvelles Pratiques, Nouveaux Enjeux." *Revue d'Histoire des Sciences* 57 (2004): 243–303.

Arnoff, E. Leonard. "Operations Research at the Case Institute of Technology." *Operations Research* 5 (1957): 289–292.

Arrow, Kenneth J. *Social Choice and Individual Values*. New York: John Wiley and Sons, 1951.

Arrow, Kenneth J. "Decision Theory and Operations Research." *Operations Research* 5 (1957): 765–774.

Arrow, Kenneth J. "The Genesis of 'Optimal Inventory Policy.'" *Operations Research* 50 (2002): 1–2.

Arrow, K. J., D. Blackwell, and M. A. Girshick. "Bayes and Minimax Solutions of Sequential Decision Problems." *Econometrica* 17 (1949): 213–244.

Arrow, Kenneth J., and Gerard Debreu. "Existence of an Equilibrium for a Competitive Economy." *Econometrica* 22 (1954): 265–290.

Arrow, Kenneth J., Theodore E. Harris, and Jacob Marschak. "Optimal Inventory Policy." *Econometrica* 19 (1951): 250–272.

Assad, Arjang A. "C. West Churchman." In *Profiles in Operations Research: Pioneers and Innovators*, ed. Arjang A. Assad and Saul I. Gass, 171–200. New York: Springer, 2011.

Assad, Arjang A., and Saul I. Gass, eds. *Profiles in Operations Research: Pioneers and Innovators*. New York: Springer, 2011.

Association of Scientific Workers. *Science and the Nation*. Harmondsworth: Penguin, 1947.

Augier, Mie. "Models of Herbert A. Simon." *Perspectives on Science* 8 (2000): 407–443.

Augustine, Dolores L. *Red Prometheus: Engineering and Dictatorship in East Germany, 1945–1990*. Cambridge, MA: MIT Press, 2007.

Austin, Brian. *Schonland: Scientist and Soldier*. Philadelphia: Institute of Physics, 2001.

Babbage, Charles. *Reflections on the Decline of Science in England and on Some of Its Causes*. London: B. Fellowes and J. Booth, 1830.

Backhouse, Roger. "Paul Samuelson, RAND and the Cowles Commission Activity Analysis Conference, 1947–1949." 2012. Unpublished draft.

Bailey, Robert A. "Application of Operations-Research Techniques to Airborne Weapons Systems Planning." *Journal of the Operations Research Society of America* 1 (1953): 187–199.

Baker, Edward K., and Robert T. Plant. "David Bendel Hertz." In *Profiles in Operations Research: Pioneers and Innovators*, ed. Arjang A. Assad and Saul I. Gass, 403–413. New York: Springer, 2011.

Balogh, Brian. *Chain Reaction: Expert Debate and Public Participation in American Commercial Nuclear Power, 1945–1975*. New York: Cambridge University Press, 1991a.

Balogh, Brian. "Reorganizing the Organizational Synthesis: Federal-Professional Relations in Modern America." *Studies in American Political Development* 5 (1991b): 119–172.

Baum, Claude. *The System Builders: The Story of SDC*. Santa Monica: Systems Development Corporation, 1981.

Beer, Stafford. "Operational Research and Accounting." *Operational Research Quarterly* 5 (1954): 1–12.

Beer, Stafford. "What Has Cybernetics to Do with Operational Research?" *Operational Research Quarterly* 10 (1959): 1–21.

Bell, G. E. "Operational Research into Air Traffic Control." *Journal of the Royal Aeronautical Society* 53 (1949a): 965–976.

Bell, G. E. "Traffic Control Research." *Flight* 55 (1949b): 292.

Bellman, Richard. "Review of *Studies in the Mathematical Theory of Inventory and Production*." *Management Science* 5 (1958): 139–141.

Bentz, Richard, William Chace, Gordon Doerfer, John Donaldson, Thomas English, James Graves, Paul Grim, et al. "Some Civil Defense Problems in the Nation's Capital

Following Widespread Thermonuclear Attack." *Operations Research* 5 (1957): 319–350.

Bernal, J. D. *The Social Function of Science*. London: G. Rutledge and Sons, 1939.

Bernal, J. D. "The Function of the Scientist in Government Policy and Administration." *Advancement of Science* 2 (1942): 14–17.

Bernal, J. D. "Lessons of the War for Science" [1945]. Reprinted in *Proceedings of the Royal Society of London. Series A, Mathematical and Physical Sciences* 342 (1975): 555–574.

Birkenhead, Earl of. *The Prof in Two Worlds: The Official Life of Prof. F. A. Lindemann, Viscount Cherwell*. London: Collins, 1961.

Blackett, P. M. S. *Military and Political Consequences of Atomic Energy*. London: Turnstile Press, 1948a.

Blackett, P. M. S. "Operational Research." *The Advancement of Science* 5 (1948b). Reprinted in Blackett 1962, 169–198.

Blackett, P. M. S. "Operational Research." *Operational Research Quarterly* 1 (1950): 3–6. Reprinted in Blackett 1962, 199–204, as "The Scope of Operational Research."

Blackett, P. M. S. "Recollections of Problems Studied, 1940–1945." *Brassey's Annual* (1953). Reprinted in Blackett 1962, 205–234.

Blackett, P. M. S. *Atomic Weapons and East-West Relations*. Cambridge, UK: Cambridge University Press, 1956.

Blackett, P. M. S. "Tizard and the Science of War," *Nature* 185 (1960). Reprinted in Blackett 1962, 101–119.

Blackett, P. M. S. "A Critique of Some Contemporary Defence Thinking." *Encounter* (April 1961a). Reprinted in Blackett 1962, 128–146.

Blackett, P. M. S. "Science and Government." *Scientific American* (April 1961b). Reprinted in Blackett 1962, 120–127.

Blackett, P. M. S. "Operational Research and Nuclear Weapons." *Journal of the Royal United Services Institute* 106 (1961c): 202–214.

Blackett, P. M. S. *Studies of War, Nuclear and Conventional*. New York: Hill and Wang, 1962.

Blackwell, David, and Albert H. Bowker. "Meyer Abraham Girshick, 1908–1955." *Annals of Mathematical Statistics* 26 (1955): 365–367.

Blackwell, David, and M. A. Girshick. *Theory of Games and Statistical Decisions*. New York: John Wiley and Sons, 1954.

Brodie, Bernard. *The Absolute Weapon: Atomic Power and World Order*. New York: Harcourt, Brace and Company, 1946.

Brooks, John. *Dreadnought Gunnery and the Battle of Jutland: The Question of Fire Control*. New York: Rutledge, 2005.

Brothers, LeRoy A. "Operations Analysis in the United States Air Force." *Journal of the Operations Research Society of America* 2 (1954): 1–16.

Brown, Andrew. *J. D. Bernal: The Sage of Science*. New York: Oxford University Press, 2005.

Bruce-Biggs, B. *Supergenius: The Mega-Worlds of Herman Kahn*. New York: North American Policy Press, 2000.

Caldwell, Peter. *Dictatorship, State Planning, and Social Theory in the German Democratic Republic*. New York: Cambridge University Press, 2003.

Camp, Glen D., Russell L. Ackoff, Wroe Alderson, Martin L. Ernst, George E. Nicholson, Jr., George P. Wadsworth, and Joseph F. McCloskey. "Report of the Education Committee." *Journal of the Operations Research Society of America* 1 (1953): 248–251.

Cardwell, D. S. L. *The Organisation of Science in England: A Retrospect*. Melbourne: Heinemann, 1957.

Cattermole, M. J. G., and A. F. Wolfe. *Horace Darwin's Shop: A History of the Cambridge Scientific Instrument Company, 1878–1968*. Bristol: Hilder, 1987.

Ceruzzi, Paul. *Internet Alley: High Technology in Tyson's Corner, 1945–2005*. Cambridge, MA: MIT Press, 2008.

Chaloupek, Gunther. "The Austrian Debate on Economic Calculation in a Socialist Society." *History of Political Economy* 22 (1990): 659–675.

Cherrier, Beatrice. "Rationalizing Human Organization in an Uncertain World: Jacob Marschak, from Ukranian Prisons to Behavioral Science Laboratories." *History of Political Economy* 42 (2010): 443–467.

Churchill, Winston S. *Their Finest Hour*. vol. 2. The Second World War. New York: Houghton Mifflin, 1949.

Churchman, C. West. *Theory of Experimental Inference*. New York: Macmillan, 1948.

Churchman, C. "The Application of Sampling to LCL Interline Settlements of Accounts on American Railroads." In *Handbook of Industrial Engineering and Management*, ed. William Grant Ireson and Eugene L. Grant, 1051–1057. Englewood Cliffs: Prentice Hall, 1955.

Churchman, C. West. *Prediction and Optimal Decision: Philosophical Issues of a Science of Values*. Englewood Cliffs: Prentice-Hall, 1961.

Churchman, C. West. "Management Science: Science of Managing and Managing of Science." *Interfaces* 24 (1994): 99–110.

Churchman, C. West, and Russell L. Ackoff. "Varieties of Unification." *Philosophy of Science* 13 (1946): 287–300.

Churchman, C. West, and Russell L. Ackoff. "An Experimental Definition of Personality." *Philosophy of Science* 14 (1947): 304–332.

Churchman, C. West, and Russell L. Ackoff. "An Approximate Measure of Value." *Journal of the Operations Research Society of America* 2 (1954): 172–181.

Churchman, C. West, Russell L. Ackoff, and E. Leonard Arnoff. *Introduction to Operations Research*. New York: John Wiley and Sons, 1957.

Churchman, C. West, Russell L. Ackoff, and Murray Wax. "Introduction." In *Measurement of Consumer Interest*, ed. C. West Churchman, Russell L. Ackoff, and Murray Wax, 1–7. Philadelphia: University of Pennsylvania Press, 1947.

Clark, Ronald W. *The Rise of the Boffins*. London: Phoenix House, 1962.

Clark, Ronald W. *Tizard*. Cambridge, MA: MIT Press, 1965.

Coffey, Thomas M. *Iron Eagle: The Turbulent Life of General Curtis LeMay*. New York: Crown, 1986.

Cohen-Cole, Jamie. *The Open Mind: Cold War Politics and the Sciences of Human Nature*. Chicago: University of Chicago Press, 2014.

Collini, Stefan. *Absent Minds: Intellectuals in Britain*. New York: Oxford University Press, 2006.

Collins, H. M., and Robert Evans. "The Third Wave of Science Studies: Studies of Expertise and Experience." *Social Studies of Science* 32 (2002): 235–296.

Collins, Harry, and Robert Evans. *Rethinking Expertise*. Chicago: University of Chicago Press, 2007.

Collins, Martin J. *Cold War Laboratory: RAND, the Air Force, and the American State, 1945–1950*. Washington, DC: Smithsonian Institution Press, 2002.

Cook, Fred J. *The Warfare State*. New York: Macmillan, 1962.

Cooke, G. W. *Agricultural Research, 1931–1981: A History of the Agricultural Research Council and a Review of Developments in Agricultural Research during the Last Fifty Years*. London: Agricultural Research Council, 1981.

Cooper, W. W. "Abraham Charnes, W. W. Cooper (et al.): A Brief History of a Long Collaboration in Developing Industrial Uses of Linear Programming." *Operations Research* 50 (2002): 35–41.

Cornford, E. C. "Dr. L. B. C. Cunningham." *Nature* 158 (1946): 408–409.

Crook, Paul. "Science and War: Radical Scientists and the Tizard-Cherwell Area Bombing Debate in Britain." *War & Society* 12 (1994): 69–101.

Crowther, J. G. *The Social Relations of Science.* London: Macmillan, 1941.

Crowther, J. G. *Statesmen of Science.* London: Cresset, 1965.

Crowther, J. G. *Science in Modern Society.* London: Cresset, 1967.

Crowther, J. G., and R. Whiddington. *Science at War.* London: HMSO, 1947.

Crowther-Heyck, Hunter. *Herbert A. Simon: The Bounds of Reason in Modern America.* Baltimore: Johns Hopkins University Press, 2005.

Crowther-Heyck, Hunter. "Patrons of the Revolution: Ideals and Institutions in Postwar Behavioral Science." *Isis* 97 (2006): 420–446.

Cunningham, L. B. C., and W. R. B. Hynd. "Random Processes in Problems of Air Warfare." *Supplement to the Journal of the Royal Statistical Society* 8 (1946): 62–97.

Daniels, H. E. "A Tribute to L. H. C. Tippett." *Journal of the Royal Statistical Society A* 145 (1982): 261–262.

Dantzig, George B. *Linear Programming and Extensions.* Princeton: Princeton University Press, 1963.

Dantzig, George B. "Reminiscences about the Origins of Linear Programming." *Operations Research Letters* 1 (1982): 43–48.

Dantzig, G., R. Fulkerson, and S. Johnson. "Solution of a Large-Scale Traveling Salesman Problem." *Journal of the Operations Research Society of America* 2 (1954): 393–410.

David, Thomas Rhodri Vivian. "British Scientists and Soldiers in the First World War (With Special Reference to Ballistics and Chemical Warfare)." PhD thesis, Imperial College London, 2009.

Dean, Burton V. "West Churchman and Operations Research: Case Institute of Technology, 1951–1957." *Interfaces* 24 (1994): 5–15.

DeGroot, Morris H. "A Conversation with David Blackwell." *Statistical Science* 1 (1986): 40–53.

DeJager, Timothy. "Pure Science and Practical Interests: The Origins of the Agricultural Research Council, 1930–1937." *Minerva* 31 (1993): 129–150.

Dell, S. "Economics and Military Operations Research." *Review of Economics and Statistics* 42 (1960): 219–222.

"The Deployment of Scientific Effort in Britain." *Nature* 161 (1948): 1–3.

Desmarais, Ralph. "Science, Scientific Intellectuals and British Culture in the Early Atomic Age." PhD thesis, Imperial College London, 2010.

Dorfman, Robert, Paul A. Samuelson, and Robert M. Solow. *Linear Programming and Economic Analysis*. New York: McGraw-Hill, 1958.

Düppe, Till, and E. Roy Weintraub. *Finding Equilibrium: Living Through the Transformation of Postwar Economics Theory*. Princeton: Princeton University Press, 2014.

Dvoretzky, A., J. Kiefer, and J. Wolfowitz. "The Inventory Problem: I. Case of Known Distributions of Demand." *Econometrica* 20 (1952a): 187–222.

Dvoretzky, A., J. Kiefer, and J. Wolfowitz. "The Inventory Problem: II. Case of Unknown Distributions of Demand." *Econometrica* 20 (1952b): 450–466.

Dyer, Davis. *TRW: Pioneering Technology and Innovation since 1900*. Boston: Harvard Business School Press, 1998.

Dyer, Davis, and Michael Aaron Dennis. *Architects of Information Advantage: The MITRE Corporation since 1958*. Montgomery: Community Communications, 1998.

Dyson, Freeman. *Disturbing the Universe*. New York: Harper & Row, 1979.

Dyson, Freeman. "A Failure of Intelligence: Operational Research in the RAF Bomber Command." *Technology Review* 109 (6) (2006): 62–71.

Easterfield, T. E. "The Special Research Unit at the Board of Trade, 1946–1949." *Journal of the Operational Research Society* 34 (1983): 565–568.

Edgerton, David. *England and the Aeroplane: An Essay on a Militant and Technological Nation*. Baskingstoke: Macmillan, 1991.

Edgerton, David. "Whatever Happened to the British Warfare State? The Ministry of Supply, 1945–1951." In *The Labour Government and Private Industry: The Experience of 1945–1951*, ed. Helen Mercer, Neil Rollings, and J. D. Tomlinson, 91–116. Edinburgh University Press, 1992.

Edgerton, David. "Tilting at Paper Tigers." *British Journal for the History of Science* 26 (1993): 67–75.

Edgerton, David. "British Scientific Intellectuals and the Relations of Science, Technology and War." In *National Military Establishments and the Advancement of Science and Technology*, ed. Paul Forman and José M. Sánchez-Ron, 1–35. Boston: Kluwer Academic, 1996a.

Edgerton, David. *Science, Technology and the British Industrial "Decline," 1870–1970*. New York: Cambridge University Press, 1996b.

Edgerton, David. "The 'White Heat' Revisited: The British Government and Technology in the 1960s." *Twentieth Century British History* 7 (1996c): 53–82.

Edgerton, David. "C. P. Snow as Anti-Historian of British Science: Revisiting the Technocratic Moment, 1959–1964." *History of Science* 43 (2005): 187–208.

Edgerton, David. *Warfare State: Britain, 1920–1970*. New York: Cambridge University Press, 2006.

"Editorial Notes [1]." *Operational Research Quarterly* 1 (1950): 1–3.

"Editorial Notes [2]." *Operational Research Quarterly* 1 (1950): 53–54.

"Editorial Notes." *Operational Research Quarterly* 4 (1953): 57–60.

Edwards, A. P. J. "Ben Lockspeiser." *Biographical Memoirs of Fellows of the Royal Society* 39 (1994): 246–261.

Edwards, Paul N. *The Closed World: Computers and the Politics of Discourse in Cold War America*. Cambridge, MA: MIT Press, 1996.

Eisenhart, Churchill, and Joan R. Rosenblatt. "W. J. Youden, 1900–1971." *Annals of Mathematical Statistics* 43 (1972): 1035–1040.

Ekbladh, David. *The Great American Mission: Modernization and the Construction of an American World Order*. Princeton: Princeton University Press, 2010.

Ellul, Jacques. *The Technological Society* [1954], trans. John Wilkinson. New York: Knopf, 1964.

Engel, Joseph H. "Operations Research for the U. S. Navy Since World War II." *Operations Research* 8 (1960): 798–809.

Engerman, David C. *Modernization from the Other Shore: American Intellectuals and the Romance of Russian Development*. Cambridge, MA: Harvard University Press, 2003.

Engerman, David C. *Know Your Enemy: The Rise and Fall of America's Soviet Experts*. New York: Oxford University Press, 2009.

Engerman, David C. "Social Science in the Cold War." *Isis* 101 (2010): 393–400.

Engerman, David C., Nils Gilman, Mark H. Haefele, and Michael E. Latham, eds. *Staging Growth: Modernization and the Global Cold War*. Amherst: University of Massachusetts Press, 2003.

Erickson, Paul Hilding. "The Politics of Game Theory: Mathematics and Cold War Culture." PhD dissertation, University of Wisconsin-Madison, 2006.

Erickson, Paul, Judy L. Klein, Lorraine Daston, Rebecca Lemov, Thomas Sturm, and Michael Gordin. *How Reason Almost Lost Its Mind: The Strange Career of Cold War Rationality*. Chicago: University of Chicago Press, 2013.

Evans, Robert. "Social Networks and Private Spaces in Economic Forecasting." *Studies in History and Philosophy of Science* 38 (2007): 686–697.

Everett, Robert R, ed. "Special Issue: SAGE (Semi-Automated Ground Environment)." *Annals of the History of Computing* 5 (1983): 319–406.

Ezrahi, Yaron. *The Descent of Icarus: Science and the Transformation of Contemporary Democracy*. Cambridge, MA: Harvard University Press, 1990.

Farren, William S., and R. V. Jones. "Henry Thomas Tizard, 1885–1959." *Biographical Memoirs of Fellows of the Royal Society* 7 (1961): 313–348.

Finkbeiner, Ann. *The Jasons: The Secret History of Science's Postwar Elite*. New York: Viking, 2006.

Flood, Merrill M. "The Objectives of TIMS." *Management Science* 2 (1956a): 178–184.

Flood, Merrill M. "The Traveling-Salesman Problem." *Operations Research* 4 (1956b): 61–75.

Forman, Paul. "Behind Quantum Electronics: National Security as Basis for Physical Research in the United States, 1940–1960." *Historical Studies in the Physical and Biological Sciences* 18 (1987): 149–229.

Fort, Adrian. *Prof: The Life of Frederick Lindemann*. London: Jonathan Cape, 2003.

Fortun, M., and S. S. Schweber. "Scientists and the Legacy of World War II: The Case of Operations Research." *Social Studies of Science* 23 (1993): 595–642.

Fourcade, Marion, and Rakesh Khurana. "From Social Control to Financial Economics: The Linked Ecologies of Economics and Business in Twentieth Century America." Harvard Business School Working Paper 11–071, 2011.

Freedman, Lawrence. *The Evolution of Nuclear Strategy*. New York: St. Martin's Press, 1981.

Gadsby, G. Neville. "The Army Operational Research Establishment." *Operational Research Quarterly* 16 (1965): 5–18.

Gass, Saul I. "The First Linear Programming Shoppe." *Operations Research* 50 (2002): 61–68.

Gass, Saul I., and Arjang A. Assad. *An Annotated Timeline of Operations Research: An Informal History*. New York: Kluwer Academic, 2005.

Gass, Saul I., and Jonathan Rosenhead. "Leonid Vital'evish Kantorovich." In *Profiles in Operations Research: Pioneers and Innovators*, ed. Arjang A. Assad and Saul I. Gass, 157–170. New York: Springer, 2011.

Gay, Hannah. *The History of Imperial College London, 1907–2007*. London: Imperial College Press, 2007.

Gerovitch, Slava. *From Newspeak to Cyberspeak: A History of Soviet Cybernetics*. Cambridge, MA: MIT Press, 2002.

Ghamari-Tabrizi, Sharon. "Simulating the Unthinkable: Gaming Future War in the 1950s and 1960s." *Social Studies of Science* 30 (2000): 163–223.

Ghamari-Tabrizi, Sharon. *The Worlds of Herman Kahn: The Intuitive Science of Thermonuclear War*. Cambridge, MA: Harvard University Press, 2005.

Gibson, Weldon B. *SRI, The Founding Years*. Los Altos: Publishing Services Center, 1980.

Gibson, Weldon B. *SRI, The Take-Off Years: The Right Moves at the Right Times*. Los Altos: Publishing Services Center, 1986.

Gieryn, Thomas F. "Boundary-Work and the Demarcation of Science from Non-Science: Strains and Interests in Professional Ideologies of Scientists." *American Sociological Review* 48 (1983): 781–795.

Gilman, Nils. *Mandarins of the Future: Modernization Theory in Cold War America*. Baltimore: Johns Hopkins University Press, 2003.

Giocoli, Nicola. "Nash Equilibrium." *History of Political Economy* 36 (2004): 639–666.

Goodeve, Charles. "Operations Research." *Nature* 161 (1948): 377–384.

Green, Alex E. S. "Finding the Japanese in March 1945." *Interfaces* 23 (5) (September/October 1993): 62–69.

Green, Alex E. S. "A Physicist with the Air Force in World War II." *Physics Today* 54 (8) (August 2001): 40–44.

Greenberg, Daniel S. "The Myth of the Scientific Elite." *Public Interest* 1 (1965): 51–62.

Greene, Benjamin P. *Eisenhower, Science Advice, and the Nuclear Test-Ban Debate*. Stanford: Stanford University Press, 2007.

Grimes, William. "David Blackwell, Scholar of Probability, Dies at 91." *New York Times*, July 17, 2010.

Guerlac, Henry E. *Radar in World War II*. 2 vols. Los Angeles: Tomash; New York: American Institute of Physics, 1987.

Gummett, Philip. *Scientists in Whitehall*. Manchester: Manchester University Press, 1980.

Habermas, Jürgen. *Technik und Wissenschaft als "Ideologie"*. Frankfurt am Main: Suhrkamp, 1968.

Habermas, Jürgen. *Toward a Rational Society: Student Protest, Science, and Politics.* Boston: Beacon, 1970.

Haines, George. *Essays on German Influence upon English Education and Science, 1850–1919.* New London: Connecticut College, 1969.

Halberstam, David. *The Best and the Brightest.* New York: Random House, 1972.

Hall, Daniel. J. G. Crowther, J. D. Bernal, V. H. Mottram, E. Charles, P. A. Gorther, and P. M. S. Blackett. *The Frustration of Science.* New York: W. W. Norton, 1935.

Harris, Arthur. *Bomber Offensive.* London: Collins, 1947.

Harrod, R. F. *The Prof: A Personal Memoir of Lord Cherwell.* London: Macmillan & Co, 1959.

Hartley, Harold, and D. Gabor. "Thomas Ralph Merton." *Biographical Memoirs of Fellows of the Royal Society* 16 (1970): 421–440.

Hauser, John R., and Glen L. Urban. "John D. C. Little." In *Profiles in Operations Research: Pioneers and Innovators*, ed. Arjang A. Assad and Saul I. Gass, 659–676. New York: Springer, 2011.

Hausrath, Alfred H. "Utilization of Negro Manpower in the Army." *Journal of the Operations Research Society of America* 2 (1954): 17–30.

Hausrath, Alfred H. *Venture Simulation in War, Business, and Politics.* New York: McGraw-Hill, 1971.

Hay, David Lowell. "Bomber Businessmen: The Army Air Forces and the Rise of Statistical Control, 1940–1945." PhD dissertation, University of Notre Dame, 1994.

Herken, Gregg. *Counsels of War.* New York: Knopf, 1985.

Herken, Gregg. *Cardinal Choices: Presidential Science Advising from the Atomic Bomb to SDI.* Stanford: Stanford University Press, 1992.

Herman, Ellen. *The Romance of American Psychology: Political Culture in the Age of Experts.* Berkeley: University of California Press, 1995.

Herman, Ellen. "Project Camelot and the Career of Cold War Psychology." In *Universities and Empire: Money and Politics in the Social Sciences during the Cold War*, ed. Christopher Simpson, 97–134. New York: New Press, 1998.

Herrmann, Cyril C., and John F. Magee. "'Operations Research' for Management." *Harvard Business Review* 31 (July 1953): 100–112.

Hertz, David B. "ORSA and TIMS Should Affiliate Rather than Merge." *Operations Research* 6 (1958): 296–297.

Hewlett, Richard G., and Francis Duncan. *Atomic Shield, 1947/1952*. vol. 2. A History of the United States Atomic Energy Commission. Washington, D.C.: U.S. Atomic Energy Commission, 1972.

Hewlett, Richard G., and Jack M. Holl. *Atoms for Peace and War, 1953–1961: Eisenhower and the Atomic Energy Commission*. Berkeley: University of California Press, 1989.

Heyck, Hunter. "Producing Reason." In *Cold War Social Science: Knowledge Production, Liberal Democracy, and Human Nature*, ed. Mark Solovey and Hamilton Cravens, 99–116. New York: Palgrave Macmillan, 2012.

Hicks, Donald, and David Smith. "Sir Owen Haddon Wansbrough-Jones: An Appreciation." *Journal of the Operational Research Society* 34 (1983): 105–109.

Hill, A. V. "Science, National and International, and the Basis of Cooperation." *Science* 93 (1941): 579–584; and *Nature* 147 (1941): 250–252.

Hill, A. V. "The Use and Misuse of Science in Government." *Advancement of Science* 2 (1942): 6–9.

Hill, A. V. *The Ethical Dilemmas of Science and Other Writings*. New York: Rockefeller Institute Press, 1960.

Hitch, Charles. "Planning Defense Production." *American Economic Review* 40 (1950): 191–198.

Hitch, Charles. "Sub-optimization in Operations Problems." *Journal of the Operations Research Society of America* 1 (1953): 87–99.

Hitch, Charles. "An Appreciation of Systems Analysis." *Journal of the Operations Research Society of America* 3 (1955): 466–481.

Hitch, Charles. "Comments by Charles Hitch." *Operations Research* 4 (1956): 426–430.

Hitch, Charles. "Operations Research and National Planning—A Dissent." *Operations Research* 5 (1957): 718–723.

Hitch, Charles. "Economics and Military Operations Research." *Review of Economics and Statistics* 40 (1958): 199–209.

Hitch, Charles. "A Further Comment on Economics and Military Operations Research." *Review of Economics and Statistics* 42 (1960): 222–223.

Hitch, Charles J., and Roland N. McKean. *The Economics of Defense in the Nuclear Age*. Cambridge, MA: Harvard University Press, 1960.

Hoffman, A. J., and P. Wolfe. "History." In *The Traveling Salesman Problem: A Guided Tour of Combinatorial Optimization*, ed. E. L. Lawler, J. K. Lenstra, A. H. G. Rinnooy Kan, and D. B. Shmoys, 1–15. New York: John Wiley & Sons, 1985.

Hogg, Ian V. *Anti-Aircraft: A History of Air Defence*. London: Macdonald and Jane's, 1978.

Holley, I. B., Jr. *Ideas and Weapons*. New Haven: Yale University Press, 1953.

Holley, I. B., Jr. "The Evolution of Operations Research and Its Impact on the Military Establishment: The Air Force Experience." In *Science, Technology, and Warfare: Proceedings of the Third Military History Symposium, USAF Academy*, ed. Monte D. Wright and Lawrence J. Paszek, 89–109. Washington, DC: U.S. Government Printing Office, 1971.

Holley, I. B., Jr. *Technology and Military Doctrine: Essays on a Challenging Relationship*. Maxwell Air Force Base: Air University Press, 2004.

Hollinger, David. "Science as a Weapon in *Kulturkämpfe* in the United States during and after World War II." *Isis* 86 (1995): 440–454.

Hore, Peter, ed. *Patrick Blackett: Sailor, Scientist, Socialist*. Portland, OR: Frank Cass, 2000.

Horkheimer, Max. *The Eclipse of Reason*. New York: Oxford University Press, 1947.

Horkheimer, Max, and Theodor W. Adorno. *Dialektik der Aufklärung: Philosophische Fragmente*. Amsterdam: Querido, 1947.

Horowitz, Irving Louis. *The Rise and Fall of Project Camelot: Studies in the Relationship between Social Science and Practical Politics*. Cambridge, MA: MIT Press, 1967.

Hounshell, David A. "The Cold War, RAND, and the Generation of Knowledge, 1946–1962." *Historical Studies in the Physical and Biological Sciences* 27 (1997): 237–268.

Hounshell, David A. "The Medium is the Message, or How Context Matters: The RAND Corporation Builds an Economics of Innovation, 1946-1962." In *Systems, Experts and Computers: The Systems Approach in Management and Engineering, World War II and After*, ed. Agatha C. Hughes and Thomas P. Hughes, 255–310. Cambridge, MA: MIT Press, 2000.

Howard, Michael. "Blackett and the Origins of Nuclear Strategy." *Journal of the Operational Research Society* 36 (1984): 89–95. Reprinted as "P. M. S. Blackett." In *Makers of Nuclear Strategy*, ed. John Bayliss and John Garnett, 153–163. London: Pinter, 1991.

Hughes, Thomas P. *American Genesis: A Century of Invention and Technological Enthusiasm, 1870–1970*. New York: Viking, 1989.

Hughes, Thomas P. *Rescuing Prometheus*. New York: Pantheon, 1998.

Hutchison, Kenneth, J. A. Gray, and Harrie Massey. "Charles Drummond Ellis." *Biographical Memoirs of Fellows of the Royal Society* 27 (1981): 199–233.

"In This Issue." *Harvard Business Review* 31 (July 1953): 7–12.

Isaac, Joel. "The Human Sciences in Cold War America." *Historical Journal (Cambridge, England)* 50 (2007): 725–746.

Isaac, Joel, and Duncan Bell, eds. *Uncertain Empire: American History and the Idea of the Cold War.* New York: Oxford University Press, 2012.

Jardini, David R. "Out of the Blue Yonder: The RAND Corporation's Diversification into Social Welfare Research." PhD dissertation, Carnegie Mellon University, 1996.

Jasanoff, Sheila. *The Fifth Branch: Science Advisers as Policymakers.* Cambridge, MA: Harvard University Press, 1990.

Johns Hopkins University Operations Research Office. *The Utilization of Negro Manpower in the Army: A 1951 Study.* McLean: Research Analysis Corporation, 1967.

Johnson, Ellis. "A Survey of Operations Research in the U.S.A." *Operational Research Quarterly* 5 (1954): 43–48.

Johnson, Ellis. "The Crisis in Science and Technology and Its Effect on Military Development." *Operations Research* 6 (1958): 11–34.

Johnson, Ellis. "The Long-Range Future of Operations Research." *Operations Research* 8 (1960): 1–23.

Johnson, H. Thomas, and Robert S. Kaplan. *Relevance Lost: The Rise and Fall of Management Accounting.* Boston: Harvard Business School Press, 1987.

Jones, R. V. "Scientists at War." *The Times*, April 6, 7, and 8, 1961.

Jones, R. V. "Lord Cherwell's Judgement in World War II." *The Oxford Magazine*, May 9, 1963, 279–286.

Jones, R. V. *Most Secret War.* London: Hamilton, 1978.

Josephson, Paul R. *Totalitarian Science and Technology.* Atlantic Highlands, NJ: Humanities Press, 1996.

Kahn, E. J. *The Problem Solvers: A History of Arthur D. Little, Inc.* Boston: Little, Brown, 1986.

Kahn, Herman. *On Thermonuclear War.* Princeton: Princeton University Press, 1960.

Kahn, Herman. *Thinking about the Unthinkable.* New York: Horizon, 1962.

Kaijser, Arne, and Joar Tiberg. "From Operations Research to Futures Studies: The Establishment, Diffusion, and Transformation of the Systems Approach in Sweden, 1945–1980." In *Systems, Experts, and Computers: The Systems Approach in Management and Engineering, World War II and After*, ed. Agatha C. Hughes and Thomas P. Hughes, 385–412. Cambridge, MA: The MIT Press, 2000.

Kaiser, David. "Cold War Requisitions, Scientific Manpower, and the Production of American Physicists after World War II." *Historical Studies in the Physical and Biological Sciences* 33 (2002): 131–159.

Kaplan, Fred. *The Wizards of Armageddon*. Stanford, CA: Stanford University Press, 1983.

Karchere, Alvin. "Review of *The Theory of Inventory Management*." *Journal of the Operations Research Society of America* 1 (1953): 314–316.

Katcher, David A. "On the Proposal for Merger of ORSA and TIMS." *Operations Research* 5 (1957): 563–564.

Katz, Bernard. "Archibald Vivian Hill." *Biographical Memoirs of Fellows of the Royal Society* 24 (1978): 71–149.

Kay, H. D. "William Kershaw Slater." *Biographical Memoirs of Fellows of the Royal Society* 17 (1971): 663–680.

Kevles, Daniel J. *The Physicists: The History of a Scientific Community in Modern America*. New York: Knopf, 1977.

Keynes, John Maynard. *A Treatise on Money*. 2 vols. New York: Harcourt Brace & Co, 1930.

Kirby, Maurice W. "Operations Research Trajectories: The Anglo-American Experience from the 1940s to the 1990s." *Operations Research* 48 (2000): 661–670.

Kirby, Maurice W. "The Intellectual Journey of Russell Ackoff: From OR Apostle to OR Apostate." *Journal of the Operational Research Society* 54 (2003a): 1127–1140.

Kirby, Maurice W. *Operational Research in War and Peace: The British Experience from the 1930s to 1970*. London: Imperial College Press, 2003b.

Kirby, Maurice W. "'A Festering Sore': The Issue of Professionalism in the History of the Operational Research Society." *Journal of the Operational Research Society* 57 (2006): 1161–1172.

Kirby, Maurice W. "Paradigm Change in Operations Research: Thirty Years of Debate." *Operations Research* 55 (2007): 1–13.

Kirby, Maurice W. "Charles Frederick Goodeve." In *Profiles in Operations Research: Pioneers and Innovators*, ed. Arjang A. Assad and Saul I. Gass, 83–94. New York: Springer, 2011.

Kirby, M. W., and M. T. Godwin. "The 'Invisible Science': Operational Research for the British Armed Forces after 1945." *Journal of the Operational Research Society* 61 (2010): 68–81.

Kirby, Maurice W., and Jonathan Rosenhead. "Russell Lincoln Ackoff." In *Profiles in Operations Research: Pioneers and Innovators*, ed. Arjang A. Assad and Saul I. Gass, 387–402. New York: Springer, 2011.

Kittel, Charles. "The Nature and Development of Operations Research." *Science* 105 (1947): 150–153.

Kjeldsen, Tinne Hoff. "A Contextualized Historical Analysis of the Kuhn-Tucker Theorem in Nonlinear Programming: The Impact of World War II." *Historia Mathematica* 27 (2000): 331–361.

Kjeldsen, Tinne Hoff. "Different Motivations and Goals in the Historical Development of the Theory of Systems of Linear Inequalities." *Archive for History of Exact Sciences* 56 (2002): 469–538.

Klein, Judy L. "Economics for a Client: The Case of Statistical Quality Control and Sequential Analysis." In *Toward a History of Applied Economics*, ed. Roger E. Backhouse and Jeff Biddle, 27–69. Durham: Duke University Press, 2000.

Klein, Judy L. "Protocols of War and the Mathematical Invasion of Policy Space." Unpublished book manuscript.

Klingaman, William K. *APL, Fifty Years of Service to the Nation: A History of the Johns Hopkins University Applied Physics Laboratory.* Laurel, MD: The Laboratory, 1993.

Kohler, Robert E. "The Management of Science: The Experience of Warren Weaver and the Rockefeller Foundation Programme in Molecular Biology." *Minerva* 14 (1976): 279–306.

Kojevnikov, Alexei. "The Phenomenon of Soviet Science." *Osiris* 23 (2008): 115–135.

Komons, Nick A. *Science and the Air Force.* Arlington: Office of Aerospace Research, 1966.

Koopman, Bernard O. "New Mathematical Methods in Operations Research." *Journal of the Operations Research Society of America* 1 (1952): 3–9.

Koopman, Bernard O. "Fallacies in Operations Research." *Operations Research* 4 (1956): 422–426.

Koopman, Bernard O. *Search and Screening: General Principles with Historical Applications.* Elmsford: Pergamon, 1980.

Koopmans, Tjalling C., ed. *Activity Analysis of Production and Allocation: Proceedings of a Conference.* New York: John Wiley and Sons, 1951.

Koopmans, Tjalling C. "Exchange Ratios between Cargoes on Various Routes (Non-Refrigerating Dry Cargoes)." In *Scientific Papers of Tjalling C. Koopmans*, 77–86. New York: Springer, 1970.

Koopmans, Tjalling C., and Stanley Reiter. "A Model of Transportation." In *Activity Analysis of Production and Allocation: Proceedings of a Conference*, ed. Tjalling C. Koopmans, 222–259. New York: John Wiley and Sons, 1951.

Kotkin, Stephen. *Magnetic Mountain: Stalinism as a Civilization*. Berkeley: University of California Press, 1995.

Kozak, Warren. *LeMay: The Life and Wars of General Curtis LeMay*. Washington, DC: Regnery, 2009.

Krige, John. *American Hegemony and the Postwar Reconstruction of Science in Europe*. Cambridge, MA: The MIT Press, 2006.

Kuklick, Bruce. *Blind Oracles: Intellectuals and War from Kennan to Kissinger*. Princeton: Princeton University Press, 2007.

Kusch, Martin. "Review of 'Bruno Latour, *We Have Never Been Modern*.'" *British Journal for the History of Science* 28 (1995): 125–126.

Lanchester, F. W. *Aircraft in Warfare: The Dawn of the Fourth Arm*. London: Constable and Company, 1916.

Latham, Michael E. *Modernization as Ideology: American Social Science and "Nation Building" in the Kennedy Era*. Chapel Hill: University of North Carolina Press, 2000.

Latham, Michael E. *The Right Kind of Revolution: Modernization, Development, and U. S. Foreign Policy from the Cold War to the Present*. Ithaca: Cornell University Press, 2011.

Lathrop, John B. "A Proposal for Merging ORSA and TIMS." *Operations Research* 5 (1957): 123–125.

Latour, Bruno. *The Pasteurization of France*, trans. Alan Sheridan and John Law. Cambridge, MA: Harvard University Press, 1988.

Latour, Bruno. *We Have Never Been Modern*, trans. Catherine Porter. Cambridge, MA: Harvard University Press, 1993.

Lear, John. "Watson-Watt's Retort to C. P. Snow." *New Scientist* 9 (1961): 537.

LeMay, Curtis E., with MacKinlay Kantor. *Mission with LeMay: My Story*. Garden City, NY: Doubleday, 1965.

Leonard, Robert. *Von Neumann, Morgenstern, and the Creation of Game Theory: From Chess to Social Science, 1900–1960*. New York: Cambridge University Press, 2010.

Levinson, Horace C. "Experiences in Commercial Operations Research." *Journal of the Operations Research Society of America* 1 (1953): 220–239.

Levinson, Horace C., and Arthur A. Brown. "Operations Research." *Scientific American* 184 (3) (March 1951): 15–21.

Light, Jennifer S. *From Warfare to Welfare: Defense Intellectuals and Urban Problems in Cold War America*. Baltimore: Johns Hopkins University Press, 2003.

Lindsay, O. J. M. "John Anthony Hardinge Giffard, 3rd Earl of Halsbury." *Biographical Memoirs of Fellows of the Royal Society* 47 (2001): 239–253.

Little, John Dutton Conant. "Use of Storage Water in a Hydroelectric System." PhD dissertation, Massachusetts Institute of Technology, 1954.

Little, John Dutton Conant. "The Use of Storage Water in a Hydroelectric System." *Journal of the Operations Research Society of America* 3 (1955): 187–197.

Llewellyn-Jones, Malcolm. "A Clash of Cultures: The Case for Large Convoys." In *Patrick Blackett: Sailor, Scientist, Socialist*, ed. Peter Hore, 138–158. Portland, OR: Frank Cass, 2000.

Lovell, Bernard. *Science and Civilization*. New York: T. Nelson, 1939.

Lovell, Bernard. "Patrick Maynard Stuart Blackett, Baron Blackett, of Chelsea." *Biographical Memoirs of Fellows of the Royal Society* 21 (1975): 1–115.

Low, David. "Does Colonel Blimp Exist?" [1937]. Reprinted in *The Complete Colonel Blimp*, ed. Mark Bryant, 41–43. London: Bellew, 1991.

Lowe, George E. "The Camelot Affair." *Bulletin of the Atomic Scientists* 22 (5) (1966): 44–48.

Lowe, R. H. "Operational Research in the Canadian Department of National Defence." *Operations Research* 8 (1960): 847–856.

Luce, R. Duncan, and Howard Raiffa. *Games and Decisions: Introduction and Critical Survey*. New York: Wiley, 1957.

Lutyens, David. "Boffins at Odds." *Daily Telegraph*, April 9, 1961.

Lyons, Gene M. *The Uneasy Partnership: Social Science and the Federal Government in the Twentieth Century*. New York: Russell Sage Foundation, 1969.

MacDougall, G. D. A. "The Prime Minister's Statistical Section." In *Lessons of the War Economy*, ed. D. N. Chester, 58–68. Cambridge, UK: Cambridge University Press, 1951.

MacDougall, G. D. A. "Machinery of Government: Some Personal Reflections." In *Policy and Politics*, ed. David Butler and A. H. Hasley, 169–181. London: Macmillan, 1978.

MacKenzie, Donald. *An Engine, Not a Camera: How Financial Models Shape Markets*. Cambridge, MA: MIT Press, 2006.

MacLane, Saunders. "The Applied Mathematics Group at Columbia in World War II." In *A Century of Mathematics in America*. vol. 3. ed. Peter Duren, 495–515. Providence: American Mathematical Society, 1989.

MacLeod, Roy M. "The Support of Victorian Science: The Endowment of Research Movement in Great Britain, 1868–1900." *Minerva* 9 (1971): 197–230.

MacLeod, Roy M. "Science for Imperial Efficiency and Social Change: Reflections on the British Science Guild, 1905–1936." *Public Understanding of Science (Bristol, England)* 3 (1994): 155–193.

Magee, John F. "The Effect of Promotional Effort on Sales." *Journal of the Operations Research Society of America* 1 (1953): 64–74.

Magee, John F. "Guides to Inventory Policy, I. Functions and Lot Sizes." *Harvard Business Review* 34 (1) (Jan. 1956a): 49–60.

Magee, John F. "Guides to Inventory Policy, II. Problems of Uncertainty." *Harvard Business Review* 34 (2) (March 1956b): 103–116.

Magee, John F. "Guides to Inventory Policy, III. Anticipating Future Needs." *Harvard Business Review* 34 (3) (May 1956c): 57–70.

Magee, John F. *Production Planning and Inventory Control*. New York: McGraw-Hill, 1958.

Magee, John F. "Operations Research at Arthur D. Little, Inc.: The Early Years." *Operations Research* 50 (2002): 149–153.

Magee, John F., and Martin L. Ernst. "Progress in Operations Research: The Challenge of the Future." In *Progress in Operations Research*. vol. 1. ed. Russell L. Ackoff, 465–491. New York: John Wiley & Sons, 1961.

Mandler, Peter. *The English National Character: The History of an Idea from Edmund Burke to Tony Blair*. New Haven: Yale University Press, 2006.

Mandler, Peter. "Deconstructing 'Cold War Anthropology.'" In *Uncertain Empire: American History and the Idea of the Cold War*, ed. Joel Isaac and Duncan Bell, 245–266. New York: Oxford University Press, 2012.

Marcuse, Herbert. *One Dimensional Man: Studies in the Ideology of Advanced Industrial Society*. Boston: Beacon, 1964.

Matusow, Alan J. *The Unraveling of America: A History of Liberalism in the 1960s*. Athens, GA: University of Georgia Press, 1984.

May, Andrew David. "The RAND Corporation and the Dynamics of American Strategic Thought, 1946–1962." PhD dissertation, Emory University, 1998.

McArthur, Charles W. *Operations Analysis in the U.S. Army Eighth Air Force in World War II*. Providence: American Mathematical Society, 1990.

McGucken, William. "On Freedom and Planning in Science: The Society for Freedom in Science, 1940–46." *Minerva* 16 (1978): 42–72.

McGucken, William. *Scientists, Society and State: The Social Relations of Science Movement in Great Britain, 1931–1947*. Columbus: Ohio University Press, 1984.

McKenna, Christopher D. *The World's Newest Profession: Management Consulting in the Twentieth Century*. New York: Cambridge University Press, 2006.

Medina, Eden. *Cybernetic Revolutionaries: Technology and Politics in Allende's Chile*. Cambridge, MA: MIT Press, 2011.

Mehrling, Perry. *Fischer Black and the Revolutionary Idea of Finance*. Hoboken: John Wiley & Sons, 2005.

Mellor, D. P. *The Role of Science and Industry*, vol. V, series 4 (Civil). In *Australia in the War of 1939–1945*, ed. Gavin Long. Canberra: Australian War Memorial, 1958.

Mercer, Alan. "Patrick Rivett." *Journal of the Operational Research Society* 56 (2005): 1119–1121.

Mills, C. *Wright. White Collar: The American Middle Classes*. New York: Oxford University Press, 1951.

Milne, David. *America's Rasputin: Walt Rostow and the Vietnam War*. New York: Hill and Wang, 2008.

Milne, E. A. "Ralph Howard Fowler, 1889–1944." *Obituary Notices of Fellows of the Royal Society* 5 (14) (1945): 60–78.

Mindell, David A. *Between Human and Machine: Feedback, Control, and Computing before Cybernetics*. Baltimore: Johns Hopkins University Press, 2002.

Miranti, Paul J. *Accountancy Comes of Age: The Development of an American Profession*. Chapel Hill: University of North Carolina Press, 1990.

Mirowski, Philip. "What Were Von Neumann and Morgenstern Trying to Accomplish?" In *Toward a History of Game Theory*, ed. E. Roy Weintraub, 113–147. Durham: Duke University Press, 1992.

Mirowski, Philip. *Machine Dreams: Economics Becomes a Cyborg Science*. New York: Cambridge University Press, 2002.

Mirowski, Philip. "Cracks, Hidden Passageways and False Bottoms: The Economics of Science and Social Studies of Economics." In *The Effortless Economy of Science?* ed. Philip Mirowski, 1–35. Durham: Duke University of Press, 2004a.

Mirowski, Philip. "The Scientific Dimensions of Social Knowledge and Their Distant Echoes in 20th-century American Philosophy of Science." *Studies in History and Philosophy of Science* 35 (2004b): 283–326.

Mirowski, Philip. "A History Best Served Cold." In *Uncertain Empire: American History and the Idea of the Cold War*, ed. Joel Isaac and Duncan Bell, 61–74. New York: Oxford University Press, 2012.

Miser, Hugh J. "Craft in Operations Research." *Operations Research* 40 (1992): 633–639.

MITRE Corporation. *MITRE: The First Twenty Years: A History of the MITRE Corporation (1958–1978)*. Bedford: MITRE Corporation, 1979.

Mood, A. M. "Review of 'Philip M. Morse and George E. Kimball, *Methods of Operations Research*.'" *Journal of the Operations Research Society of America* 1 (1953): 306–308.

Mood, A. M. "Diversification of Operations Research." *Operations Research* 13 (1965): 169–178.

Morgan, Mary S. *The World in the Model: How Economists Work and Think*. New York: Cambridge University Press, 2012.

Morgenstern, Oskar. "Abraham Wald, 1902–1950." *Econometrica* 19 (1951): 361–367.

Morrell, Jack, and Arnold Thackray. *Gentlemen of Science: Early Years of the British Association for the Advancement of Science*. New York: Oxford University Press, 1981.

Morse, Philip M. "Must We Always Be Gadgeteers?" *Physics Today* 3 (12) (December 1950): 4–5.

Morse, Philip M. "Trends in Operations Research." *Journal of the Operations Research Society of America* 1 (1953): 159–165.

Morse, Philip M. "Where Is the New Blood?" *Journal of the Operations Research Society of America* 3 (1955): 383–387.

Morse, Philip M. *In at the Beginnings: A Physicist's Life*. Cambridge, MA: MIT Press, 1977.

Morse, Philip M., and George E. Kimball. "Methods of Operations Research," Operations Evaluation Group, Report 54. 1946. http://www.cna.org/sites/default/files/research/1100005400.pdf.

Morse, Philip M., and George E. Kimball. *Methods of Operations Research*. New York: MIT Press and John Wiley & Sons, 1951.

Morton, N. W. "A Brief History of the Development of Canadian Military Operational Research." *Operations Research* 4 (1956): 187–192.

Mumford, Lewis. *Technics and Civilization*. New York: Harcourt, Brace and Company, 1934.

Mumford, Lewis. *The Myth of the Machine*. 2 vols. New York: Harcourt, Brace, Jovanovich, 1967 and 1970.

Nash, John. "Equilibrium Points in n-Person Games." *Proceedings of the National Academy of Sciences of the United States of America* 36 (1950): 49–50.

Nash, John. "Non-Cooperative Games." *Annals of Mathematics* 54 (1951): 286–295.

Nassar, Sylvia. *A Beautiful Mind: The Life of Mathematical Genius and Nobel Laureate John Nash.* New York: Simon and Schuster, 1998.

Needell, Allan A. *Science, Cold War, and the American State: Lloyd V. Berkner and the Balance of Professional Ideals.* Amsterdam: Harwood Academic, 2000.

Noble, David. *America by Design: Science, Technology, and the Rise of Corporate Capitalism.* New York: Oxford University Press, 1979.

Nye, Mary Jo. *Blackett: Physics, War and Politics in the Twentieth Century.* Cambridge, MA: Harvard University Press, 2004.

Oikonomou, Alexandros-Panagiotis. "The Hidden Persuaders: Government Scientists and Defence in Post-war Britain." PhD thesis, Imperial College London, 2010.

Olby, Robert. "Social Imperialism and State Support for Agricultural Research in Edwardian Britain." *Annals of Science* 48 (1991): 509–526.

"Operational Research Club." *Operational Research Quarterly* 1 (1950): 36.

Operations Evaluation Group. "Search and Screening," Report 56. 1946. http://www.cna.org/sites/default/files/research/1100005600.pdf.

O'Rand, Angela M. "Mathematizing Social Science in the 1950s: The Early Development and Diffusion of Game Theory." In *Toward a History of Game Theory*, ed. E. Roy Weintraub, 177–204. Durham: Duke University Press, 1992.

Ortolano, Guy. "The Literature and the Science of 'Two Cultures' Historiography." *Studies in History and Philosophy of Science* 39 (2008): 143–150.

Ortolano, Guy. *The Two Cultures Controversy: Science, Literature and Cultural Politics in Postwar Britain.* New York: Cambridge University Press, 2009.

Overy, Richard. *The Bombing War: Europe 1939–1945.* London: Allen Lane, 2013.

Owens, Larry. "Mathematicians at War: Warren Weaver and the Applied Mathematics Panel, 1942–1945." In *The History of Modern Mathematics, Vol. II: Institutions and Applications*, ed. David E. Rowe and John McCleary, 287–305. Boston: Academic Press, 1989.

Pawle, Gerald. *The Secret War, 1939–1945.* London: Harrap, 1956.

Pennycuick, K. "Army Operational Research in the Far East 1952/54." *Operational Research Quarterly* 5 (1954): 120–129.

Peyton, John. *Solly Zuckerman: A Scientist out of the Ordinary.* London: John Murray, 2001.

Pickering, Andy. "Cyborg History and the World War II Regime." *Perspectives on Science* 3 (1995): 1–48.

Pickering, Andy. *The Cybernetic Brain: Sketches of Another Future*. Chicago: University of Chicago Press, 2009.

Platt, William J. "Industrial Economics and Operations Research at the Stanford Research Institute." *Journal of the Operations Research Society of America* 2 (1954): 411–418.

Pocock, J. W. "Operations Research and the Management Consultant." *Journal of the Operations Research Society of America* 1 (1953): 137–144.

Ponturo, John. "Analytical Support for the Joint Chiefs of Staff: The WSEG Experience, 1948–1976," Institute for Defense Analyses Study S-507. 1979. http://www.dod.mil/pubs/foi/joint_staff/jointStaff_jointOperations/20.pdf.

Porter, Theodore M. *Trust in Numbers: The Pursuit of Objectivity in Science and Public Life*. Princeton: Princeton University Press, 1995.

Porter, Theodore M. "Positioning Social Science in Cold War America." In *Cold War Social Science: Knowledge Production, Liberal Democracy, and Human Nature*, ed. Mark Solovey and Hamilton Cravens, ix–xv. New York: Palgrave Macmillan, 2012.

Poundstone, William. *Prisoner's Dilemma*. New York: Doubleday, 1992.

Pratt, T. H. "A Rose by Any Other Name: An Outline of Operational Analysis in Admiralty Headquarters, 1947–1970." *Journal of Naval Science* 7 (1981): 1–9, 104–113, 161–169, 218–227.

Pugh, George E. "Operations Research for the Secretary of Defense and the Joint Chiefs of Staff." *Operations Research* 8 (1960): 839–846.

Quade, E. S., ed. *Analysis for Military Decisions*. Chicago: Rand McNally, 1964a.

Quade, E. S. "The Selection and Use of Strategic Air Bases: A Case Study." In *Analysis for Military Decisions*, ed. E. S. Quade, 24–63. Chicago: Rand McNally, 1964b.

Quade, E. S., and W. I. Boucher, eds. *Systems Analysis and Policy Planning: Applications in Defense*. New York: American Elsevier, 1968.

Rand, Graham K. "Berwyn Hugh Patrick Rivett." In *Profiles in Operations Research: Pioneers and Innovators*, ed. Arjang A. Assad and Saul I. Gass, 477–492. New York: Springer, 2011.

Rau, Erik Peter. "Combat Scientists: The Emergence of Operations Research in the United States during World War II." PhD dissertation, University of Pennsylvania, 1999.

Rau, Erik Peter. "The Adoption of Operations Research in the United States during World War II." In *Systems, Experts and Computers: The Systems Approach in*

Management and Engineering, World War II and After, ed. Agatha C. Hughes and Thomas P. Hughes, 57–92. Cambridge, MA: MIT Press, 2000.

Rau, Erik Peter. "Technological Systems, Expertise, and Policy Making: The British Origins of Operational Research." In *Technologies of Power: Essays in Honor of Thomas Parke Hughes and Agatha Chipley Hughes*, ed. Michael Thad Allen and Gabrielle Hecht. Cambridge, MA: MIT Press, 2001.

Redmond, Kent C., and Thomas M. Smith. *From Whirlwind to MITRE: The R&D Story of the SAGE Air Defense Computer*. Cambridge, MA: The MIT Press, 2000.

Rees, Mina. "The Mathematical Sciences and World War II." *American Mathematical Monthly* 87 (1980): 607–621.

Rees, Mina. "Warren Weaver, 1894–1978." *Biographical Memoirs, National Academy of Sciences* 57 (1987): 493–529.

Richardson, F. D. "Charles Frederick Goodeve." *Biographical Memoirs of Fellows of the Royal Society* 27 (1981): 307–353.

Rider, Robin E. "Operations Research and Game Theory: Early Connections." In *Toward a History of Game Theory*, ed. E. Roy Weintraub, 225–239. Durham: Duke University Press, 1992.

Riker, William H. "The Entry of Game Theory into Political Science." In *Toward a History of Game Theory*, ed. E. Roy Weintraub, 207–224. Durham: Duke University Press, 1992.

Riordan, Michael, and Lillian Hoddeson. *Crystal Fire: The Birth of the Information Age*. New York: W. W. Norton & Company, 1997.

Robin, Ron. *The Making of the Cold War Enemy: Culture and Politics in the Military-Intellectual Complex*. Princeton: Princeton University Press, 2001.

Rohde, Joy. *Armed with Expertise: The Militarization of American Social Science Research during the Cold War*. Ithaca: Cornell University Press, 2013.

Rose, Hilary, and Steven Rose. *Science and Society*. London: Allen Lane, The Penguin Press, 1969.

Rosenberg, David Alan. "Reality and Responsibility: Power and Process in the Making of United States Nuclear Strategy, 1945–68." *Journal of Strategic Studies* 9 (1986): 35–52.

Rosenhead, Jonathan. "Operational Research at the Crossroads: Cecil Gordon and the Development of Post-War OR." *Journal of the Operational Research Society* 40 (1989): 3–28.

Rosenhead, Jonathan. "Anthony Stafford Beer." In *Profiles in Operations Research: Pioneers and Innovators*, Arjang A. Assad and Saul I. Gass, 593–612. New York: Springer, 2011.

Ross, Donald. 1980. "Twenty Years of Research at the SACLANT ASW Research Centre." SACLANTCEN Special Report M-93. http://www.cmre.nato.int/employment/current-vacancies/doc_download/84-twenty-years-of-research-at-the-saclant-asw-research-centre-1959-1979.

Rosser, J. Barkley. "Mathematics and Mathematicians in World War II." *Notices of the American Mathematical Society* 29 (1982): 509–515.

Rostow, W. W. *Stages of Economic Growth: A Non-Communist Manifesto*. Cambridge, UK: Cambridge University Press, 1960.

Salveson, Melvin E. "The Institute of Management Sciences: A Prehistory and Commentary on the Occasion of TIMS' 40th Anniversary." *Interfaces* 27 (1997): 74–78.

Samuelson, P. A. "A Note on the Pure Theory of Consumer's Behavior." *Economica* 5 (1938): 61–71.

Samuelson, P. A. "Consumption Theory in Terms of Revealed Preference." *Economica* 15 (1948): 243–253.

Samuelson, P. A. "Abstract of a Theorem Concerning Substitutability in Open Leontief Models." In *Activity Analysis of Production and Allocation: Proceedings of a Conference*, ed. Tjalling C. Koopmans, 142–146. New York: John Wiley and Sons, 1951.

Sapolsky, Harvey M. *Science and the Navy: The History of the Office of Naval Research*. Princeton: Princeton University Press, 1990.

Savage, Leonard J. *The Foundations of Statistics*. New York: John Wiley and Sons, 1954.

Scarf, Hebert E. "The Optimality of (S,s) Policies in the Dynamic Inventory Problem." In *Mathematical Methods in the Social Sciences, 1959*, ed. Kenneth J. Arrow, Samuel Karlin, and Patrick Suppes, 196–202. Stanford: Stanford University Press, 1960.

Schaffer, Simon. "Astronomers Mark Time: Discipline and the Personal Equation." *Science in Context* 2 (1988): 115–145.

Schelling, Thomas C. *The Strategy of Conflict*. Cambridge, MA: Harvard University Press, 1960.

Schweber, S. S. "The Mutual Embrace of Science and the Military: ONR and the Growth of Physics in the United States After World War II." In *Science, Technology, and the Military*, ed. Everett Mendelsohn, Merrett Rowe Smith, and Peter Weingart, 1–45. Boston: Kluwer Academic, 1988.

Science in War. 1940. London: Penguin.

Sent, Esther-Mirjam. "Herbert A. Simon as a Cyborg Scientist." *Perspectives on Science* 8 (2000): 380–406.

Shah, Hemant. *The Production of Modernization: Daniel Lerner, Mass Media, and the Passing of Traditional Society*. Philadelphia: Temple University Press, 2011.

Shapin, Steven. "Lowering the Tone in the History of Science: A Noble Calling." In *Never Pure: Historical Studies of Science as if It Was Produced by People with Bodies, Situated in Time, Space, Culture, and Society, and Struggling for Credibility and Authority*, ed. Steven Shapin, 1–14. Baltimore: Johns Hopkins University Press, 2010.

Shapin, Steven, and Simon Schaffer. *Leviathan and the Air-Pump: Hobbes, Boyle, and the Experimental Life*. Princeton: Princeton University Press, 1985.

Shapley, Deborah. *Promise and Power: The Life and Times of Robert McNamara*. Boston: Little, Brown, 1993.

Sherry, Michael S. *The Rise of American Air Power: The Creation of Armageddon*. New Haven: Yale University Press, 1987.

Shrader, Charles R. *History of Operations Research in the United States Army, Volume 1: 1942–1962*. Washington, DC: United States Army, 2006.

Shrader, Charles R. *History of Operations Research in the United States Army, Volume II: 1961–1973*. Washington, DC: United States Army, 2008.

Shrader, Charles R. *History of Operations Research in the United States Army, Volume III: 1973–1995*. Washington, DC: United States Army, 2009.

Shubik, Martin. "What Is an Application and When Is Theory a Waste of Time?" *Management Science* 33 (1987): 1511–1522.

Shubik, Martin. "Game Theory and Operations Research: Some Musings 50 Years Later." *Operations Research* 50 (2002): 192–196.

Shurkin, Joel N. *Broken Genius: The Rise and Fall of William Shockley, Creator of the Electronic Age*. New York: Macmillan, 2006.

Simon, Herbert A. *Administrative Behavior: A Study of Decision-Making Processes in Administrative Organizations*. New York: The Macmillan Company, 1947.

Simon, Herbert A. "Effects of Technological Change in a Linear Model." In *Activity Analysis of Production and Allocation: Proceedings of a Conference*, ed. Tjalling C. Koopmans, 260–276. New York: John Wiley and Sons, 1951.

Simon, Herbert A. "On the Application of Servomechanism Theory in the Study of Production Control." *Econometrica* 20 (1952): 247–268.

Simon, Herbert A. *The Sciences of the Artificial*. Cambridge, MA: The MIT Press, 1969.

Simpson, Christopher. *Science of Coercion: Communication Research and Psychological Warfare, 1945–1960*. New York: Oxford University Press, 1994.

Smeed, R. J. "Some Factors Influencing the Road Behaviour of Vehicle Drivers." *Operational Research Quarterly* 3 (1952): 60–67.

Smith, Bruce L. R. *The RAND Corporation: Case Study of a Nonprofit Advisory Corporation.* Cambridge, MA: Harvard University Press, 1966.

Smith, M. Nicholas, Jr. "Comments." *Journal of the Operations Research Society of America* 2 (1954): 181–187.

Smith, M. Nicholas, Jr., S. Stanley Walters, Franklin C. Brooks, and David H. Blackwell. "The Theory of Value and the Science of Decision: A Summary." *Journal of the Operations Research Society of America* 1 (1953): 103–113.

Snow, C. P. *The Two Cultures and the Scientific Revolution.* New York: Cambridge University Press, 1959.

Snow, C. P. *Science and Government.* Cambridge, MA: Harvard University Press, 1961.

Solovey, Mark. "Project Camelot and the 1960s Epistemological Revolution: Rethinking the Politics-Patronage-Social Science Nexus." *Social Studies of Science* 31 (2001): 171–206.

Solovey, Mark. *Shaky Foundations: The Politics-Patronage-Social Science Nexus in Cold War America.* New Brunswick, NJ: Rutgers University Press, 2013.

Solovey, Mark, and Hamilton Cravens. *Cold War Social Science: Knowledge Production, Liberal Democracy, and Human Nature.* New York: Palgrave Macmillan, 2012.

Solow, Herbert. "Operations Research." *Fortune* 43 (4) (April 1951): 105–122.

Spufford, Francis. *Red Plenty.* London: Faber, 2010.

Stigler, George J. "The Cost of Subsistence." *Journal of Farm Economics* 27 (1945): 303–314.

Stuart, Alan. "Sir Maurice Kendall, 1907–1983." *Journal of the Royal Statistical Society. Series A (General)* 147 (1984): 120–122.

Sumida, Jon Tetsuro. *In Defence of Naval Supremacy: Finance, Technology, and British Naval Policy, 1889–1914.* London: Unwin Hyman, 1989.

Sumida, Jon Tetsuro. "The Quest for Reach: The Development of Long-Range Naval Gunnery in the Royal Navy, 1901–1912." In *Tooling for War: Military Transformations in the Industrial Age,* ed. Stephen Chiabotti, 49–96. Chicago: Imprint, 1996.

Thiesmeyer, Lincoln R., and John E. Burchard. *Combat Scientists.* Boston: Little, Brown, 1947.

Thomas, William. "The Heuristics of War: Scientific Method and the Founders of Operations Research." *British Journal for the History of Science* 40 (2007): 251–274.

Thomas, William. "Operations Research vis-à-vis Management at the Massachusetts Institute of Technology and Arthur D. Little." *Business History Review* 86 (2012): 99–122.

Tidman, Keith R. *The Operations Evaluation Group: A History of Naval Operations Analysis*. Annapolis: Naval Institute Press, 1984.

Tippett, L. H. C. "Operational Research at the Shirley Institute." *Operational Research Quarterly* 1 (1950): 19–24.

Tomlinson, Rolfe C., ed. *OR Comes of Age: A Review of the Work of the Operational Research Branch of the National Coal Board, 1948–1969*. London: Tavistock, 1971.

Trefethen, Florence N. "A History of Operations Research." In *Operations Research for Management*, ed. Joseph F. McCloskey and Florence N. Trefethen, 3–35. Baltimore: Johns Hopkins University Press, 1954.

Trundle, G. T., Jr. "Your Inventory a Graveyard?" *Factory Management and Maintenance* 94 (1936): 45.

Turner, Frank M. "Public Science in Britain, 1880–1919." *Isis* 71 (Dec. 1980): 589–608.

United States Army Air Forces. *Operations Analysis in World War II*. Philadelphia: Stevenson Brothers, 1948.

Vazsonyi, Andrew. "The Uses of Mathematics in Production and Inventory Control." *Management Science* 1 (1954): 70–85.

Vazsonyi, Andrew. "The Use of Mathematics in Production and Inventory Control—II (Theory of Scheduling)." *Management Science* 1 (1955): 207–223.

Vazsonyi, Andrew. "Operations Research in Production Control—A Progress Report." *Operations Research* 4 (1956): 19–31.

Vazsonyi, Andrew. *Which Door Has the Cadillac: Adventures of a Real-Life Mathematician*. New York: Writers Club, 2002.

Von Neumann, John, and Oskar Morgenstern. *Theory of Games and Economic Behavior*. Princeton: Princeton University Press, 1944.

Waddington, C. H. "Operational Research." *Nature* 161 (1948): 404.

Waddington, C. H. *OR in World War 2: Operational Research against the U-Boat*. London: Elek Science, 1973.

Wakelam, Randall. *The Science of Bombing: Operational Research in RAF Bomber Command*. Toronto: University of Toronto Press, 2009.

Wald, Abraham. *Sequential Analysis*. New York: John Wiley and Sons, 1947.

Wald, Abraham. *Statistical Decision Functions*. New York: John Wiley and Sons, 1950.

Wald, A., and J. Wolfowitz. "Optimum Character of the Sequential Probability Ratio Test." *Annals of Mathematical Statistics* 19 (1948): 326–339.

Wallis, W. Allen. "The Statistical Research Group, 1942–1945." *Journal of the American Statistical Association* 75 (1980): 320–330.

Wang, Zuoyue. *In Sputnik's Shadow: The President's Science Advisory Committee and Cold War America*. New Brunswick, NJ: Rutgers University Press, 2008.

Wansbrough-Jones, O. H. In "Operational Research in War and Peace." *Advancement of Science* 4 (1948): 321–323.

War Office. *Text Book of Anti-Aircraft Gunnery*. 2 vols. London: HMSO, 1924–1925.

Waring, Stephen P. *Taylorism Transformed: Scientific Management Theory since 1945*. Chapel Hill: University of North Carolina Press, 1991.

Waring, Stephen P. "Cold Calculus: The Cold War and Operations Research." *Radical History Review* 63 (1995): 28–51.

Watson-Watt, Robert. In "Operational Research in War and Peace." *Advancement of Science* 4 (1948): 320–321.

Watson-Watt, Robert. *Three Steps to Victory*. London: Oldhams, 1957.

Watson-Watt, Robert. "The Truth about Churchill's Aide." *The Saturday Review*, March 4, 1961.

Weaver, Warren. *Free Science*. Oxford: Potter Press, 1945.

Weaver, Warren. *Scene of Change: A Lifetime in American Science*. New York: Scribner, 1970.

Weida, Nancy C. "Andrew Vazsonyi." In *Profiles in Operations Research: Pioneers and Innovators*, ed. Arjang A. Assad and Saul I. Gass, 273–291. New York: Springer, 2011.

Weintraub, E. Roy, ed. *Toward a History of Game Theory*. Durham: Duke University Press, 1992.

Weintraub, E. Roy, ed. *MIT and the Transformation of American Economics*. Durham: Duke University Press, 2014.

Werskey, Gary. *The Visible College*. London: Allen Lane, 1978.

Westwick, Peter J. *The National Labs: Science in an American System, 1947–1974*. Cambridge, MA: Harvard University Press, 2003.

Whitin, Thomson M. *The Theory of Inventory Management*. Princeton: Princeton University Press, 1953.

Whitson, W. L. "The Growth of Operations Research in the U.S. Army." *Operations Research* 8 (1960): 809–824.

Whyte, William H. *The Organization Man*. New York: Simon and Schuster, 1956.

Wilkie, Tom. *British Science and Politics since 1945*. Oxford: Basil Blackwell, 1991.

Wilks, S. S. "Research on Consumer Products as a Counterpart of Wartime Research." In *Measurement of Consumer Interest*, ed. C. West Churchman, Russell L. Ackoff, and Murray Wax, 135–138. Philadelphia: University of Pennsylvania Press, 1947.

Willard, Hal. "A-Attack to Kill 80%, 11 Boy Scientists Predict." *Washington Post*, December 28, 1956.

Williams, E. C. "Review of *Methods of Operations Research*." *Operational Research Quarterly* 2 (1951): 49–50.

Williams, E. C. "The Origin of the Term 'Operational Research' and the Early Development of the Military Work." *Operational Research Quarterly* 19 (1968): 111–113.

Williams, J. D. *The Compleat Strategyst, Being a Primer on the Theory of Games of Strategy*. New York: McGraw-Hill, 1954.

Wilson, Thomas. *Churchill and the Prof*. London: Cassell, 1995.

Winner, Langdon. *Autonomous Technology: Technics-out-of-Control as a Theme in Political Thought*. Cambridge, MA: MIT Press, 1977.

Wohlstetter, Albert. "The Delicate Balance of Terror." *Foreign Affairs* 37 (1959): 211–234.

Wohlstetter, Albert. "Scientists, Seers and Strategy." *Foreign Affairs* 41 (1963): 466–478.

Wohlstetter, Albert. "Analysis and Design of Conflict Systems." In *Analysis for Military Decisions*, ed. E. S. Quade, 103–148. Chicago: Rand McNally, 1964a.

Wohlstetter, Albert. "Strategy and the Natural Scientists." In *Sciences and National Policy-Making*, ed. Robert Gilpin and Christopher Wright, 174–239. New York: Columbia University Press, 1964b.

Wohlstetter, Albert. "Sin and Games in America." In *Game Theory and Related Approaches to Social Behavior*, ed. Martin Shubik, 209–225. New York: John Wiley and Sons, 1964c.

Wolf, William. *Boeing B-29 Superfortress: The Ultimate Look, From Drawing Board to VJ-Day*. Atglen: Schiffer, 2005.

Wolfe, Audra J. *Competing with the Soviets: Science, Technology, and the State in Cold War America*. Baltimore: Johns Hopkins University Press, 2013.

Wolfowitz, J. "Abraham Wald, 1902–1950." *Annals of Mathematical Statistics* 23 (1952): 1–13.

Wong, Stanley. *The Foundations of Paul Samuelson's Revealed Preference Theory: A Study by the Method of Rational Reconstruction.* Boston: Routledge and Kegan Paul, 1978.

Wood, Marshall K., and Murray A. Geisler. "Development of Dynamic Models for Program Planning." In *Activity Analysis of Production and Allocation: Proceedings of a Conference,* ed. Tjalling C. Koopmans, 189–215. New York: John Wiley and Sons, 1951.

Yates, F. "Operational Research." *Nature* 161 (1948): 609.

Young, Stephanie Caroline. "Power and the Purse: Defense Budgeting and American Politics, 1947–1972." PhD dissertation, University of California, Berkeley, 2010.

Zacks, Shelemyahu. "Jacob Wolfowitz, 1910–1981." *Biographical Memoirs, National Academy of Sciences* 82 (2003): 372–384.

Zimmerman, Carroll L. *Insider at SAC: Operations Analysis under General LeMay.* Manhattan, KS: Sunflower University Press, 1988.

Zimmerman, David. *Top Secret Exchange: The Tizard Mission and the Scientific War.* Buffalo: McGill-Queens University Press, 1996.

Zimmerman, David. *Britain's Shield: Radar and the Defeat of the Luftwaffe.* Stroud: Sutton, 2001.

Zuckerman, Solly. In "Operational Research in War and Peace." *Advancement of Science* 4 (1948): 323–326.

Zuckerman, Solly. "Judgment and Control in Modern Warfare." *Foreign Affairs* 40 (1962): 196–212.

Zuckerman, Solly. *Scientists and War: The Impact of Science on Military and Civil Affairs.* New York: Harper & Row, 1967.

Zuckerman, Solly. *From Apes to Warlords.* New York: Harper & Row, 1978.

Zuckerman, Solly. *Monkeys, Men and Missiles: An Autobiography, 1946–1988.* London: Collins, 1988.

Index